St. Olaf College

AUG 19 1985

Science Library

COMPUTATIONAL METHODS FOR PARTIAL DIFFERENTIAL EQUATIONS

ELLIS HORWOOD SERIES IN MATHEMATICS AND ITS APPLICATIONS

Series Editor: Professor G. M. BELL, Chelsea College, University of London

Statistics and Operational Research
Editor: B. W. CONOLLY, Chelsea College, University of London

Baldock, G. R. & Bridgeman, T.	Mathematical Theory of Wave Motion
de Barra, G.	Measure Theory and Integration
Berry, J. S., Burghes, D. N., Huntley, I. D., James, D. J. G. & Moscardini, A. O.	Teaching and Applying Mathematical Modelling
Burghes, D. N. & Borrie, M.	Modelling with Differential Equations
Burghes, D. N. & Downs, A. M.	Modern Introduction to Classical Mechanics and Control
Burghes, D. N. & Graham, A.	Introduction to Control Theory, including Optimal Control
Burghes, D. N., Huntley, I. & McDonald, J.	Applying Mathematics
Burghes, D. N. & Wood, A. D.	Mathematical Models in the Social, Management and Life Sciences
Butkovskiy, A. G.	Green's Functions and Transfer Functions Handbook
Butkovskiy, A. G.	Structure of Distributed Systems
Chorlton, F.	Textbook of Dynamics, 2nd Edition
Chorlton, F.	Vector and Tensor Methods
Dunning-Davies, J.	Mathematical Methods for Mathematicians, Physical Scientists and Engineers
Eason, G., Coles, C. W. & Gettinby, G.	Mathematics and Statistics for the Bio-sciences
Exton, H.	Handbook of Hypergeometric Integrals
Exton, H.	Multiple Hypergeometric Functions and Applications
Exton, H.	q-Hypergeometric Functions and Applications
Faux, I. D. & Pratt, M. J.	Computational Geometry for Design and Manufacture
Firby, P. A. & Gardiner, C. F.	Surface Topology
Gardiner, C. F.	Modern Algebra
Gasson, P. C.	Geometry of Spatial Forms
Goodbody, A. M.	Cartesian Tensors
Goult, R. J.	Applied Linear Algebra
Graham, A.	Kronecker Products and Matrix Calculus: with Applications
Graham, A.	Matrix Theory and Applications for Engineers and Mathematicians
Griffel, D. H.	Applied Functional Analysis
Hanyga, A.	Mathematical Theory of Non-linear Elasticity
Hoskins, R. F.	Generalised Functions
Hunter, S. C.	Mechanics of Continuous Media, 2nd (Revised) Edition
Huntley, I. & Johnson, R. M.	Linear and Nonlinear Differential Equations
Jaswon, M. A. & Rose, M. A.	Crystal Symmetry: The Theory of Colour Crystallography
Johnson, R. M.	Linear Differential Equations and Difference Equations: A Systems Approach
Kim, K. H. & Roush, F. W.	Applied Abstract Algebra
Kosinski, W.	Field Singularities and Wave Analysis in Continuum Mechanics
Marichev, O. I.	Integral Transforms of Higher Transcendental Functions
Meek, B. L. & Fairthorne, S.	Using Computers
Muller-Pfeiffer, E.	Spectral Theory of Ordinary Differential Operators
Nonweiler, T. R. F.	Computational Mathematics: An Introduction to Numerical Analysis
Oldknow, A. & Smith, D.	Learning Mathematics with Micros
Ogden, R. W.	Non-linear Elastic Deformations
Rankin, R.	Modular Forms
Ratschek, H. & Rokne, Jon	Computer Methods for the Range of Functions
Scorer, R. S.	Environmental Aerodynamics
Smith, D. K.	Network Optimisation Practice: A Computational Guide
Srivastava, H. M. & Karlsson, P. W.	Multiple Gaussian Hypergeometric Series
Srivastava, H. M. & Manocha, H. L.	A Treatise on Generating Functions
Sweet, M. V.	Algebra, Geometry and Trigonometry in Science, Engineering and Mathematics
Temperley, H. N. V. & Trevena, D. H.	Liquids and Their Properties
Temperley, H. N. V.	Graph Theory and Applications
Thom, R.	Mathematical Models of Morphogenesis
Thomas, L. C.	Games Theory and Applications
Townend, M. Stewart	Mathematics in Sport
Twizell, E. H.	Computational Methods for Partial Differential Equations
Wheeler, R. F.	Rethinking Mathematical Concepts
Willmore, T. J.	Total Curvature in Riemannian Geometry
Willmore, T. J. & Hitchin, N.	Global Riemannian Geometry

COMPUTATIONAL METHODS FOR PARTIAL DIFFERENTIAL EQUATIONS

E. H. TWIZELL, B.Sc., M.A., Ph.D.
Lecturer in Numerical Analysis
Department of Mathematics and Statistics, Brunel University

ELLIS HORWOOD LIMITED
Publishers · Chichester

Halsted Press: a division of
JOHN WILEY & SONS
New York · Chichester · Brisbane · Toronto

First published in 1984 by
ELLIS HORWOOD LIMITED
Market Cross House, Cooper Street, Chichester, West Sussex, England PO19 1EB

The publisher's colophon is reproduced from James Gillison's drawing of the ancient Market Cross, Chichester.

Distributors:
Australia, New Zealand, South-east Asia:
Jacaranda-Wiley Ltd., Jacaranda Press,
JOHN WILEY & SONS INC.,
G.P.O. Box 859, Brisbane, Queensland 4001, Australia

Canada:
JOHN WILEY & SONS CANADA LIMITED
22 Worcester Road, Rexdale, Ontario, Canada.

Europe, Africa:
JOHN WILEY & SONS LIMITED
Baffins Lane, Chichester, West Sussex, England.

North and South America and the rest of the world:
Halsted Press: a division of
JOHN WILEY & SONS
605 Third Avenue, New York, N.Y. 10016, U.S.A.

© 1984 E. H. Twizell/Ellis Horwood Limited

British Library Cataloguing in Publication Data
Twizell, E. H.
Computational methods for partial differential equations. —
(Ellis Horwood series in mathematics and its applications)
1. Differential equations, Partial — Numerical solutions
I. Title
515.3'53 QA374
Library of Congress Card No. 84-19752

ISBN 0-85312-383-7 (Ellis Horwood Limited)
ISBN 0-470-27511-1 (Halsted Press)

Typeset by Ellis Horwood Limited.
Printed in Great Britain by Unwin Brothers of Woking.

COPYRIGHT NOTICE —
All Rights Reserved. No part of this publication may be reproduced, stored in a retrieval system, or transmitted, in any form or by any means, electronic, mechanical, photocopying, recording or otherwise, without the permission of Ellis Horwood Limited, Market Cross House, Cooper, Street, Chichester, West Sussex, England PO19 1EB

Table of Contents

Preface 9

Chapter 1 Introductory Mathematics
    1.1   Introductory matrix algebra 11
    1.2   Solution of linear algebraic systems. 17
    1.3   Solution of non-linear equations. 25
    1.4   Lagrangian interpolation. 28
    1.5   Some expansions. 31
    1.6   Classification of second order partial differential equations 37

Chapter 2 Elliptic Equations: Finite Difference Methods
    2.1   The Dirichlet, Neumann and Robbins problems. 42
    2.2   Alternating direction implicit (ADI) methods 54
    2.3   Irregular boundaries; an elastic membrane problem 61
    2.4   Nonrectangular coordinates. 67

Chapter 3 Elliptic Equations: Finite Element Methods
    3.1   Introduction. 81
    3.2   Fundamental concepts 83
    3.3   Variational formulations. 87
    3.4   Methods of finite element solution 91
    3.5   Basis functions for different elements 93
    3.6   Convergence and error analysis. 101
    3.7   The Dirichlet problem 101
    3.8   Rational basis functions 107

Chapter 4 Hyperbolic Equations
    4.1   The method of characteristics for second order equations 116
    4.2   A specimen problem 122
    4.3   Propagation of discontinuities in initial conditions. 126

	4.4	Finite element solutions 128
	4.5	Low order finite difference methods 132
	4.6	Analyses of finite difference methods 139
	4.7	Higher order methods 149
	4.8	Two space variables 154
	4.9	First order equations 159
	4.10	The role of characteristics 161
	4.11	Central difference methods for first order equations 166
	4.12	Backward difference methods for first order equations 176

Chapter 5 Parabolic Equations

	5.1	Introduction 200
	5.2	Low order finite difference methods 202
	5.3	Analyses of the methods 207
	5.4	Extrapolations 216
	5.5	Extension to two-space dimensions 220
	5.6	Higher order approximations in time 227
	5.7	Non-constant coefficients 248
	5.8	Diffusion-convection equations 252

Bibliography and References 266

Index 273

To my wife Anne
and our son Linus

Preface

This book has been written for undergraduate and graduate students in mathematics who are embarking on a first course in the numerical solution of partial differential equations. The book is also intended for undergraduate and graduate students and research workers in the physical and biomedical sciences, and in the engineering disciplines, who need to acquire knowledge of computational techniques for solving the partial differential equations which arise in the mathematical modelling of physical, biomedical and engineering systems.

It is not assumed that the reader has previous knowledge of the theory of partial differential equations, though first courses in calculus and matrix algebra are prerequisites. It is assumed that the reader has access to a high-speed computer and can write programs in a high level scientific language.

The book begins with a chapter devoted to the revision of topics from undergraduate calculus and matrix algebra syllabuses which are relevant to the following chapters. Emphasis throughout the book is on the numerical solution of linear partial differential equations by finite difference methods. Time dependent problem are transformed into systems of ordinary differential equations before computing the solution and the concepts of A-stability and L-stability are discussed. Hyperbolic equtions are also solved by characteristic methods and there is a background chapter on finite element methods for elliptic equations.

I would like to thank Mrs P. Denham, Mrs B. Yates and Mrs R. E. Clarke of Brunel University for typing the manuscript so professionally and Mrs V. R. Kessler also of Brunel University for help in the general preparation of the typescript. It gives me pleasure to acknowledge the contributions made by a number of my students, in particular Dr P. Smith, Dr A. Q. M. Khaliq, Dr S. I. A. Tirmizi, Miss Isla E. Finlayson and Messrs P. J. Aston, N. J. Dickens and R. J. Harvey. I am grateful to the Institute of Mathematics and Its Applications and to SIAM Publications for permission to use copyright material, and to my colleague Mr G. D. Smith, whose own successful book I have used for many years, for his useful comments on the chapter on parabolic equations.

E. H. Twizell
March 1984

1

Introductory Mathematics

1.1 INTRODUCTION MATRIX ALGEBRA

1.1.1 Some definitions

In the numerical solution of partial differential equations, much of the theory requires the reader to have a knowledge of elementary matrix algebra. In particular, the eigenvalues of a matrix play an important role in establishing many theories in the analysis of computational methods for partial differential equations. The reader will see many times that the computed solution of a linear partial difference equation is obtained by solving a linear system of the form

$$A\mathbf{u} = \mathbf{b}, \tag{1.1}$$

where A is a square matrix of order N; that is, the matrix has N rows and N columns. The vectors $\mathbf{u}$ and $\mathbf{b}$ each have N elements or components; the vector $\mathbf{u}$ is the vector of unknown functional values which is to be determined, and the vector $\mathbf{b}$ is known.

The solution of (1.1), which may be written in the form

$$\mathbf{u} = A^{-1}\mathbf{b}, \tag{1.2}$$

is known to be unique whenever the inverse matrix A^{-1} exists, for then A^{-1} is unique. The solution $\mathbf{u}$ is therefore unique provided the determinant of A does not vanish, that is, provided the matrix A is *non-singular*.

It is clear that solving the linear system (1.1) is equivalent to inverting the matrix A. Such linear systems associated with partial differential equations generally have orders N which make them too large for solution by the most elementary method of finding A^{-1}, in which the adjoint matrix is formed. Consequently, a numerical method is usually chosen; the choice of method depends on the form or structure of the matrix A and is particularly important when A is large.

The success of many computational methods for solving linear systems of the form (1.1) depends on the eigenvalues of a matrix being sufficiently small.

1.1.2 Eigenvalues and eigenvectors

Let **x** be a non-zero column vector with N elements such that, for the square matrix A of order N, $A\mathbf{x} = \lambda \mathbf{x}$ for some scalar λ. Then λ is said to be an *eigenvalue* of A and **x** the associated *eigenvector*. The *eigenvalue problem* is to determine all possible eigenvalues and eigenvectors which satisfy this relation. Clearly the eigenvector **x** corresponding to the eigenvalue λ is not unique as any scalar multiple of **x** will satisfy the relation with the same λ.

It is a simple matter to derive an equation from which λ may be determined. The relation $A\mathbf{x} = \lambda \mathbf{x}$ may be written in the form

$$(A - \lambda I)\mathbf{x} = \mathbf{0}, \qquad (1.3)$$

where **0** is the *zero vector* of order N, all the elements of which are zero, and a necessary and sufficient condition for (1.3) to have a non-trivial solution is given by

$$f(\lambda) \equiv \det(A - \lambda I) = 0. \qquad (1.4)$$

Equation (1.4) is the *characteristic equation* of A and $f(\lambda)$ itself is called the characteristic polynomial of A. This polynomial is of degree N in λ and thus has N roots λ_i, with $i = 1, \ldots, N$, which may or may not be distinct or real; the eigenvalues of A, which in older texts are sometimes called the *characteristic roots* of the matrix, may or may not be distinct or real. In the case when A has real elements, complex eigenvalues occur in complex conjugate pairs.

Three very useful theorems, relating to the eigenvalues of a matrix A, in the numerical solution of partial differential equations, are as follows:

Theorem 1.1 (The *Cayley-Hamilton Theorem*) Every square matrix satisfies its own characteristic equation.

Theorem 1.2 (*Geršgorin's First Theorem*) The eigenvalue of the square matrix A which has largest modulus, cannot exceed the largest sum of the moduli of the elements along any row or down any column of A.

Theorem 1.3 (*Geršgorin's Circle Theorem* or *Brauer's Theorem*) Every eigenvalue of the matrix A lies inside at least one of the discs in the complex plane with centre a_{ii} and radius

$$r_i = \sum_{j \neq i} |a_{ij}|, \, i = 1, \ldots, N.$$

Sec. 1.1]   Introduction Matrix Algebra   13

The *spectral radius* of a square matrix A, denoted by $\rho(A)$, is given by

$$\rho(A) = \max_{i} |\lambda_i|$$

where $\lambda_i (i = 1, \ldots, N)$ are the eigenvalues of A. The spectral radius is the radius of the smallest disc in the complex plane, with centre at the origin, which contains all the eigenvalues of A. It follows from Theorems 1.2, and 1.3 that

$$\rho(A) \leqslant \min \left[\max_{i} \sum_{j=1}^{N} |a_{ij}|, \max_{j} \sum_{i=1}^{N} |a_{ij}| \right]. \quad (1.5)$$

1.1.3 Norms

Let A be a square matrix of order N; a *norm* of A is a non-negative real number denoted by $\|A\|$ which satisfies the conditions

(i) $\|A\| \geqslant 0$ with equality only when A is the null matrix,   (1.6)
(ii) $\|cA\| = |c| \|A\|$ for any complex scalar c,   (1.7)
(iii) $\|A+B\| \leqslant \|A\| + \|B\|$ (the *triangle inequality*),   (1.8)
(iv) $\|AB\| \leqslant \|A\| \|B\|$ (the *Schwarz inequality*)   (1.9)

where B is also a square matrix of order N. The equivalent properties of a vector norm will be discussed in Chapter 3.

In its broadest sense, a norm serves as a measure of the 'size' or 'magnitude' of a given matrix. Some of the important norms in matrix linear algebra are defined as follows:

(i) The *Euclidean norm*, denoted by $\|A\|_E$, is given by

$$\|A\|_E = \left[\sum_{i=1}^{N} \sum_{j=1}^{N} |a_{ij}|^2 \right]^{1/2} \quad (1.10)$$

(ii) The *spectral norm* or the L_2 *norm*, denoted by $\|A\|_S$ or $\|A\|_2$ is given by

$$\|A\|_S = \|A\|_2 = \max_{r} [|\lambda_r(\bar{A}^T A)|]^{1/2}. \quad (1.11)$$

(The elements of the matrix $\bar{A}$ are the complex conjugates of those of A.) This norm is arguably the most important in analyzing numerical methods for the solution of partial differential equations and, in keeping with many other texts, will be denoted by $\|A\|$, that is, with no subscript. When A is symmetric $\rho(A) = \|A\|$ and when A is non-symmetric $\rho(A) \leqslant \|A\|$.

(iii) The L_1 *norm*, denoted by $\|A\|_1$, is given by

$$\|A\|_1 = \max_{j} \sum_{i=1}^{N} |a_{ij}|. \quad (1.12)$$

(iv) The L_∞ norm, denoted by $\|A\|_\infty$, is given by

$$\|A\|_\infty = \max_i \sum_{j=1}^{N} |a_{ij}|. \qquad (1.13)$$

The L_1 norm is clearly the maximum sum of the moduli of the elements in the individual columns of the matrix A and the L_∞ norm is the maximum sum of the moduli of the elements in the individual rows.

1.1.4 Computation of eigenvalues and eigenvectors

In the numerical solution of partial differential equations, the solution is often found by solving a linear system, the coefficient matrix of which is large and sparse. The analysis of the method being used usually involves, at some stage, knowledge of the eigenvalues of the matrix and, in particular, success of the numerical method often requires the eigenvalue with greater modulus not to exceed a certain value.

One numerical method for finding the maximum modulus eigenvalue is the so-called *Power Method*. This method starts with an initial approximation $\mathbf{x}^{(0)} \neq \mathbf{0}$ to the eigenvector associated with the (unknown) eigenvalue of largest modulus, and, after a series of iterations, converges to give the largest modulus eigenvalue and its associated eigenvector. Suppose that the element of largest modulus of A is in row i and column j of the matrix, then two common choices of $\mathbf{x}^{(0)}$ are the zero vector with the zero in position i, or the zero in position j, replaced by $+1$. A third common choice is to let $\mathbf{x}^{(0)} = (1,1,\ldots,1)^T$.

An algorithm for the method is as follows:

1. choose $\mathbf{x}^{(0)}$, the tolerances ϵ_λ and ϵ_x, and set $r = 0$;
2. compute $\mathbf{y}^{(1)} = A\mathbf{x}^{(0)}$ and let $\lambda^{(1)}$ be the element of $\mathbf{y}^{(1)}$ of largest modulus (note: $\lambda^{(1)}$ is *not* the modulus, it is the algebraic value of this element);
3. let $\mathbf{x}^{(1)} = \mathbf{y}^{(1)}/\lambda^{(1)}$ be the normalised form of $\mathbf{y}^{(1)}$ (the element of $\mathbf{x}^{(1)}$ of greatest magnitude is now $+1$), then $\lambda^{(1)}$ is a first approximation to the algebraic value of the eigenvalue of largest modulus and $\mathbf{x}^{(1)}$ is the associated eigenvector;
4. let $r = r + 1$;
5. compute $\mathbf{y}^{(r+1)}$ from the relation $\mathbf{y}^{(r+1)} = A\mathbf{x}^{(r)}$ and let $\lambda^{(r+1)}$ be the algebraic value of the element of $\mathbf{y}^{(r+1)}$ having greatest modulus;
6. compute $\mathbf{x}^{(r+1)} = \mathbf{y}^{(r+1)}/\lambda^{(r+1)}$;
7. compare corresponding elements of $\mathbf{x}^{(r)}$, $\mathbf{x}^{(r+1)}$ and $\lambda^{(r)}$, $\lambda^{(r+1)}$. If $|x_m^{(r+1)} - x_m^{(r)}| < \epsilon_x$ for *all* elements $m = 1,\ldots,N$, and $|\lambda^{(r+1)} - \lambda^{(r)}| < \epsilon_\lambda$, the algorithm has converged and the process is terminated, otherwise return to step 4.

Following termination of the algorithm the final value of λ and the final vector $\mathbf{x}$ are, respectively, the eigenvalue of largest modulus and the associated eigenvector, to the prescribed tolerance ϵ. Provided the linear system of algebraic equations associated with the matrix A is not ill-conditioned, the power method will converge to the desired results.

Apart from the case of large, sparse matrices, the power method has limited appeal, more modern and more sophisticated methods being preferred. The method is believed to be the first iterative method for solving the numerical eigenvalue problem. The many other numerical methods for finding the eigenvalues of a matrix are described in numerous specialist texts (Forsythe and Moler (1967) and Goult *et al.* (1974)) and in general numerical analysis texts (Gerald and Wheatley (1984), Burden *et al.* (1981), and Johnson and Riess (1982)).

As a numerical example, consider the matrix

$$A = \begin{bmatrix} -5 & 1 \\ 30 & -4 \end{bmatrix}$$

and find the largest modulus eigenvalue and the corresponding eigenvector, correct to two decimal places.

Table 1.1

r	$y_1^{(r)}$	$y_2^{(r)}$	$\lambda^{(r)}$	$x_1^{(r)}$	$x_2^{(r)}$
0	—	—	—	1.000	0.000
1	−5.000	30.000	30.000	−0.167	1.000
2	1.835	−9.010	−9.010	−0.204	1.000
3	2.020	−10.120	−10.120	−0.200	1.000
4	2.000	−10.000	−10.000	−0.200	1.000
5	2.000	−10.000	−10.000		

Taking $\mathbf{x}^{(0)} = (1,0)^T$, the computation proceeds as in Table 1.1. The largest modulus eigenvalue is therefore -10.00 to two decimal places and the corresponding eigenvector is $(-0.20, 1.00)^T$. The reader will notice that a better choice of initial vector would probably have been $\mathbf{x}^{(0)} = (0,1)^T$; this is because for $r = 1, 2, \ldots$ the element of largest modulus in $\mathbf{x}^{(r)}$ occurred as the second element.

It is easy to verify that the characteristic equation is $f(\lambda) \equiv \lambda^2 + 9\lambda - 10 = 0$ and the eigenvalues of A are $\lambda = -10, \lambda = +1$.

1.1.5 Stiffness

Significant computational difficulties can occur when solving a system of N first order ordinary differential equations of the linear form

$$\mathbf{y}'(t) = A\mathbf{y}(t) + \boldsymbol{\phi}(t) \; ; \mathbf{y}(t_0) = \mathbf{y}_0, \tag{1.14}$$

when the real parts of the N eigenvalues of the matrix A are negative and widely varying in magnitude. Such problems will be met in Chapter 5.

Assuming that the initial function $\mathbf{y}_0$ may be written as a linear combination of the normalized eigenvectors $\mathbf{x}_i (i=1, \ldots, N)$ of the matrix A, say

$$\mathbf{y}_0 = \sum_{i=1}^{N} \alpha_i \mathbf{x}_i,$$

then the theoretical solution of (1.14) at time $t = nl$ may be written in the form

$$\mathbf{y}(t) = \sum_{i=1}^{N} \alpha_i \{\exp(l\lambda_i)\}^n \mathbf{x}_i + \psi(t)$$

where l is some convenient time step, n is an integer, $\lambda_i (i=1, \ldots, N)$ is an eigenvalue of A, and ψ depends on ϕ; the first part of $\mathbf{y}(t)$ is called the *transient* part of the solution and $\psi(t)$ the *steady state* part. It is clear that if $\text{Re}(\lambda_i) < 0$, $\mathbf{y}(t) \to \psi(t)$ as $t, n \to \infty$. In problems where $\text{Re}(\lambda_i) < 0$, and are widely varying in magnitude, some components of the transient part of the solution decay rapidly while others decay slowly, and unless extreme care is taken, the round-off error associated with the transient terms can swamp the solution and produce useless results.

This is the problem of stiffness and (1.14) is called a *stiff system*. A formal definition is as follows:

Definition 1.1 The linear system (1.14) is said to be *stiff* if the eigenvalues $\lambda_r (r=1, \ldots, N)$ of the matrix A satisfy

(i) $\text{Re}(\lambda_r) < 0$ for all $r = 1, 2, \ldots, N$;

(ii) $\max_r \{|\text{Re}(\lambda_r)|\} \gg \min_r \{|\text{Re}(\lambda_r)|\}$.

Definition 1.2 The ratio $\max_r \{|\text{Re}(\lambda_r)|\} / \min_r \{|\text{Re}(\lambda_r)|\}$ is called the *stiffness ratio*.

Definition 1.3 The non-linear system $y' = \mathbf{f}(t, \mathbf{y})$; $\mathbf{y}(t_0) = \mathbf{y}_0$ is aid to be *stiff in an interval I of t* if, for $t \in I$, the eigenvalues $\lambda_r(t), (r=1, 2, \ldots, N$ where N is the number of equations in the system) of the Jacobian $\partial \mathbf{f} / \partial \mathbf{y}$ satisfy

(i) $\text{Re}[\lambda_r(t)] < 0$ for all $r = 1, 2, \ldots, N$;

(ii) $\max_r \{|\text{Re}[\lambda_r(t)]|\} \gg \min_r \{|\text{Re}[\lambda_r(t)]|\}$.

Clearly the stiffness ratio of a non-linear system varies with t.

1.2 SOLUTION OF LINEAR ALGEBRAIC SYSTEMS

1.2.1 Direct methods

Direct methods of solving a linear system of algebraic equations, written in matrix form as

$$A\mathbf{x} = \mathbf{b} \tag{1.15}$$

where $A = [a_{ij}]$ $(i,j = 1, \ldots, N)$ is a square matrix of order N, and $\mathbf{x} = (x_1, x_2, \ldots, x_N)^T$ and $\mathbf{b} = (b_1, b_2, \ldots, b_N)^T$ are vectors with N components, are those methods for which one application of an elimination or decomposition process leads to the numerical solution.

Direct methods are very reliable, though for some of the simpler problems in the numerical solution of partial differential equations their use is not recommended. The reason for this is quite simple, as will be seen in later chapters: the matrix A may well be *sparse*, that is most of its elements are zero, and a direct method of solution may produce *fill* with up to half of its elements becoming non-zero.

The most commonly used direct methods are described and analyzed in most general numerical analysis texts, see, for example, Gerald and Wheatley (1984), Burden *et al.* (1981), Johnson and Riess (1982) and Morris (1983). These methods include the method of *Gaussian elimination* (sometimes referred to as the method of *pivotal condensation*) and the *Doolittle, Crout* and *Choleski decomposition* methods, the last being used only for systems of the form (1.15) in which A is symmetric.

All of these direct methods are expensive to use: the Gauss, Doolittle and Grout methods each involve $\frac{1}{3}N^3 + N^2 - \frac{1}{3}N$ multiplications or divisions, where N is the order of A, and $\frac{1}{3}N^3 + \frac{1}{2}N^2 - \frac{5}{6}N$ additions or subtractions. The Choleski method for symmetric coefficient matrices is slightly less expensive requiring $\frac{1}{6}N^3 + \frac{3}{2}N^2 + \frac{1}{3}N$ multiplications or divisions and $\frac{1}{6}N^3 + \frac{7}{6}N^2 - N$ additions or subtractions; it also requires the computation of N square roots.

1.2.2 Iterative methods

When the linear system is small enough to be economically stored in the memory of a computer, it is generally most efficient to use a direct method to obtain the solution. Large sparse systems, on the other hand, which are frequently associated with the numerical solution of partial differential equations, can generally be solved more efficiently use an *iterative method*, provided that full advantage is taken of the sparseness of the coefficient matrix.

An iterative technique to solve a linear system of the form (1.15) starts with an initial approximation $\mathbf{x}^{(0)}$ to the solution $\mathbf{x}$ and generates a sequence of vectors $\mathbf{x}^{(0)}, \mathbf{x}^{(1)}, \mathbf{x}^{(2)}, \ldots$ which converges to $\mathbf{x}$. Such iterative methods convert the given system $A\mathbf{x} = \mathbf{b}$ into an equivalent system of the form $\mathbf{x} = B\mathbf{x} + \mathbf{c}$ where B is a square matrix of order N and $\mathbf{c}$ is a vector with N components.

The choice of $\mathbf{x}^{(0)}$ is largely arbitrary (the zero vector is a common choice) and the sequence of iterates is generated from

$$\mathbf{x}^{(k+1)} = B\mathbf{x}^{(k)} + \mathbf{c}; k = 0, 1, 2, \ldots . \qquad (1.16)$$

A sufficient condition for the convergence of (1.16) is

$$\|B\| < 1. \qquad (1.17)$$

The iteration procedure is continued until some criterion for stopping is satisfied; two such criteria are

$$\|\mathbf{x}^{(k+1)} - \mathbf{x}^{(k)}\| < \epsilon \text{ for some } k \qquad (1.18)$$

and

$$\frac{\|\mathbf{x}^{(k+1)} - \mathbf{x}^{(k)}\|}{\|\mathbf{x}^{(k)}\|} < \epsilon \qquad (1.19)$$

where $\epsilon > 0$ is some *tolerance*. The usual norm used is the L_∞ norm for vectors given by

$$\|\mathbf{x}\| = \max_r |x_r|; r = 1, 2, \ldots, N \qquad (1.20)$$

The absolute error criterion (1.18) is suitable for use when the variables $x_i (i=1, \ldots, N)$ are all of the same order of magnitude, while the relative error criterion (1.19) is suitable for use when the x_i have widely differing orders of magnitude.

The two iteration methods commonly used are the *Jacobi method* and the *Gauss–Seidel method*. In the Gauss-Seidel method an iterate is used as soon as it is calculated whereas in the Jacobi method a new set of iterates $x_m^{(k+1)} (m=1, \ldots, N; k = 0, 1, 2, \ldots)$ is not used until the iteration procedure has completed its sweep through the system.

A general comparison of the two methods produces the following observations.

(i) the Gauss-Seidel method invariably converges when the Jacobi method does,
(ii) the Gauss-Seidel method may converge when the Jacobi method diverges,
(iii) generally, when both methods converge, the Gauss-Seidel method converges more quickly than the Jacobi method.

These observations are no more than general guidelines; in fact there are linear systems for which the Jacobi method converges while the Gauss-Seidel method does not (see Varga, 1962; p. 74).

Acceleration of converge can usually be obtained by the introduction of *acceleration parameters* to the right-hand side of (1.16) which takes the form

$$\mathbf{x}^{(k+1)} = (D+\omega L)^{-1} [(1-\omega)D - \omega U]\mathbf{x}^{(k)} + \omega(D+\omega L)^{-1} \mathbf{b}. \qquad (1.21)$$

Equation (1.21) describes the method of *successive over-relaxation* (SOR). In (1.21), ω is the acceleration parameter and D, L, U are, respectively, diagonal,

Solution of Linear Algebraic Systems

lower-triangular and upper-triangular matrices such that $D + L + U = A$. The parameter ω generally lies in the interval $1 < \omega < 2$; when $\omega = 1$, (1.21) is the Gauss-Siedel method.

For more details of iterative and SOR methods, descriptions of algorithms for their implementation and program listings, the reader is referred again to the texts by Gerald and Wheatley (1984), Burden et al. (1981), Johnson and Riess (1982) and Morris (1983).

When the matrix of coefficients in (1.15) is symmetric, the method of *conjugate gradients* may be used to determine $\mathbf{x}$. This method, which uses the best features of both direct and iterative methods needs a *residual vector* $\mathbf{r} = \mathbf{b} - A\mathbf{x}$ and a scalar quantity $s = \mathbf{r}^T\mathbf{r}$ which is the sum of the squares of the components of $\mathbf{r}$. The aim of the method is clearly to minimise s.

The iterates $\mathbf{x}^{(k)}$ ($k = 0, 1, 2, \ldots$) are determined from

$$\mathbf{x}^{(k+1)} = \mathbf{x}^{(k)} + \alpha_k \mathbf{w}^{(k)}$$

where α_k is a scalar and $\mathbf{w}^{(k)}$ is a vector to be determined. The scalar α_k is chosen to make $s_{k+1} = \mathbf{r}^{(k)T}\mathbf{r}^{(k)}$ a minimum, in which case

$$\alpha = \mathbf{w}^{(k)T}\mathbf{r}^{(k)}/\mathbf{w}^{(k)T}A\mathbf{w}^{(k)}$$

(see, for example, Greig (1980)). In determining $\mathbf{w}^{(k)}$, the combination of $\mathbf{r}^{(k)}$ and $\mathbf{w}^{(k-1)}$ given by

$$\mathbf{w}^{(k)} = \mathbf{r}^{(k)} + \beta_{k-1}\mathbf{w}^{(k-1)}$$

is chosen. The successive constants β_k can be chosen so that the process terminates in exactly N steps. To this end, $\mathbf{w}^{(k)}$ is made *conjugate* (that is orthogonal with respect to the matrix A) to $\mathbf{w}^{(k-1)}$, that is $\mathbf{w}^{(k)T}A\mathbf{w}^{(k-1)} = 0$. The corresponding value of β is thus given by

$$\beta_{k-1} = -\mathbf{r}^{(k)T}A\mathbf{w}^{(k-1)}/\mathbf{w}^{(k-1)T}A\mathbf{w}^{(k-1)}.$$

The full computational algorithm is given by

$$\beta_{k-1} = -\mathbf{r}^{(k)T}A\mathbf{w}^{(k-1)}/\mathbf{w}^{(k-1)T}A\mathbf{w}^{(k-1)},$$

$$\mathbf{w}^{(k)} = \mathbf{r}^{(k)} + \beta_{k-1}\mathbf{w}^{(k-1)},$$

$$\alpha_k = \mathbf{r}^{(k)T}\mathbf{w}^{(k)}/\mathbf{w}^{(k)T}A\mathbf{w}^{(k)},$$

$$\mathbf{x}^{(k+1)} = \mathbf{x}^{(k)} + \alpha_k\mathbf{w}^{(k)},$$

$$\mathbf{r}^{(k+1)} = \mathbf{b} - A\mathbf{x}^{(k+1)} = \mathbf{r}^{(k)} - \alpha_k A\mathbf{w}^{(k)}$$

with $\mathbf{x}^{(0)}$ arbitrary and $\mathbf{w}^{(-1)} = \mathbf{0}$ so that $\mathbf{w}^{(0)} = \mathbf{r}^{(0)}$.

It is well beyond the intended scope of the present text to undergo a study of the advanced numerical methods for solving equations of the form (1.15). Such methods include band and envelope methods, the minimum degree ordering algorithm for use when A is sparse, quotient tree methods for finite difference and finite element problems, one-way dissection methods for finite element problems, and nested dissection methods. For a detailed study of these methods the reader is referred to an advanced text, such as that by George and Liu (1981) which contains a number of FORTRAN subroutines.

1.2.3 Special forms of the coefficient matrix

Frequently in the numerical solution of partial differential equations in one space variable, A takes the form of the tridiagonal matrix

$$M = \begin{bmatrix} d_1 & e_1 & & & & 0 \\ c_2 & d_2 & e_2 & & & \\ & \ddots & \ddots & \ddots & & \\ 0 & & & c_{N-1} & d_{N-1} & e_{N-1} \\ & & & & c_N & d_N \end{bmatrix} ; \qquad (1.22)$$

indeed, it usually occurs that the c_i are all equal to each other as are the d_i and the e_i.

The LU decomposition of such a matrix may be carried out by decomposing M, which is usually positive definite, into

$$M = LU = \begin{bmatrix} 1 & & & & 0 \\ l_2 & 1 & & & \\ & l_3 & 1 & & \\ & & \ddots & \ddots & \\ 0 & & & l_N & 1 \end{bmatrix} \begin{bmatrix} u_1 & e_1 & & & 0 \\ & u_2 & e_2 & & \\ & & \ddots & \ddots & \\ & & & & e_{N-1} \\ 0 & & & & u_N \end{bmatrix} \qquad (1.23)$$

and using the following algorithm which is based on the Doolittle method:

(i) let $u_1 = d_1$;
(ii) let $l_i = c_i/u_{i-1}$ and then $u_i = d_i - l_i e_{i-1}$ for $i = 2, \ldots, N$.

The system (1.15) can now be written as the pair of equations

$$L\mathbf{y} = \mathbf{b}, \qquad (1.24)$$

$$U\mathbf{x} = \mathbf{y}. \qquad (1.25)$$

The intermediate vector $\mathbf{y}$ is found by the method of forward substitution from (1.24) giving

$$y_1 = b_1; y_i = b_i - l_i y_{i-1}, i = 2, \ldots, N. \qquad (1.26)$$

Sec. 1.2] Solution of Linear Algebraic Systems 21

The vector **y** is then used in (1.25) from which **x** is found by back substitution giving

$$x_N = y_N/u_N; \quad x_i = (y_i - e_i x_{i+1})/u_i, \quad i = N-1, \ldots, 2, 1. \quad (1.27)$$

This algorithm requires only $5N-4$ multiplications or divisions and $3N-3$ additions or subtractions and obviously has a considerable advantage in CPU time over the full elimination or decomposition methods of Gauss, Doolittle and Crout which do not take into account the tridiagonality of M in (1.22). This advantage is particularly marked for large N.

In the numerical solution of partial differential equations in two space variables (see, for instance, section 5.5 of Chapter 5 where the Lawson-Morris and Peaceman-Rachford methods are described), the solution is found by solving a linear system of the form (1.15) where, now, the matrix A is of order N^2 and takes one of the forms

$$A = \begin{bmatrix} M & & & 0 \\ & M & & \\ & & \ddots & \\ 0 & & & M \end{bmatrix} \quad (1.28)$$

or

$$A = \begin{bmatrix} D & E & & & & 0 \\ C & D & E & & & \\ & & \ddots & \ddots & \ddots & \\ & & & & C & D & E \\ 0 & & & & & C & D \end{bmatrix} \quad (1.29)$$

In (1.28), A is block diagonal with tridiagonal blocks, M being the tridiagonal matrix of order N given by (1.22). The solution of (1.15) with A given by (1.28) and (1.22) is easily obtained by adapting the decomposition algorithm described by equations (1.23) through (1.27), the vectors **x** and **b** now being of order N^2 also. The adapted algorithm is as follows

(i) let $u_1 = d_1$;
(ii) let $l_i = c_i/u_{i-1}$ and then $u_i = d_i - l_i e_{i-1}$ for $i = 2, \ldots, N$;
(iii) for each $k = 0, 1, \ldots, N-1$ let

$$y_{kN+1} = b_{kN+1} \text{ and } y_{kN+i} = b_{kN+i} - l_i y_{kN+i-1} \text{ for } i = 2, \ldots, N;$$

(iv) for each $k = N-1, \ldots, 1, 0$ compute

$$x_{(k+1)N} = y_{(k+1)N}/u_{(k+1)N} \text{ and } x_{kN+i} = (y_{kN+i} - e_i y_{kN+i+1})/u_i,$$
$$i = N-1, \ldots, 2, 1$$

In (1.29), A is block tridiagonal with diagonal blocks. The diagonal matrices C, D, E are each of order N and usually take the forms

$$C = \begin{bmatrix} c & & & 0 \\ & c & & \\ & & \ddots & \\ 0 & & & c \end{bmatrix}, \quad D = \begin{bmatrix} d & & & 0 \\ & d & & \\ & & \ddots & \\ 0 & & & d \end{bmatrix}, \quad E = \begin{bmatrix} e & & & 0 \\ & e & & \\ & & \ddots & \\ 0 & & & e \end{bmatrix}$$

(1.30)

where $c \neq 0$, $d \neq 0$, $e \neq 0$. The tridiagonal solver described by (1.23) to (1.27) is adapted as follows:

(i) let $u_1 = d$;
(ii) let $l_i = c/u_{i-1}$ and $u_i = d - l_i e$ for $i = 2, \ldots, N$;
(iii) for $j = 2, 3, \ldots, N$ let

$$y_{(j-1)N+i} = b_{(j-1)N+i} - l_j b_{(j-2)N+i} \text{ for each } i = 1, 2, \ldots, N;$$

(iv) for $j = N^2, N^2-1, \ldots, N(N-1) + 1$ compute

$$x_j = y_j/u_N;$$

(v) for $j = N-1, \ldots, 2, 1$ compute

$$x_{(j-1)N+i} = (y_{(j-1)N+i} - e x_{jN+i})/u_j \text{ for each } i = N, \ldots, 2, 1.$$

In subsection 5.6.2 of Chapter 5, obtaining the solution of the partial differential equation in one space variable requires the matrix A (of order N) in (1.15) to take the quindiagonal form

$$Q = \begin{bmatrix} g & d & e & & & 0 \\ f & c & d & e & & \\ p & f & c & d & e & \\ & \ddots & \ddots & \ddots & \ddots & \ddots \\ 0 & & p & f & c & d & e \\ & & & p & f & c & g \end{bmatrix}.$$

(1.31)

The solution vector $\mathbf{x}$ (of order N) may be obtained using the following decomposition algorithm for which it is easy to write a computer program:

(i) let $v_1 = g$, $r_1 = d$, $z_2 = f/v_1$, $r_2 = d - e z_2$, $v_2 = c - r_1 z_2$;
(ii) for $i = 3, 4, \ldots, N-2$ let

$$q_i = p/v_{i-2}, z_i = (f - q_i r_{i-2})/v_{i-1},$$
$$v_i = c - r_{i-1} z_i - e q_i, r_i = d - p z_i;$$

(iii) let $q_{N-1} = p/v_{N-3}, q_N = p/v_{N-2}$,
$$z_{N-1} = (f - q_{N-1}r_{N-3})/v_{N-2},$$
$$r_{N-1} = d - ez_{N-1}, v_{N-1} = c - eq_{N-1} - r_{N-2}z_{N-1},$$
$$z_N = (f - q_N r_{N-1})/v_{N-1}, v_N = g - eq_N - r_{N-1}z_N;$$

(iv) let $y_1 = b_1, y_2 = b_2 - y_1 z_2$;
(v) let $y_i = b_i - q_i y_{i-2} - y_{i-1} z_i$ for $i = 3, 4, \ldots, N$;
(vi) let $x_N = x_N/v_N, x_{N-1} = (y_{N-1} - r_{N-1} x_N)/v_{N-1}$;
(vii) for $i = N-2, \ldots, 2, 1$ compute

$$x_i = (y_i - ex_{i+2} - r_i x_{i+1})/v_i.$$

This algorithm requires $11N-16$ multiplications or divisions and $8N-13$ additions or subtractions and is consequently economical in CPU time.

In subsection 5.6.3 of Chapter 5, the numerical methods described require the solution of a system of the form (1.15) where, again, A is of order N and takes one of the forms

$$A = \begin{bmatrix} Q & & 0 \\ & Q & \\ & & \ddots \\ 0 & & Q \end{bmatrix} \quad (1.32)$$

or

$$A = \begin{bmatrix} G & D & D & & & \\ F & C & D & E & 0 & \\ P & F & C & D & E & \\ & \ddots & \ddots & \ddots & \ddots & \\ 0 & P & F & C & D & E \\ & & P & F & C & G \end{bmatrix} \quad (1.33)$$

In (1.32), A is block diagonal with quindiagonal blocks, Q being the quindiagonal matrix or order N given by (1.31). The solution of (1.15) with A being given by (1.32) and (1.31) is easily obtained by adapting the quindiagonal solver already described. The vectors $\mathbf{x}$ and $\mathbf{b}$ are each of order N^2 once more and the algorithm is as follows:

(i) let $v_1 = g, r_1 = d, z_2 = f/v_1, r_2 = d - ez_2, v_2 = c - r_1 z_2$;
(ii) for $i = 3, 4, \ldots, N-2$ let

$$q_i = p/v_{i-2}, z_i = (f - q_i r_{i-2})/v_{i-1},$$
$$v_i = c - r_{i-1} z_i - eq_i, r_i = d - pz_i;$$

(iii) let
$$q_{N-1} = p/v_{N-3}, q_N = p/v_{N-2},$$
$$z_{N-1} = (f - q_{N-1}r_{N-3})/v_{N-2},$$
$$r_{N-1} = d - ez_{N-1}, v_{N-1} = c - eq_{N-1} - r_{N-2}z_{N-1},$$
$$z_N = (f - q_N r_{N-2})/v_{N-1}, v_N = g - eq_N - r_{N-1}z_N;$$

(iv) for each $k = 0, 1, \ldots, N-1$ let
$$y_{kN+1} = b_{kN+1}, y_{kN+2} = b_{kN+2} - y_{kN+1}z_2 \text{ and}$$
$$y_{kN+i} = b_{kN+i} - q_i y_{kN+i-2} - y_{kN+i-1}z_i \text{ for } i = 3, 4, \ldots, N;$$

(v) for each $k = N-1, \ldots, 1, 0$ compute
$$x_{(k+1)N} = y_{(k+1)N}/v_N, x_{(k+1)N-1} = (y_{(k+1)N-1} - r_{N-1}x_{(k+1)N})/v_{N-1}$$
and
$$x_{kN+i} = (y_{kN+i} - ex_{kN+i+2} - r_i x_{kN+i+1})/v_i \text{ for } i = N-2, \ldots, 2, 1.$$

The matrix A in (1.33) is block quindiagonal with diagonal blocks. The diagonal matrices C, D, E, P, G are each of or order N; C, D, E usually take the forms given in (1.30) while P and G usually take the forms

$$P = \begin{bmatrix} p & & 0 \\ & p & \\ & & \ddots \\ 0 & & & p \end{bmatrix}, \quad G = \begin{bmatrix} g & & 0 \\ & g & \\ & & \ddots \\ 0 & & & g \end{bmatrix}. \tag{1.34}$$

The quindiagonal solver for A given by (1.33) is adapted as follows:

(i) let $v_1 = g, r_1 = d, z_2 = f/v_1, r_2 = d - ez_2, v_2 = c - r_1 z_2;$
(ii) let $q_i = p/v_{i-2}, z_i = (f - q_i r_{i-2})/v_{i-1},$
$$v_i = c - r_{i-1}z_i - eq_i, r_i = d - ez_i \text{ for } i = 3, 4, \ldots, N-2;$$

(iii) let $q_{N-1} = p/v_{N-3}, q_N = p/v_{N-2},$
$$z_{N-1} = (f - q_{N-1}r_{N-3})/v_{N-2}, r_{N-1} = d - ez_{N-1},$$
$$v_{N-1} = c - eq_{N-1} - r_{N-2}z_{N-1}, z_N = (f - q_N r_{N-2})/v_{N-1},$$
$$v_N = g - eq_N - r_{N-1}z_N;$$

(iv) for each $k = 1, 2, \ldots, N$ let
$$y_k = b_k, y_{N+k} = b_{N+k} - y_k z_2;$$

(v) for each $k = 3, \ldots, N$ let
$$y_{(k-1)N+j} = b_{(k-1)N+j} - q_k y_{(k-3)N+j} - y_{(k-2)N+j}z_k$$
$$\text{for } j = 1, \ldots, N;$$

(vi) for each $k = 0, 1, \ldots, N-1$

$$x_{N^2-k} = y_{N^2-k}/v_N,$$
$$x_{N^2-N-k} = (y_{N^2-N-k} - r_{N-1}x_{N^2-k})/v_{N-1};$$

(vii) for each $k = N-2, \ldots, 2, 1$

$$x_{kN-j} = (y_{kN-j} - r_k x_{(k+1)N} - ex_{(k+2)N-j})/v_k \text{ for } j = 0, 1, \ldots, N-1.$$

1.3 SOLUTION OF NON-LINEAR EQUATIONS.

1.3.1 Fixed-point iteration methods

The methods to be discussed in section 1.3 are concerned with finding the roots or zeros of the single non-linear equation

$$f(x) = 0. \tag{1.35}$$

In this subsection the methods will involve rewriting (1.35) in the form

$$x = g(x) \tag{1.36}$$

and finding $\alpha \in [a, b]$ which satisfies

$$f(\alpha) = 0 \text{ and } \alpha = g(\alpha); \tag{1.37}$$

α is one of the zeros of the non-linear equation (1.35) and is known as a *fixed point* of g in $[a, b]$.

The following theorem gives sufficient conditions for the existence and uniqueness of a fixed point:

Theorem 1.4 If $g(x)$ is continuous for $x \in [a, b]$ and $g(x) \in [a, b]$ for all $x \in [a, b]$, then g has a fixed point in $[a, b]$. Further, suppose $g'(x)$ exists on (a, b) and

$$|g'(x)| \leq k < 1 \text{ for all } x \in [a, b] \tag{1.38}$$

Then g has a unique fixed point α in $[a, b]$.

In order to locate the fixed point α, an initial approximation x_0 to α is chosen and the sequence $x_0, x_1, x_2, \ldots$ is generated by the *one-point iteration formula*

$$x_{n+1} = g(x_n); n = 0, 1, 2, \ldots. \tag{1.39}$$

If this sequence converges then, remembering that g is continuous,

$$\alpha = \lim_{n \to \infty} x_{n+1} = \lim_{n \to \infty} g(x_n) = g\left(\lim_{n \to \infty} x_n\right) = g(\alpha).$$

and a solution to (1.36) is obtained. Iterating using (1.39) to obtain α is known as *fixed point iteration*.

The fact that the conditions of Theorem 1.4 are not necessary are illustrated by the following simple example:

Example Let $f(x) = x^2 - x - 2$; then clearly $f(x) = 0$ has zeros at $x = +2$ and $x = -1$. Let $x = g(x) = x^2 - 2$, so that $g'(x) = 2x$. Using the one-point iteration formula $x_{n+1} = x_n^2 - 2$ with $x_0 = 1$ gives $x_1 = -1$ so that convergence to the zero at $x = -1$ is immediate. However, $|g'(1)| = 2 \not< 1$ so that (1.38) is violated yet convergence has occurred.

In practice it is obviously desirable to find a procedure which will guarantee that the function g not only converges to a fixed point but does so as quickly as possible. This entails finding an interval $[a, b]$ which satisfies the following theorem:

Theorem 1.5 Let g be continuous on $[a, b]$ and suppose that $g(x) \in [a, b]$ for all $x \in [a, b]$. Further, let $g'(x)$ exist on (a, b) with

$$|g'(x)| \leq k < 1 \text{ for all } x \in (a, b). \tag{1.40}$$

If x_0 is any number in $[a, b]$, then the sequence defined by

$$x_{n+1} = g(x_n); n = 0, 1, 2, \ldots$$

will converge to the unique fixed point α in $[a, b]$.

Using Theorem 1.5, precise error estimates can be found, as indicated by the following lemmas:

Lemma 1.6 If g satisfies the hypotheses of Theorem 1.5, a bound for the error involved in using x_n to approximate α is given by

$$|x_n - \alpha| \leq k^n \max \{x_0 - a, b - x_0\}; n = 0, 1, 2, \ldots . \tag{1.41}$$

Lemma 1.7 If g satisfies the hypotheses of Theorem 1.5, then

$$|x_n - \alpha| < \frac{k^n}{1-k} |x_0 - x_1|; n = 0, 1, 2, \ldots \tag{1.42}$$

Lemmas 1.6 and 1.7 show that the rate of convergence depends on the value of k, the rate of convergence depending upon the factor $k^n/(1-k)$. Clearly, the smaller k can be made, the faster the convergence.

If the sequence $x_0, x_1, x_2, \ldots$ tends to the limit α in such a way that

$$\lim_{x_n \to \alpha} \frac{x_{n+1} - \alpha}{(x_n - \alpha)^p} = C \tag{1.43}$$

for some constant $C \neq 0$ and for $p \geq 1$, then C is called the *asymptotic* error constant of the method and p is the *order of convergence*; p and C are connected by the following theorem:

Theorem 1.8 The order of convergence of a convergent one point iteration function is p if and only if

$$g'(\alpha) = g''(\alpha) = \ldots = g^{(p-1)}(\alpha) = 0 \text{ and } g^{(p)}(\alpha) \neq 0. \tag{1.44}$$

The asymptotic error constant has the value $C = g^{(p)}(\alpha)/p!$

1.3.2 The Newton-Raphson method

This method (sometimes referred to simply as *Newton's method*) is one of the most powerful and well known numerical techniques for finding a zero of (1.35).

Suppose that the function $f(x)$ is twice continuously differentiable for $x \in [a, b]$. Let $x^* \in [a, b]$ be an approximation to the zero α such that $f'(x^*) \neq 0$ and $|\alpha - x^*|$ is sufficiently small that $|\alpha - x^*|^2$ and higher powers are negligible.

Consider now the second degree Taylor polynomial for $f(x)$ about x^*; this is given by

$$f(x) = f(x^*) + (x-x^*)f'(x^*) + \tfrac{1}{2}(x-x^*)^2 f''(r(x)) \tag{1.45}$$

and is considered further in section 1.4. In (1.45), $r(x)$ lies between x and x^*.

Since $f(\alpha) = 0$, equation (1.45) gives

$$0 = f(x^*) + (\alpha - x^*)f'(x^*) + \tfrac{1}{2}(\alpha - x^*)^2 f''(r(x)), \tag{1.46}$$

Neglecting the second order term in (1.46) and solving for α gives

$$\alpha \cong x^* - f(x^*)/f'(x^*). \tag{1.47}$$

Equation (1.47) suggests the iteration formula

$$x_{n+1} = x_n - f(x_n)/f'(x_n); n = 0, 1, 2, \ldots \tag{1.48}$$

which is known as the Newton-Raphson method.

Newton's method is a special one-point iteration formula; this may be seen by writing

$$g(x) = x - f(x)/f'(x). \tag{1.49}$$

It follows from (1.49) and Theorem 1.8 that the Newton-Raphson method is second order convergent, since it was assumed that $f'(x) \neq 0$ for $x \in [a, b]$. In the case of α being a multiple zero of multiplicity m, $f'(\alpha) = 0$ and (1.48) yields a sequence of iterates which is only first order convergent. Newton's method for a root of multiplicity m is modified to

$$x_{n+1} = x_n - mf(x_n)/f'(x_n); n = 0, 1, 2, \ldots . \tag{1.50}$$

It is important to choose a good initial approximation x_0 for use in (1.48) otherwise the sequence of iterates may diverge from α. The following converge theorem illustrates the importance of the choice of x_0 for the Newton-Raphson method:

Theorem 1.9 Let $f(x)$ be twice continuously differentiable for x in some closed interval $[a, b]$. If $\alpha \in [a, b]$ is such that $f(\alpha) = 0$ and $f'(\alpha) \neq 0$, then there exists a number $\delta > 0$ such that (1.48) generates a sequence $x_1, x_2, x_3, \ldots$ which converges to α for any initial value $x_0 \in [\alpha-\delta, \alpha+\delta]$.

1.3.3 Stopping the iterations

The stopping criteria discussed in section 1.2 for the numerical solution of linear systems are easily adapted for use with single non-linear equations of the form (1.35).

A tolerance $\epsilon > 0$ is selected in advance and iterations using (1.39) or (1.48) are continued until one of the stopping criteria

$$|x_{k+1} - x_k| < \epsilon \tag{1.51}$$

or

$$\frac{|x_{k+1} - x_k|}{|x_k|} < \epsilon, x_k \neq 0 \tag{1.52}$$

is satisfied for some $k = 0, 1, 2, \ldots$.

1.4 LAGRANGIAN INTERPOLATION

In section 2.3 of Chapter 2 the displacements of points outside the boundary of an elastic membrane must be estimated in terms of the displacements at points in the membrane and on its boundary. This is most easily achieved by Lagrangian interpolation and, as one displacement is required in terms of two others, linear Lagrangian interpolation is all that is required.

Suppose that two distinct points $(x_0, f(x_0))$, $(x_1, f(x_1))$, with $x_0 < x_1$, are given and that it is required to find the interpolating polynomial $p(x)$ of first degree which passes through these points. The polynomial $p(x)$ is thus an approximation to the function $f(x)$ in the interval $x_0 \leq x \leq x_1$; clearly $p(x_0) = f(x_0)$ and $p(x_1) = f(x_1)$.

Sec. 1.4] Lagrangian Interpolation

Suppose next that the polynomial has the form

$$p(x) = a + bx, \tag{1.53}$$

where the parameters a, b are to be determined; the *method of undertermined coefficients*, so named for a reason which will become clear to the reader, will be used to find a, b. The polynomial $p(x)$ passes through the two given points so that

$$f(x_0) = a + b\, x_0,$$
$$f(x_1) = a + b\, x_1.$$

It is then a very simple exercise to show that

$$a = \frac{x_1 f(x_0) - x_0 f(x_1)}{x_1 - x_0}, \quad b = \frac{f(x_0) - f(x_1)}{x_0 - x_1}$$

and substituting these values of a, b into (1.53) gives

$$p(x) = \frac{x - x_1}{x_0 - x_1} f(x_0) + \frac{x - x_0}{x_1 - x_0} f(x_1) \tag{1.54}$$

which is the form of $p(x)$ known as the *Lagrange interpolating polynomial* of degree 1.

The principle may easily be extended to find the second degree polynomial through the three known points $(x_i, f(x_i))$, $i = 0, 1, 2$, for which $x_0 < x_1 < x_2$. The polynomial is assumed to have the form

$$p(x) = a + bx + cx^2$$

where a, b, c are parameters. Using the method of undetermined coefficients, a, b, c are found and it may then be shown that the Lagrange interpolating polynomial of degree 2 takes the form

$$p(x) = \frac{(x - x_1)(x - x_2)}{(x_0 - x_1)(x_0 - x_2)} f(x_0) + \frac{(x - x_0)(x - x_2)}{(x_1 - x_0)(x_1 - x_2)} f(x_1)$$

$$+ \frac{(x - x_0)(x - x_1)}{(x_2 - x_0)(x_2 - x_1)} f(x_2). \tag{1.55}$$

The Lagrangian interpolating polynomial of degree N passes through the $N+1$ data points $(x_i, f(x_i))$ for $i = 0, 1, \ldots, N$, with $x_0 < x_1 < \ldots < x_N$. Its form is a generalisation of (1.54), (1.55) and is given by

$$p(x) = \sum_{i=0}^{N} \left[\prod_{\substack{j=0 \\ j \neq i}}^{N} \frac{(x - x_j)}{(x_i - x_j)} f(x_i) \right]. \tag{1.56}$$

One of the most useful features of (1.56) is that the given data points need not be equally spaced with regard to the $x_i (i = 0, 1, \ldots, N)$.

The next step is to calculate a bound on the error involved in approximating a function by an interpolating polynomial, for it is as important to know how much a computed value is in error as it is to have the value itself. The error bound which will now be found is restricted to those functions having derivatives with known bounds. This is the case when trigonometric, exponential or logarithmic functions are to be interpolated using tabulated values.

Consider the function

$$g(x) = f(x) - p(x) - \frac{f(X) - p(X)}{(X-x_0)(X-x_1)\ldots(X-x_N)} \cdot (x-x_0)(x-x_1)\ldots(x-x_N), \quad (1.57)$$

where X is some point in the open interval (x_0, x_N) at which the function value $f(X)$ is to be approximated by the polynomial value $p(X)$ calculated from (1.56). Assume that $f(x)$ possesses $N+1$ continuous derivatives in (x_0, x_N) and that $p(x)$ is of degree at most N.

It is clear that $g(x_i) = 0$ $(i = 0, 1, \ldots, N)$ and that $g(X) = 0$; that is $g(x)$ vanishes $N+2$ times in the closed interval $[x_0, x_N]$. It follows, therefore, that $g'(x)$ vanishes $N+1$ times, $g''(x)$ vanishes N times, etc., in $[x_0, x_N]$. In particular $g^{(N+1)}(x)$ must vanish once in $[x_0, x_N]$; that is, there exists a number r in the interval $[x_0, x_N]$ such that $g^{(N+1)}(r) = 0$.

Differentiating $g(x)$ $N+1$ times gives

$$g^{(N+1)}(x) = f^{(N+1)}(x) - \frac{f(X)-p(X)}{(X-x_0)(X-x_1)\ldots(X-x_N)} \cdot (N+1)! \quad (1.58)$$

since $p^{(N+1)}(x) = 0$, $p(x)$ being a polynomial of degree less than $N+1$. Putting $x = r$ in (1.58) gives

$$0 = f^{(N+1)}(r) - \frac{f(X)-p(X)}{(X-x_0)(X-x_1)\ldots(X-x_N)} \cdot (N+1)!$$

which leads to

$$f(X) - p(X) = (X-x_0)(X-x_1)\ldots(X-x_N) \frac{f^{(N+1)}(r)}{(N+1)!}. \quad (1.59)$$

In order to use (1.59) it is necessary to know the value of $f^{(N+1)}(r)$ or to be able to estimate its value. Often, it is possible to place an upper bound on $|f^{(N+1)}(r)|$ and thus on the modulus of the error $f(X) - p(X)$.

As an example, consider the function $f(x) = \sin x$ and pass an interpolating polynomial $p(x)$ of degree six through the tabulated values of $\sin x_i$, where $x_i = i\pi/12$ and $i = 0, 1, \ldots, 6$; the error in the computed value $p(5\pi/24)$, for instance, is calculated as follows.

For the function $f(x) = \sin x$, $f^{(N+1)}(x) = f^{(vii)}(x) = -\cos x$ so that $|f^{(vii)}(r)| \leq 1$ for any r in the interval $[0, \frac{1}{2}\pi]$. Now,

$$f(X) - p(X) = (X - x_0)(X - x_1)\ldots(X - x_6) \cdot \frac{f^{(vii)}(r)}{7!},$$

and for $X = 5\pi/24$,

$$|f(5\pi/24) - p(5\pi/24)| = \left| \left(\frac{5\pi}{24} - 0\right)\left(\frac{5\pi}{24} - \frac{\pi}{12}\right)\ldots\left(\frac{5\pi}{24} - \frac{\pi}{2}\right) \right| \cdot$$

$$\frac{f^{(vii)}(r)}{7!} \leq \frac{1}{5040} \cdot 5 \cdot 3 \cdot 1 \cdot 1 \cdot 3 \cdot 5 \cdot 7 \cdot \frac{\pi^7}{24^7}.$$

This gives the error bound

$$|f(5\pi/24) - p(5\pi/24)| \leq 2.06 \times 10^{-7}.$$

1.5 SOME EXPANSIONS

The key formula in expressing the derivatives which appear in partial differential equations in terms of function values, is *Taylor's formula*. Suppose, first of all, that $u = u(x)$ is a function of one independent variable x. If this function is continuous and has $n + 1$ continuous derivatives on some closed interval I containing the point $x = a$ (that is, $u(x) \in C^{n+1}(I)$), then, for each $x \in I$,

$$u(x) = u(a) + \frac{(x-a)}{1!} u'(a) + \frac{(x-a)^2}{2!} u''(a) + \ldots + \frac{(x-a)^n}{n!} u^{(n)}(a) + \quad (1.60)$$

$$R_n(x, a).$$

This is known as Taylor's formula and the term $R_n(x, a)$ is usually referred to as the *remainder term*. The remainder term is usually written in one of the following three forms, where r depends on n and x:

(i) *The Lagrange form*

$$R_n(x, a) = \frac{(x-a)^{n+1}}{(n+1)!} u^{(n+1)}(r), \quad r \text{ between } a \text{ and } x;$$

(ii) *The Cauchy form*

$$R_n(x, a) = \frac{(x-a)(x-r)^n}{n!} u^{(n+1)}(r), \quad r \text{ between } a \text{ and } x;$$

(iii) *The integral form*

$$R_n(x, a) = \int_a^x \frac{(x-t)^n}{n!} u^{(n+1)}(t) dt.$$

The polynomial $P_n(x, a)$, given by

$$P_n(x, a) = u(a) + \frac{(x-a)}{1!} u'(a) + \frac{(x-a)^2}{2!} u''(a) + \ldots + \frac{(x-a)^n}{n!} u^{(n)}(a), \quad (1.61)$$

is known as the *Taylor polynomial* of degree n of $u(x)$ at a, and the coefficients $u(a), u'(a)/1!, u''(a)/2!, \ldots$ are called the *Taylor coefficients of $u(x)$ at a*. Hence

$$u(x) = P_n(x, a) + R_{n+1}(x, a).$$

The Taylor polynomial $P_n(x, a)$ is therefore an approximation to $u(x)$, the error in this approximation being $R_{n+1}(x, a)$.

In the special case $a = 0$, Taylor's formula (1.60) becomes

$$u(x) = u(0) + \frac{x}{1!} u'(0) + \frac{x^2}{2!} u''(0) + \ldots + \frac{x^n}{n!} u^{(n)}(0) + R_n(x). \quad (1.62)$$

In this form, the formula is known as *Maclaurin's formula* and the remainder term $R_n(x) \equiv R_n(x, 0)$ takes one of the following three forms, where r depends on n and x:

(i) *The Lagrange form*

$$R_n(x) = \frac{x^{n+1}}{(n+1)!} u^{(n+1)}(r), \, r \text{ between 0 and } x;$$

(ii) *The Cauchy form*

$$R_n(x) = \frac{x(x-r)^n}{n!} u^{(n+1)}(r), \, r \text{ between 0 and } x;$$

(iii) *The integral form*

$$R_n(x) = \int_0^x \frac{(x-t)^n}{n!} u^{(n+1)}(t) \, dt.$$

Some Expansions

Often, Taylor's and Maclaurin's formulas are used to express $u(x)$ as a power series. To this end, the series

$$u(x) = u(a) + \frac{(x-a)}{1!} u'(a) + \frac{(x-a)^2}{2!} u''(a) + \ldots =$$

$$\sum_{k=0}^{\infty} \frac{(x-a)^k}{k!} u^{(k)}(a) \qquad (1.63)$$

is called the *Taylor series* of $u(x)$ at a or the *Taylor expansion* of $u(x)$ in powers of $x - a$. Similarly, the series

$$u(x) = u(0) + \frac{x}{1!} u'(0) + \frac{x^2}{2!} u''(0) + \ldots = \sum_{k=0}^{\infty} \frac{x^k}{k!} u^{(k)}(0) \qquad (1.64)$$

is called the *Maclaurin series* of $u(x)$ or the *Maclaurin expansion* of $u(x)$.

In the numerical solution of differential equations it is often more convenient to consider the Taylor series in a slightly different form. In (1.63) x, a are replaced simultaneously by $x + h$, x; $h > 0$ is called an increment in x and the Taylor series becomes

$$u(x + h) = u(x) + \frac{h}{1!} u'(x) + \frac{h^2}{2!} u''(x) + \ldots = \sum_{k=0}^{\infty} \frac{h^k}{k!} u^{(k)}(x). \qquad (1.65)$$

Taking only the first two terms of the right-hand side of (1.65) gives $u(x + h) = u(x) + hu'(x) + 0(h^2)$, where the expression $0(h^2)$ indicates that the error $R_2(x)$ has *principal part* proportional to h^2. It follows that

$$\frac{du}{dx} = \frac{u(x+h) - u(x)}{h} + 0(h), \qquad (1.66)$$

which is called a *first order, forward difference* approximant to du/dx. It is easy to show that a *first order, backward difference* approximant to du/dx is

$$\frac{du}{dx} = \frac{u(x) - u(x-h)}{h} + 0(h). \qquad (1.67)$$

The Taylor expansion for $u(x - h)$, which is used in (1.67), is found by replacing h with $-h$ in (1.65) to give

$$u(x - h) = u(x) - \frac{h}{1!} u'(x) + \frac{h^2}{2!} u''(x) - \ldots = \sum_{k=0}^{\infty} (-1)^k \frac{x^k}{k!} u^{(k)}(x).$$

$$(1.68)$$

Taking terms up to and including h^3 on the right-hand sides of (1.65), (1.68) and subtracting gives

$$u(x+h) - u(x-h) = 2hu'(x) + O(h^3)$$

so that

$$\frac{du}{dx} = \frac{u(x+h) - u(x-h)}{2h} + O(h^2). \tag{1.69}$$

This is a *second order, central difference* approximant to du/dx.

Taking terms up to and including h^4 on the left-hand sides of (1.65), (1.68) and adding gives

$$u(x+h) + u(x-h) = 2u(x) + h^2 u''(x) + O(h^4),$$

the terms in h, h^3 having vanished. This gives

$$\frac{d^2 u}{dx^2} = \frac{u(x+h) - 2u(x) + u(x-h)}{h^2} + O(h^2) \tag{1.70}$$

which is a *second order, central difference* approximant to $d^2 u/dx^2$.

Suppose, next, that $u = u(x, t)$ is a function of two independent variables x, t and suppose that x is given an increment $h > 0$ but that t is fixed. The Taylor series given by (1.65) becomes

$$u(x+h, t) = u(x, t) + \frac{h}{1!}\frac{\partial u(x, t)}{\partial x} + \frac{h^2}{2!}\frac{\partial^2 u(x, t)}{\partial x^2} + \ldots \tag{1.71}$$

while (1.68) becomes

$$u(x-h, t) = u(x, t) - \frac{h}{1!}\frac{\partial u(x, t)}{\partial x} + \frac{h^2}{2!}\frac{\partial^2 u(x, t)}{\partial x^2} - \ldots \tag{1.72}$$

Taking only the first two terms of the right-hand side of (1.71) gives $u(x+h, t) = u(x, t) + h\partial u/\partial x + O(h^2)$, from which it follows that

$$\frac{\partial u}{\partial x} = \frac{u(x+h, t) - u(x, t)}{h} + O(h). \tag{1.73}$$

Similarly, it may be shown from (1.72) that

$$\frac{\partial u}{\partial x} = \frac{u(x, t) - u(x-h, t)}{h} + O(h). \tag{1.74}$$

The expressions for $\partial u/\partial x$ in (1.73), (1.74) are first order approximations.

Some Expansions

Taking terms up to and including h^3 in (1.71), (1.72), and subtracting, gives

$$u(x+h, t) - u(x-h, t) = 2h \frac{\partial u}{\partial x} + O(h^3)$$

from which it follows that

$$\frac{\partial u}{\partial x} = \frac{u(x+h, t) - u(x-h, t)}{2h} + O(h^2) \tag{1.75}$$

which is a second order central difference replacement for $\partial u/\partial x$.

Taking terms up to and including h^4 in (1.71), (1.72), and adding, gives

$$u(x+h, t) + u(x-h, t) = 2u(x, t) + h^2 \frac{\partial^2 u}{\partial x^2} + O(h^4)$$

from which it follows that

$$\frac{\partial^2 u}{\partial x^2} = \frac{u(x+h, t) - 2u(x, t) + u(x-h, t)}{h^2} + O(h^2) \tag{1.76}$$

which is a second order approximation of the second derivative $\partial^2 u/\partial x^2$.

Suppose now that x is fixed and that t is given an increment l or $-l$. The Taylor series (1.71), (1.72) become

$$u(x, t+l) = u(x, t) + \frac{l}{1!} \frac{\partial u(x, t)}{\partial t} + \frac{l^2}{2!} \frac{\partial^2 u(x, t)}{\partial t^2} + \cdots,$$

$$u(x, t-l) = u(x, t) - \frac{l}{1!} \frac{\partial u(x, t)}{\partial t} + \frac{l^2}{2!} \frac{\partial^2 u(x, t)}{\partial t^2} - \cdots,$$

respectively, and it is easy to show that

$$\frac{\partial u}{\partial t} = \frac{u(x, t+l) - u(x, t)}{l} + O(l), \tag{1.77}$$

$$\frac{\partial u}{\partial t} = \frac{u(x, t) - u(x, t-l)}{l} + O(l), \tag{1.78}$$

$$\frac{\partial u}{\partial t} = \frac{u(x, t+l) - u(x, t-l)}{2l} + O(l^2), \tag{1.79}$$

$$\frac{\partial^2 u}{\partial t^2} = \frac{u(x, t+l) - 2u(x, t) + u(x, t-l)}{l^2} + O(l^2). \tag{1.80}$$

The approximations given by (1.73)–(1.80) will be used frequently throughout the book.

36   **Introductory Mathematics**   [Ch. 1

The dependent variable $u = u(x, t)$ will arise in Chapters 4, 5 where time-dependent problems will be discussed, the variable x denoting a space dimension and the variable t denoting time. Steady-state problems will be discussed in Chapter 2 where $u = u(x, y)$ will be a function of two space variables x, y. Formulas (1.77), (1.78), (1.79), (1.80) can easily be adapted to give the derivatives of u with respect to y. Assuming that the increment in y is k, these derivatives are

$$\frac{\partial u}{\partial y} = \frac{u(x, y + k) - u(x, y)}{k} + O(k), \tag{1.81}$$

$$\frac{\partial u}{\partial y} = \frac{u(x, y) - u(x, y - k)}{k} + O(k), \tag{1.82}$$

$$\frac{\partial u}{\partial y} = \frac{u(x, y + k) - u(x, y - k)}{2k} + O(k^2), \tag{1.83}$$

$$\frac{\partial^2 u}{\partial y^2} = \frac{u(x, y + k) - 2u(x, y) + u(x, y - k)}{k^2} + O(k^2). \tag{1.84}$$

Occasionally in Chapter 2, the mixed second derivative $\partial^2 u/\partial x \partial y$ will be required. In order to obtain this, Taylor's series in which both x, y receive increments is required. There are clearly four such forms and they are given by

$$u(x + h, y + k) = u(x, y) + \frac{1}{1!}\left(h\frac{\partial u(x, y)}{\partial x} + k\frac{\partial u(x, y)}{\partial y}\right)$$
$$+ \frac{1}{2!}\left(h^2\frac{\partial^2 u(x, y)}{\partial x^2} + 2hk\frac{\partial^2 u(x, y)}{\partial x \partial y} + k^2\frac{\partial^2 u(x, y)}{\partial y^2}\right) + \ldots, \tag{1.85}$$

$$u(x + h, y - k) = u(x, y) + \frac{1}{1!}\left(h\frac{\partial u(x, y)}{\partial x} - k\frac{\partial u(x, y)}{\partial y}\right)$$
$$+ \frac{1}{2!}\left(h^2\frac{\partial^2 u(x, y)}{\partial x^2} - 2hk\frac{\partial^2 u(x, y)}{\partial x \partial y} + k^2\frac{\partial^2 u(x, y)}{\partial y^2}\right) + \ldots, \tag{1.86}$$

$$u(x - h, y + k) = u(x, y) + \frac{1}{1!}\left(-h\frac{\partial u(x, y)}{\partial x} + k\frac{\partial u(x, y)}{\partial y}\right)$$
$$+ \frac{1}{2!}\left(h^2\frac{\partial^2 u(x, y)}{\partial x^2} - 2hk\frac{\partial^2 u(x, y)}{\partial x \partial y} + k^2\frac{\partial^2 u(x, y)}{\partial y^2}\right) + \ldots, \tag{1.87}$$

$$u(x - h, y - k) = u(x, y) + \frac{1}{1!}\left(-h\frac{u(x, y)}{\partial x} - k\frac{\partial u(x, y)}{\partial y}\right)$$
$$+ \frac{1}{2!}\left(h^2\frac{\partial^2 u(x, y)}{\partial x^2} + 2hk\frac{\partial^2 u(x, y)}{\partial x \partial y} + k^2\frac{\partial^2 u(x, y)}{\partial y^2}\right) + \ldots. \tag{1.88}$$

Sec. 1.6] Classification of Second Order Partial Differential Equations

It is easy to show that

$$u(x+h, y+k) - u(x+h, y-k) - u(x-h, y+k) + u(x-h, y-k)$$
$$= 4hk \frac{\partial^2 u}{\partial x \partial y} + 0(h+k)^4, \tag{1.89}$$

terms involving third order derivatives having vanished in this linear combination of the four functional values. It follows from (1.89) that

$$\frac{\partial^2 u}{\partial x \partial y} = \frac{u(x+h, y+k) - u(x+h, y-k) - u(x-h, y+k) + u(x-h, y-k)}{4hk}$$
$$+ 0\left(\frac{(h+k)^4}{hk}\right). \tag{1.90}$$

In the special case where $k = h$, (1.90) becomes

$$\frac{\partial^2 u}{\partial x \partial y} = \frac{u(x+h, y+h) - u(x+h, y-h) - u(x-h, y+h) + u(x-h, y-h)}{4h^2}$$
$$+ 0(h^2). \tag{1.91}$$

The approximations (1.90), (1.91) are both second order replacements for $\partial^2 u / \partial x \partial y$.

1.6 CLASSIFICATION OF SECOND ORDER PARTIAL DIFFERENTIAL EQUATIONS

The general quasi-linear partial differential equation of the second order in two independent variables x, t has the form

$$a \frac{\partial^2 u}{\partial x^2} + b \frac{\partial^2 u}{\partial x \partial t} + c \frac{\partial^2 u}{\partial t^2} = e, \tag{1.92}$$

where $u = u(x, t)$ and a, b, c, e are functions of x, t, u, $\partial u / \partial x$, $\partial u / \partial t$ but not of the second order derivatives. In this section, the abbreviations p, q, r, s, w will be used to denote $\partial u / \partial x$, $\partial u / \partial t$, $\partial^2 u / \partial x^2$, $\partial^2 u / \partial x \partial t$, $\partial^2 u / \partial t^2$, respectively. Equation (1.92) can therefore be written in the alternative form

$$ar + bs + cw = e. \tag{1.93}$$

Suppose that the values of u, p, q are known at all points (x, t) on some smooth curve in the (x, t) plane, then these values must satisfy the total differential formula

$$du = \frac{\partial u}{\partial x} dx + \frac{\partial u}{\partial t} dt = p \, dx + q \, dt. \tag{1.94}$$

Similarly, p and q satisfy the formulas

$$dp = \frac{\partial p}{\partial x} dx + \frac{\partial p}{\partial t} dt = r\, dx + s\, dt, \qquad (1.95)$$

$$dq = \frac{\partial q}{\partial x} dx + \frac{\partial q}{\partial t} dt = s\, dx + w\, dt. \qquad (1.96)$$

Equations (1.93), (1.95), (1.96) form a system of three equations in r, s, w which will not have a unique solution at any point at which the determinant of the coefficients of r, s, w vanishes. That is, wherever

$$\begin{vmatrix} a & b & c \\ dx & dt & 0 \\ 0 & dx & dt \end{vmatrix} = 0 \qquad (1.97)$$

holds. Expanding the determinant of (1.97) gives the equation

$$a\left(\frac{dt}{dx}\right)^2 - b\,\frac{dt}{dx} + c = 0 \qquad (1.98)$$

which has two real distinct, one real repeated, or conjugate complex roots in dt/dx according as the discriminant $b^2 - 4ac$ is greater than, equal to, or less than zero. In a domain in the (x, t) plane in which u, p, q are defined, and throughout which equation (1.98) has two, one or no real roots, the differential equation (1.92) is said to be *hyperbolic, parabolic* or *elliptic* respectively.

In the case of elliptic equations which are steady-state, the time variable t is replaced by a second space derivative, y say, and (1.92) takes the form

$$a\frac{\partial^2 u}{\partial x^2} + b\frac{\partial^2 u}{\partial x \partial y} + c\frac{\partial^2 u}{\partial y^2} = e\,; \quad b^2 - 4ac < 0. \qquad (1.99)$$

Examples. Classify the following equations as hyperbolic, parabolic, or elliptic, assuming that x, t, y, u are all real

1. $\dfrac{\partial^2 u}{\partial x^2} + u\dfrac{\partial^2 u}{\partial x \partial y} + (1+u^2)\dfrac{\partial^2 u}{\partial y^2} = 0.$

   Here $a = 1$, $b = u$, $c = 1+u^2$, $e = 0$ so that

   $$b^2 - 4ac = u^2 - 4(1+u^2) = -4 - 3u^2 < 0.$$

   Hence the equation is always elliptic.

2. $\dfrac{\partial^2 u}{\partial x^2} + 2\dfrac{\partial^2 u}{\partial x \partial t} + (1-x^2-t^2)\dfrac{\partial^2 u}{\partial t^2} = x^2 + t^2.$

Here $a = 1$, $b = 2$, $c = 1-x^2-t^2$, $e = x^2+t^2$ so that

$$b^2 - 4ac = 4 - 4(1-x^2-t^2) = 4(x^2+t^2).$$

Hence the equation is hyperbolic except at the point $(0, 0)$ in the (x, t) plane where it is parabolic.

3. $\dfrac{\partial^2 u}{\partial x^2} + x^3 \dfrac{\partial^2 u}{\partial t^2} = 0.$

Here $a = 1$, $b = 0$, $c = x^3$, $e = 0$ so that

$$b^2 - 4ac = 0 - 4x^3 = -4x^3.$$

Hence the equation is (a) hyperbolic if $x < 0$,
(b) parabolic if $x = 0$, (c) elliptic if $x > 0$.

PROBLEMS

1. Prove that any similarity transformation $C^{-1}AC$ applied to a matrix A leaves the eigenvalue of A unchanged.
2. Prove that for any matrix A, $\|A\|_1 = \|A^T\|_\infty$.
3. Prove that for any square matrix A, $\|A\|_E = [\text{trace}\,(AA^T)]^{1/2}$.
4. Prove that the trace of any square matrix A is equal to the algebraic sum of its eigenvalues.
5. Given a square matrix A of order N whose eigenvalues are distinct and are given by $\lambda_1, \lambda_2, \ldots, \lambda_N$, with corresponding eigenvectors $x_1, x_2, \ldots, x_N$. Prove that $\lambda_1^2, \lambda_2^2, \ldots, \lambda_N^2$ are the eigenvalues of the matrix A^2 and write down the corresponding eigenvectors.
   Generalise these results to the matrices A^k ($k = 1, 2, \ldots, K$) and write down the eigenvalues of the matrix polynomial $f(A) = a_0 I + a_1 A + \ldots + a_K A^K$, where $a_0, a_1, \ldots, a_K$ are real constants.
6. Determine the characteristic equation of the matrix

$$A = \begin{bmatrix} 1 & 2 \\ 3 & 4 \end{bmatrix}.$$

Starting with the vector $(0, 1)^T$ use an iterative method to determine the dominant eigenvalue of A and its asssciated eigenvector. Work to three decimal places giving all answers to two decimal places.

7. A first order initial value problem is given by $\mathbf{y}' = A\mathbf{y}$, $\mathbf{y}(0) = \mathbf{y}_0$, where

$$A = \begin{bmatrix} -42 & 38 & -40 \\ 38 & -42 & 40 \\ 80 & -80 & -80 \end{bmatrix}.$$

Find the stiffness ratio of this system.

8. Given the system of two first order ordinary differential equations

$$y_1' = -y_1 - y_1 y_2 + 294 y_2 \, ; y_1(0) = 1,$$

$$y_2' = \frac{y_1}{98} - \frac{y_1 y_2}{98} - 3y_2 \, ; y_2(0) = 0.$$

Show that, initially, the eigenvalues λ_1, λ_2 of the Jacobian matrix satisfy the characteristic equation

$$98\lambda^2 + 393\lambda + 2 = 0$$

and find the initial stiffness ratio.

9. Use the method of Gaussian elimination to solve the system

$$\begin{aligned} x_1 + x_2 - x_3 &= 3, \\ -x_1 - 2x_2 + x_3 &= -5, \\ -2x_1 + x_2 - 3x_3 &= 0. \end{aligned}$$

10. Use Doolittle's method to solve the linear system

$$\begin{aligned} 2x_1 + x_2 + x_3 &= 2, \\ x_1 - 2x_2 + x_3 &= -2, \\ 3x_1 - 3x_2 - x_3 &= 1. \end{aligned}$$

11. Use the method of Gauss–Seidel to solve the system

$$\begin{aligned} 10x_1 + x_2 + x_3 &= 1, \\ x_1 + x_2 + 10x_3 &= 1, \\ x_1 - 10x_2 + x_3 &= -1, \end{aligned}$$

giving the answers to two decimal places.

12. (i) Use the method of undetermined coefficients to derive the second degree Lagrange interpolating polynomial given by (1.55).

(ii) Verify equation (1.58).

13. Classify the following into elliptic, hyperbolic or parabolic types

(a) $\dfrac{\partial^2 u}{\partial x^2} = \dfrac{1}{k^2} \dfrac{\partial^2 u}{\partial t^2}$; k a non-zero, real constant,

(b) $x^2 \dfrac{\partial^2 u}{\partial x^2} + y^2 \dfrac{\partial^2 u}{\partial y^2} = 0$; x, y real,

(c) $\dfrac{\partial u}{\partial x} = \dfrac{k}{t^2} \dfrac{\partial}{\partial t} \left(t^2 \dfrac{\partial u}{\partial x} \right)$; k a real constant.

14. Use the one-point iteration formula $x = g(x)$ to find three fixed points of the equation $f(x) \equiv x - \tan x$ other than $x = 0$.
15. Prove that the one-point iteration function

$$x_{n+1} = \frac{1}{2}\left(x_n + \frac{2}{x_n}\right);\ n = 0, 1, \ldots$$

converges to $\sqrt{2}$ for any x_0.

16. Use Newton's method to show that, to a relative accuracy of 0.5%, there is a zero of the equation $f(x) \equiv 2e^{-x} - \sin x = 0$ at $x = 0.9210$.
17. Use the Newton-Raphson method to show that there is a root of the equation $f(x) \equiv e^x - 3x^2 = 0$ at $x = -0.45896$ to five decimal places (take $x_0 = -0.5$).

2

Elliptic Equations: Finite Difference Methods

2.1 THE DIRCHLET, NEUMANN AND ROBBINS PROBLEMS

Elliptic partial differential equations are usually associated with steady-state or equilibrium problems in which there is more than one independent variable. In this chapter problems with two independent space variables x, y will be considered. The domain of integration will be some region R in the (x, y) plane bounded by a closed curve ∂R. Together with the elliptic equation in R, boundary conditions everywhere on ∂R will be specified. These may be of the form of function values of the dependent variable u, or the value of the outward normal derivative, or a mixture of both.

It was noted in Chapter 1 that the general quasi-linear partial differential equation of the second order given by

$$a \frac{\partial^2 u}{\partial x^2} + b \frac{\partial^2 u}{\partial x \partial y} + c \frac{\partial^2 u}{\partial x^2} = e \qquad (2.1)$$

is elliptic if $b^2 - 4ac < 0$. In (2.1) the terms a, b, c, e are functions of u, x, y, $\partial u/\partial x$, $\partial u/\partial y$ but not of the second derivatives. Three classes of elliptic problem arise, corresponding to the three classes of boundary condition. One easy example of each will be considered in sub-section 2.1.1, 2.1.2 and 2.1.3. These examples have been chosen to make the reader familiar with finite difference replacements. Much more difficult problems do arise in the real world, the increase in difficulty usually arising as a result of an irregular or internal boundary. One such problem will be considered in detail in section 2.3.

For time-dependent problems further difficulties are encountered when there are discontinuities between initial conditions and boundary conditions. This type of problem is considered in detail, using finite difference methods, in Chapters 4 and 5.

2.1.1 The Dirichlet problem

The Dirichlet boundary problem requires the solution $u(x, y)$ of (2.1) in R which has the given value

$$u = f(x, y) \qquad (2.2)$$

[Sec. 2.1] The Dirchlet, Neumann and Robbins Problems 43

on ∂R. The simplest elliptic form of (2.1) has $a = c = 1$ and $b = e = 0$ giving $b^2 - 4ac = -4 < 0$; equation (2.1) is thus

$$\frac{\partial^2 u}{\partial x^2} + \frac{\partial^2 u}{\partial y^2} = 0, \quad (x, y) \in R \tag{2.3}$$

which is known as *Laplace's equation*. The simplest region R is the unit square bounded by the lines $x = 0$, $x = 1$, $y = 0$, $y = 1$ in the (x, y) plane: these lines form ∂R on which the value of u is specified.

A square *mesh* is now superimposed on R and the value of u calculated at each point of the mesh. It will be seen that the choice of a fine mesh leads to more accurate values of the solution than a coarse mesh. Usually, the mesh consists of equally spaced straight lines parallel to the x, y axes, though it will be seen later in this chapter 4 that a characteristic mesh may be used. The word *grid* is used in the literature as often as the word mesh, and the word *node* or the phrase *nodal point* are often used to describe a mesh point.

A uniform mesh consisting of M lines parallel to the x-axis and M lines parallel to the y-axis discretizes $R \cup \partial R$ into M^2 mesh points in R together with $4M + 2$ mesh points on ∂R. Defining h to be the constant distance between the mesh lines, it follows that $(M+1)h = 1$ and that the (x, y) coordinates of the mesh points are $x_k = kh$, $y_m = mh$ where $k, m = 0, 1, \ldots, M, M+1$.

The theoretical value of the function at the mesh point (kh, mh) is $u(kh, mh)$; this will be denoted by $u_{k,m}$. The computed value of $u_{k,m}$, determined using some numerical procedure or algorithm, will be denoted by $U_{k,m}$.

The computed solution at the M^2 mesh points in R is obtained by applying some numerical approximation to Laplace's equation (2.3) at every mesh point. This results in a system of equations the solution of which gives the value of U at every mesh point; the solution to (2.3) with the associated boundary condition of the form (2.2), is thus found implicitly.

Applying Taylor's theorem with two independent variables, discussed in Chapter 1, gives

$$u_{k+1,m} = \left(u + h\frac{\partial u}{\partial x} + \frac{1}{2}h^2\frac{\partial^2 u}{\partial x^2} + \frac{1}{6}h^3\frac{\partial^3 u}{\partial x^3} + \frac{1}{24}h^4\frac{\partial^4 u}{\partial x^4} + \ldots\right)_{k,m}, \tag{2.4}$$

$$u_{k-1,m} = \left(u - h\frac{\partial u}{\partial x} + \frac{1}{2}h^2\frac{\partial^2 u}{\partial x^2} - \frac{1}{6}h^3\frac{\partial^3 u}{\partial x^3} + \frac{1}{24}h^4\frac{\partial^4 u}{\partial x^4} - \ldots\right)_{k,m}, \tag{2.5}$$

$$u_{k,m+1} = \left(u + h\frac{\partial u}{\partial y} + \frac{1}{2}h^2\frac{\partial^2 u}{\partial y^2} + \frac{1}{6}h^3\frac{\partial^3 u}{\partial y^3} + \frac{1}{24}h^4\frac{\partial^4 u}{\partial y^4} + \ldots\right)_{k,m}, \tag{2.6}$$

$$u_{k,m-1} = \left(u - h\frac{\partial u}{\partial y} + \frac{1}{2}h^2\frac{\partial^2 u}{\partial y^2} - \frac{1}{6}h^3\frac{\partial^3 u}{\partial y^3} + \frac{1}{24}h^4\frac{\partial^4 u}{\partial y^4} - \ldots\right)_{k,m}, \tag{2.7}$$

where it is assumed that u is sufficiently often differentiable with respect to both x and y. Adding (2.4), (2.5), (2.6) and (2.7) leads to

$$u_{k+1,m} + u_{k,m+1} + u_{k-1,m} + u_{k,m-1} - 4u_{k,m}$$

$$- h^2 \left(\frac{\partial^2 u}{\partial x^2} + \frac{\partial^2 u}{\partial y^2} \right)_{k,m} = \frac{1}{12} h^4 \left(\frac{\partial^4 u}{\partial x^4} + \frac{\partial^4 u}{\partial y^4} \right)_{k,m} + \ldots . \qquad (2.8)$$

From (2.3), it is clear that the term in h^2 in (2.8) vanishes. The left-hand side then yields the *five point* formula

$$U_{k+1,m} + U_{k,m+1} + U_{k-1,m} + U_{k,m-1} - 4U_{k,m} = 0 \qquad (2.9)$$

which is a finite difference replacement of Laplace's equation (2.3). After dividing by h^2, the right-hand side of (2.8) gives the *local truncation error* of (2.9); the *principal part* of this error is

$$\frac{1}{12} h^2 \left(\frac{\partial^4 u}{\partial x^4} + \frac{\partial^4 u}{\partial y^4} \right)_{k,m}.$$

This error is said to be of order h^2, written as $0(h^2)$.

Equation (2.9) is now applied to all M^2 interior mesh points in the order

$$(x_1, y_m), (x_2, y_m), \ldots, (x_M, y_m)$$

for each $m = 1, 2, \ldots, M$. It is noted that, when (2.9) is applied to the $4M-4$ interior mesh points adjacent to the boundary ∂R, the known values of the boundary conditions are involved. Since, either x or y is constant along each side of the square, the boundary conditions may be written

$$u = \begin{cases} f_0(x) \text{ along } y = 0 \text{ with } 0 \leqslant x \leqslant 1, \\ f_1(x) \text{ along } y = 1 \text{ with } 0 \leqslant x \leqslant 1, \\ g_0(y) \text{ along } x = 0 \text{ with } 0 \leqslant y \leqslant 1, \\ g_1(y) \text{ along } x = 1 \text{ with } 0 \leqslant y \leqslant 1, \end{cases} \qquad (2.10)$$

Applying (2.9) to the interior mesh points gives a linear system of M^2 equations in the function values $U_{k,m}$ ($k, m = 1, 2, \ldots, M$). It is convenient to write this system in matrix form as

$$A\mathbf{U} = \mathbf{b}, \qquad (2.11)$$

Sec. 2.1] The Dirchlet, Neumann and Robbins Problems

where A is an M^2-square matrix which may be written in block tridiagonal form as

$$A = \begin{bmatrix} B & I & & & O \\ I & B & I & & \\ & \ddots & \ddots & \ddots & \\ & & I & B & I \\ O & & & I & B \end{bmatrix} \qquad (2.12)$$

In (2.12) I is the identity matrix of order M and B is the tridiagonal matrix

$$B = \begin{bmatrix} -4 & 1 & & & O \\ 1 & -4 & 1 & & \\ & \ddots & \ddots & \ddots & \\ & & 1 & -4 & 1 \\ O & & & 1 & -4 \end{bmatrix} \qquad (2.13)$$

The eigenvalues of A are given by

$$\lambda_{i,j} = -4\left(\sin^2 \frac{i\pi}{2(M+1)} + \sin^2 \frac{j\pi}{2(M+1)}\right); \; (i,j = 1, 2, \ldots, M).$$

The vector $\mathbf{U}$ is of order $M^2 \times 1$; and is given by

$$\mathbf{U} = [U_{1,1}, U_{2,1}, \ldots, U_{M,1}; U_{1,2}, \ldots, U_{M,2}; \ldots; U_{1,M}, U_{2,M}, \ldots, U_{M,M}]^T, \qquad (2.14)$$

T denoting transpose, the elements being the unknown values of the function u. The vector $\mathbf{b}$ is the vector of boundary condition and is also or order $M^2 \times 1$; its elements are

$$\mathbf{b} = [-f_0(x_1) - g_0(y_1), -f_0(x_2), \ldots, -f_0(x_{M-1}), -f_0(x_M) - g_1(y_1);$$
$$-g_0(y_2), 0, \ldots, 0, -g_1(y_2); \ldots; -g_0(y_{M-1}), 0, \ldots, 0, -g_1(y_{M-1});$$
$$-f_1(x_1) - g_0(y_M), -f_1(x_2), \ldots, -f_1(x_{M-1}), -f_1(x_M) - g_1(x_M)]^T. \qquad (2.15)$$

It is noted that, in (2.15), there are $4M-4$ non-zero elements.

The matrix A, as given in (2.12), is non-singular and the soluton of (2.3) with boundary conditions (2.1) is unique and is given by

$$\mathbf{U} = A^{-1}\mathbf{b}.$$

The solution may be computed using one of the direct or iterative methods named in Chapter 1. The matrix A is sparse and an interative method will require less storage as only the non-zero elements, not more than five in each row, need be stored.

In recent years a large research literature has grown on methods for the numerical solution of large sparse systems of the form (2.11). Such methods are applicable to the Dirichlet problem when the solution is to be computed using a very fine mesh, and to the numerical computation of heat flow in three space dimension, time-dependent models such as that of Twizell and Smith (1981). For full details of methods for solving large sparse systems the reader is referred to the works in Barker (1977).

2.1.2 The Neumann problem

The Neumann problem requires the solution $u(x, y)$ of (2.1) over $R \cup \partial R$ with the boundary condition given in the form

$$\frac{\partial u}{\partial n} = \phi(x, y)$$

on ∂R. Here, $\partial/\partial n$ denotes differentiation along the normal to ∂R directed away from R, and it is assumed that

$$\int_{\partial R} \phi(x, y) \, dr = 0,$$

where dr is an element of length along ∂R.

To illustrate the Neumann problem, Laplace's equation (2.3) will be solved on $R \cup \partial R$, where R is the unit square with boundary ∂R consisting of the lines $x = 0$, $x = 1$, $y = 0$, $y = 1$, as for the Dirichlet problem. Along each side of the square, one of x or y is constant and the derivative boundary conditions may be written

$$-\frac{\partial u}{\partial x} = \phi(0, y) \text{ along } x = 0,$$

$$\frac{\partial u}{\partial x} = \phi(1, y) \text{ along } x = 1,$$

$$-\frac{\partial u}{\partial y} = \phi(x, 0) \text{ along } y = 0,$$

$$\frac{\partial u}{\partial y} = \phi(x, 1) \text{ along } y = 1,$$

(2.16)

where the directions of $\partial u/\partial x$ and $\partial u/\partial y$ are those of increasing x and y; a minus sign is attached, where appropriate, to ensure that normal derivatives are all directed outwards.

Sec. 2.1]   **The Dirchlet, Neumann and Robbins Problems**   47

Superimposing the same uniform mesh on $R \cup \partial R$ as for the Dirichlet problem, the solution to (2.3) is to be found at $(M+2)^2$ points altogether, M^2 in R and $4M+4$ on ∂R.

The five point formula (2.9) is now applied to all $(M+2)^2$ points of the mesh. Applying (2.9) to the $4M+4$ points on ∂R involves the points $(kh, -h)$, $(kh, 1+h)$, $(-h, mh)$, $(1+h, mh)$, where $k, m = 0, 1, \ldots, M+1$, which lie outside ∂R. The derivative boundary conditions are used to eliminate these exterior points. Equation (2.9) has principal truncation error of order h^4, and replacing the derivatives in (2.16) with approximations which are first order accurate (that is, error = $O(h^2)$), leads to approximations to u at the $4M+4$ exterior points which are one order of accuracy fewer than (2.9). The replacements for the derivatives in (2.16) are

$$\frac{\partial u}{\partial x} = [u(x+h, y) - u(x-h, y)]/2h + O(h^2),$$
$$\frac{\partial u}{\partial y} = [u(x, y+h) - u(x, y-h)]/2h + O(h^2).$$
(2.17)

Using these replacements in (2.14), and using the notations $\phi_{0,m} \equiv \phi(0, mh)$, $\phi_{M+1,m} \equiv \phi(1, mh)$, $\phi_{k,0} \equiv \phi(kh, 0)$, $\phi_{k,M+1} \equiv \phi(kh, 1)$ where $k, m = 0, 1, \ldots, M+1$, leads to

$$U_{-1,m} = U_{1,m} + 2h\phi_{0,m} + O(h^3),$$

$$U_{M+2,m} = U_{M,m} + 2h\phi_{M+1,m} + O(h^3),$$

$$U_{k,-1} = U_{k,1} + 2h\phi_{k,0} + O(h^3),$$

$$U_{k,M+2} = U_{k,M} + 2h\phi_{k,M+1} + O(h^3)$$

which may be used when (2.9) is applied to points along $x=0, x=1, y=0, y=1$ respectively. In this way, the boundary conditions (2.16) are incorporated into the five point scheme (2.9).

The $(M+2)^2$ linear equations which arise from the application of (2.9) to the mesh points of $R \cup \partial R$ may be written in the matrix form

$$A\mathbf{U} = 2h\boldsymbol{\phi},$$
(2.18)

where, now, A is a matrix of order $(M+2)^2$ given in block form as

$$A = \begin{bmatrix} B & 2I & & & \\ I & B & I & & O \\ & & \ddots & \ddots & \ddots \\ & O & & I & B & I \\ & & & & 2I & B \end{bmatrix}.$$
(2.19)

The matrix B for the Neumann problem is tridiagonal of order $M+2$ and is given by

$$B = \begin{bmatrix} -4 & 2 & & & & \\ 1 & -4 & 1 & & 0 & \\ & 1 & -4 & 1 & & \\ & & \ddots & \ddots & \ddots & \\ & & & 1 & -4 & 1 \\ & 0 & & & 2 & -4 \end{bmatrix};$$

the matrix I in (2.19) is the identify matrix of order $M+2$.

The vector $\mathbf{U}$ is the vector of computed values of the solution at the $(M+2)^2$ mesh points of $R \cup \partial R$; it is given by

$$\mathbf{U} = [U_{0,0}, U_{1,0}, \ldots, U_{M+1,0}; U_{0,1}, U_{1,1}, \ldots, U_{M+1,1}; \ldots ;$$

$$U_{0,M}, U_{1,M}, \ldots, U_{M+1,M}; U_{0,M+1}, U_{1,M+1}, \ldots, U_{M+1,M+1}]^T. \quad (2.20)$$

The vector $\boldsymbol{\phi}$ is the vector of boundary conditions; it is of order $(M+1)^2 \times 1$ and is given by

$$\boldsymbol{\phi} = [2\phi_{0,0}, \phi_{1,0}, \ldots, \phi_{M,0}, 2\phi_{M+1,0}; \phi_{0,1}, 0, \ldots, 0, \phi_{M+1,1}; \ldots ;$$

$$\phi_{0,M}, 0, \ldots, 0, \phi_{M+1,M}; 2\phi_{0,M+1}, \phi_{1,M+1}, \ldots, \phi_{M,M+1}, 2\phi_{M+1,M+1}]^T.$$
(2.21)

The solution of the matrix equation (2.18) is

$$\mathbf{U} = 2hA^{-1}\boldsymbol{\phi}. \qquad (2.22)$$

However the eigenvalues of the matrix A are given by

$$\lambda_{i,j} = -4\left(\sin^2 \frac{i\pi}{2(M+1)} + \sin^2 \frac{j\pi}{2(M+1)}\right),$$

where $i,j = 0, 1, \ldots, M+1$, and it is clear that $\lambda_{0,0} = 0$. The matrix A is therefore singular and the Neumann problem does not have a unique solution. The system (2.18) has only M^2+4M+3 linearly independent equations, it being possible to write the remaining equation as a linear combination of the others. The solution of (2.18) thus involves an arbitrary constant, this always being a characteristic of the solution of the Neumann problem. The system is solved by putting, say, $U_{M+1,M+1}$ equal to some arbitrary constant and solving, say, the first M^2+4M+3 equations for the other $U_{k,m}$ in terms of $U_{M+1,M+1}$.

2.1.3 The Robbins problem
The Robbins problem requires the solution $u(x, y)$ of (2.1) over $R \cup \partial R$ with boundary conditions which are a mixture of function values and first derivatives.

Sec. 2.1]   The Dirchlet, Neumann and Robbins Problems   49

To illustrate the Robbins problem, Laplace's equation (2.3) will again be solved on $R \cup \partial R$ where R is the unit square bounded by the lines $x, y = 0$ and $x, y = 1$. Along each side of ∂R, one of x or y is constant and the boundary conditions associated with the Robbins problem may be expressed as

$$-\frac{\partial u}{\partial x} + p_0 u = g_0(y), 0 \leq y \leq 1 \quad (2.23)$$

$$\frac{\partial u}{\partial x} + p_1 u = g_1(y), 0 \leq y \leq 1 \quad (2.24)$$

$$-\frac{\partial u}{\partial y} + q_0 u = f_0(x), 0 \leq x \leq 1 \quad (2.25)$$

$$\frac{\partial u}{\partial y} + q_1 u = f_1(x), 0 \leq x \leq 1 \quad (2.26)$$

along $x=0$, $x=1$, $y=0$, $y=1$, respectively. In (2.23), (2.24), (2.25), (2.26) the directions of $\partial u/\partial x$ and $\partial u/\partial y$ are those of positive increasing x and y and attaching a minus a sign to the derivatives in equations (2.23), (2.25) ensures that each normal derivative to ∂R is directed away from the interior of R. The quantities p_0, p_1, q_0, q_1 are real valued and may be positive, negative or zero.

The region $R \cup \partial R$ is covered by the same mesh used in the discussion of the Neumann problem, so that there are, once more, M^2 interior mesh points and $4M+4$ boundary points.

The differential equation (2.3) is replaced by the five-point finite difference scheme (2.9) which is applied to all points of the mesh. Equation (2.9) may be used unaltered when applied to the M^2 interior points of the mesh but, as for the Neumann problem, it requires some modification when applied to the $4M+4$ boundary points where the mixed boundary conditions are used to eliminate points outside $R \cup \partial R$.

The derivative terms in the boundary conditions associated with the Robbins problem will be approximated by the finite difference replacements given in (2.17). The error in each equation of (2.17) is $O(h^2)$. This will have the effect of reducing the accuracy in the computed solution vector $\mathbf{U}$ by one order.

Of the boundary points, the four corner points are dealt with in a different way to the $4M$ others, as follows:

Applying (2.9) to the corner point with coordinates $(0, 0)$ gives

$$U_{1,0} + U_{0,1} + U_{-1,0} + U_{0,-1} - 4U_{0,0} + O(h^4) = 0. \quad (2.27)$$

50   **Elliptic Equations: Finite Difference Methods**   [Ch. 2

The boundary condition (2.23) applies along the line $x = 0$; at the point $(0, 0)$ this gives

$$-\frac{1}{2h}(U_{1,0} - U_{-1,0}) + p_0 U_{0,0} = g_0(0) + O(h^2)$$

leading to

$$U_{-1,0} = -2hp_0 U_{0,0} + U_{1,0} + 2hg_0(0) + O(h^3) \ . \tag{2.28}$$

The boundary condition (2.25) applies along $y=0$; at $(0, 0)$ this gives

$$-\frac{1}{2h}(U_{0,1} - U_{0,-1}) + q_0 U_{0,0} = f_0(0) + O(h^2)$$

leading to

$$U_{0,-1} = -hq_0 U_{0,0} + U_{0,1} + 2hf_0(0) + O(h^3). \tag{2.29}$$

Substituting for $U_{-1,0}$ and $U_{0,-1}$ from (2.28) and (2.29) in (2.27) gives, after dividing by 2,

$$-[h(p_0 + q_0) + 2]U_{0,0} + U_{1,0} + U_{0,1} = -h[f_0(0) + g_0(0)] \tag{2.30}$$

in which the error is $O(h^3)$.

At the corner with coordinates $(1, 0)$, equation (2.9) gives

$$U_{M+2,0} + U_{M+1,1} + U_{M,0} + U_{M+1,-1} - 4U_{M+1,0} + O(h^4) = 0. \tag{2.31}$$

The boundary condition (2.25) applies along the boundary line $y=0$; at the corner point $(1, 0)$ this gives

$$-\frac{1}{2h}(U_{M+1,1} - U_{M+1,-1}) + q_0 U_{M+1,0} = f_0(1) + O(h^2)$$

and thus

$$U_{M+1,-1} = -2hq_0 U_{M+1,0} + U_{M+1,1} + 2hf_0(1) + O(h^3). \tag{2.32}$$

The boundary condition (2.24) applies along $x=1$; at $(1, 0)$ this gives

$$\frac{1}{2h}(U_{M+2,0} - U_{M,0}) + p_1 U_{M+1,0} = g_1(0) + O(h^2)$$

and thus

$$U_{M+2,0} = U_{M,0} - 2hp_1 U_{M+1,0} + 2hg_1(0) + O(h^3). \tag{2.33}$$

Sec. 2.1] The Dirchlet, Neumann and Robbins Problems

Substituting for $U_{M+2,0}$ and $U_{M+1,-1}$ from (2.33) and (2.32) in (2.31) gives

$$U_{M,0} - [h(p_1 + q_0) + 2] U_{M+1,0} + U_{M+1,1} = -h[f_0(1) + g_1(0)]$$
(2.34)

which has error $O(h^3)$.

At the corner point with coordinates (1, 1), equation (2.9) becomes

$$U_{M+2,M+1} + U_{M+1,M+2} + U_{M,M+1} + U_{M+1,M} - 4U_{M+1,M+1} + O(h^4) = 0.$$
(2.35)

The boundary condition (2.24) applies along $x=1$; at (1, 1) this becomes

$$\frac{1}{2h}(U_{M+2,M+1} - U_{M,M+1}) + p_1 U_{M+1,M+1} = g_1(1) + O(h^2)$$

giving

$$U_{M+2,M+1} = U_{M,M+1} - 2hp_1 U_{M+1,M+1} + 2hg_1(1) + O(h^3).$$
(2.36)

The boundary condition (2.26) applies along the boundary line $y=1$; at the corner point (1, 1) this becomes

$$\frac{1}{2h}(U_{M+1,M+2} - U_{M+1,M}) + q_1 U_{M+1,M+1} = f_1(1) + O(h^2)$$

giving

$$U_{M+1,M+2} = U_{M+1,M} - 2hq_1 U_{M+1,M+1} + 2hf_1(1) + O(h^3).$$
(2.37)

Substituting for $U_{M+2,M+1}$ and $U_{M+1,M+2}$ from (2.36) and (2.37) in (2.35) gives

$$U_{M+1,M} + U_{M,M+1} - [h(p_1+q_1) + 2] U_{M+1,M+1} = -h[f_1(1) + g_1(1)]$$
(2.38)

in which error is $O(h^3)$.

The last corner point has coordinates (0, 1). At this point (2.9) becomes

$$U_{1,M+1} + U_{0,M+2} + U_{-1,M+1} + U_{0,M} - 4U_{0,M+1} + O(h^4) = 0.$$
(2.39)

The boundary conditions along the boundary lines $y=1$ and $x=0$ are (2.26) and (2.23) respectively. At (0, 1) equation (2.26) becomes

$$\frac{1}{2h}(U_{0,M+2} - U_{0,M}) + q_1 U_{0,M+1} = f_1(0) + O(h^2)$$

which yields

$$U_{0,M+2} = U_{0,M} - 2hq_1 U_{0,M+1} + 2hf_1(0) + O(h^3).$$
(2.40)

At (0, 1) equation (2.23) becomes

$$-\frac{1}{2h}(U_{1,M+1} - U_{-1,M+1}) + p_0 U_{0,M+1} = g_0(1) + O(h^2)$$

yielding

$$U_{-1,M+1} = -2hp_0 U_{0,M+1} + U_{1,M+1} + 2hg_0(1) + O(h^3). \quad (2.41)$$

Substituting for $U_{0,M+2}$ and $U_{-1,M+1}$ from (2.40) and (2.41) in (2.39) leads to

$$U_{0,M} - [h(p_0 + q_1) + 2]U_{0,M+1} + U_{1,M+1} = -h[f_1(0) + g_0(1)] \quad (2.42)$$

which has error $O(h^3)$.

The $4M$ boundary points which are not corner points are dealt with as follows:

Along the boundary line $x = 0$, the five point scheme (2.9) becomes

$$U_{0,m-1} + U_{1,m} + U_{0,m+1} + U_{-1,m} - 4U_{0,m} + O(h^4) = 0, \quad (2.43)$$

for $m = 1, 2, \ldots, M$. Equation (2.23) is the boundary condition which applies along $x = 0$; it becomes

$$-\frac{1}{2h}(U_{1,m} - U_{-1,m}) + p_0 U_{0,m} = g_0(mh) + O(h^2), (m=1, 2, \ldots, M)$$

giving

$$U_{-1,m} = -2hp_0 U_{0,m} + U_{1,m} + 2hg_0(mh) + O(h^3) \quad (2.44)$$

for $m = 1, 2, \ldots, M$. Substituting this value of $U_{-1,M}$ in (2.43) gives

$$\frac{1}{2}U_{0,m-1} - (hp_0 + 2)U_{0,m} + U_{1,m} + \frac{1}{2}U_{0,m+1} = -hg_0(mh), \quad (2.45)$$

for $m = 1, 2, \ldots, M$, in which the error is $O(h^3)$.

Along the boundary line $x = 1$, the five-point scheme (2.9) becomes

$$U_{M+2,m} + U_{M+1,m+1} + U_{M,m} + U_{M+1,m-1} - 4U_{M+1,m} + O(h^4) = 0, \quad (2.46)$$

for $m = 1, 2, \ldots, M$. The boundary condition which applies along $x = 1$ is given by (2.24); it becomes

$$\frac{1}{2h}(U_{M+2,m} - U_{M,m}) + p_1 U_{M+1,m} = g_1(mh) + O(h^2), (m = 1, 2, \ldots, M)$$

giving

$$U_{M+2,m} = U_{M,m} - 2hp_1 U_{M+1,m} + 2hg_1(mh) + O(h^3), \quad (2.47)$$

Sec. 2.1] The Dirchlet, Neumann and Robbins Problems 53

where $m = 1, 2, \ldots, M$. Substituting for $U_{M+2,m}$ for (2.47) in (2.46) gives

$$\frac{1}{2} U_{M+1,m-1} + U_{M,m} - (hp_1 + 2)U_{M+1,m} + \frac{1}{2} U_{M+1,m+1} = -hg_1(mh), \quad (2.48)$$

for $m = 1, 2, \ldots, M$, which has error $O(h^3)$.

Along the boundary line $y = 0$, equation (2.9) gives

$$U_{k+1,0} + U_{k,1} + U_{k-1,0} + U_{k,-1} - 4U_{k,0} + O(h^4) = 0 \quad (2.49)$$

for $k = 1, 2, \ldots, M$. Equation (2.25) gives the boundary condition which applies along $y = 0$; for $k = 1, 2, \ldots, M$ this equation becomes

$$-\frac{1}{2h}(U_{k,1} - U_{k,-1}) + q_0 U_{k,0} = f_0(kh) + O(h^2)$$

leading to

$$U_{k,-1} = -2hq_0 U_{k,0} + U_{k,1} + 2hf_0(kh) + O(h^3), \quad (2.50)$$

where $k = 1, 2, \ldots, M$. Substituting of $U_{k,-1}$ from (2.50) in (2.49) gives

$$\frac{1}{2} U_{k-1,0} - (hq_0 + 2)U_{k,0} + \frac{1}{2} U_{k+1,0} + U_{k,1} = -hf_0(kh), \quad (2.51)$$

in which $k = 1, 2, \ldots, M$ and in which the error is $O(h^3)$.

The last boundary line is the line $y = 1$. Along this part of the boundary, the five point formula (2.9) becomes

$$U_{k+1,M+1} + U_{k,M+2} + U_{k-1,M+1} + U_{k,M} - 4U_{k,M+1} + O(h^4) = 0 \quad (2.52)$$

for $k = 1, 2, \ldots, M$. The boundary condition along $y = 1$, equation (2.26), gives

$$\frac{1}{2h}(U_{k,M+2} - U_{k,M}) + q_1 U_{k,M+1} = f_1(kh) + O(h^2), \ (k = 1, 2, \ldots, M)$$

which yields

$$U_{k,M+2} = U_{k,M} - 2hq_1 U_{k,M+1} + 2hf_1(kh) + O(h^3), \quad (2.53)$$

for $k = 1, 2, \ldots, M$. Substituting for $U_{k,M+2}$ from (2.53) in (2.52) gives

$$U_{k,M} + \frac{1}{2} U_{k-1,M+1} - (hq_1 + 2)U_{k,M+1} + \frac{1}{2} U_{k+1,M+1} = -hf_1(kh), \quad (2.54)$$

where $k = 1, 2, \ldots, M$ and the error is $O(h^3)$.

A linear system of the form

$$AU = hg \qquad (2.55)$$

may now be formed, the solution of which, the vector **U**, is the solution of (2.3) over $R \cup \partial R$ with boundary conditions given by equations (2.23), (2.24), (2.25) and (2.26). The elements of **U** are arranged in the order given by (2.20).

The rows of matrix A are compiled by applying the appropriate finite difference equation to each mesh point (kh, mh), $(k, m = 0, 1, \ldots, M+1)$, in $R \cup \partial R$ arranged in the order implied by (2.20). The finite difference schemes to be used in compiling the rows of A are summarised as follows:

  row 1, equation (2.30) is used;
  rows 2 to $M+1$, equation (2.51) is used with $k = 1, 2, \ldots, M$ respectively;
  row $M+2$, equation (2.34) is used;
  rows $i(M+2) + 1$, $(i = 1, 2, \ldots, M)$, equation (2.45) is used with $m = 1, 2, \ldots, M$ respectively;
  rows $i(M+2) + j$, $(j = 2, 3, \ldots, M+1$ for each $i = 1, 2, \ldots, M)$ equation (2.9) divided by 2 is used with $k = 1, 2, \ldots, M$ for each $m = 1, 2, \ldots, M$;
  rows $(i+1)(m+2)$, $(i=1, 2, \ldots, M)$, equation (2.48) is used with $m = 1, 2, \ldots, M$ respectively;
  row $(M+1)(M+2) + 1$, equation (2.42) is used;
  rows $(M+1)(M+2) + j$, $(j = 2, 3, \ldots, M+1)$, equation (2.54) is used with $k = 1, 2, \ldots, M$ respectively;
  row $(M+2)^2$, equation (2.38) is used.

The vector **g** in (2.55) is the vector of boundary conditions with $(M+2)^2$ elements given by

$$[-f_0(0)-g_0(0), -f_0(h), \ldots, -f_0(Mh), -f_0(1)-g_1(0); -g_0(h), 0, \ldots, 0, -g_1(h);$$

$$\ldots; -g_0(Mh), 0, \ldots, 0, -g_1(Mh);$$

$$-f_1(0)-g_0(1), -f_1(h), \ldots, -f_1(Mh), -f_1(1)-g_1(1)]^T. \qquad (2.56)$$

The matrix A in (2.55) is sparse and an iterative technique may be used to find the solution vector **U**. One method of solution is to use an alternating direction implicit (ADI) method as described in the next section.

2.2 ALTERNATING DIRECTION IMPLICIT (ADI) METHODS

Alternating Direction Implicit (ADI) methods were introduced by Peaceman and Rachford (1955) for the numerical solution of elliptic and parabolic partial differential equations. In the case of elliptic equations the method uses an

Sec. 2.2] Alternating Direction Implict (ADI) Methods

iterative procedure which, if it converges at all, converges to the solution of the partial differential equation.

In solving Laplace's equation (2.3), for example, over the mesh superimposed on the unit square described in section 2.1, one of the second order derivatives, say $\partial^2 u/\partial y^2$, is approximated at the rth iterate ($r = 0, 1, \ldots$) by the finite difference replacement

$$\frac{\partial^2 u}{\partial y^2} = [u(x, y-h) - 2u(x, y) + u(x, y+h)]/h^2 + O(h^2) \qquad (2.57)$$

and the other second order derivative by the replacement

$$\frac{\partial^2 u}{\partial x^2} = [u(x-h, y) - 2u(x, y) + u(x+h, y)]/h^2 + O(h^2) \qquad (2.58)$$

at an intermediate iterate $r + \frac{1}{2}$. In keeping with Peaceman and Rachford (1955), this suggests the iteration scheme

$$2U_{k,m}^{(r+\frac{1}{2})} - 2U_{k,m}^{(r)} = [U_{k-1,m}^{(r+\frac{1}{2})} - 2U_{k,m}^{(r+\frac{1}{2})} + U_{k+1,m}^{(r+\frac{1}{2})} + U_{k,m-1}^{(r)} - 2U_{k,m}^{(r)}$$

$$+ U_{k,m+1}^{(r)}]/h^2 + O(h^2) \qquad (2.59)$$

at the point (kh, mh), where $U_{k,m}^{(r)}$ denotes the value of $U_{k,m}$ at the rth iterate.

In proceeding from the intermediate iterate $r + \frac{1}{2}$ to the iterate $r + 1$, $\partial^2 u/\partial y^2$ is now replaced by (2.57) at the iterate $r + 1$ and $\partial^2 u/\partial x^2$ is again replaced by (2.58) at the intermediate iterate. At the mesh point (kh, mh) this suggests the iteration scheme

$$2U_{k,m}^{(r+1)} - 2U_{k,m}^{(r+\frac{1}{2})} = [U_{k-1,m}^{(r+\frac{1}{2})} - 2U_{k,m}^{(r+\frac{1}{2})} + U_{k+1,m}^{(r+\frac{1}{2})} + U_{k,m-1}^{(r+1)} - 2U_{k,m}^{(r+1)}$$

$$+ U_{k,m+1}^{(r+1)}]/h^2 + O(h^2). \qquad (2.60)$$

The value of $U_{k,m}^{(r+1)}$ is found in terms of $U_{k,m}^{(r)}$ by eliminating the intermediate iterates from (2.59) and (2.60). The resulting iteration procedure is said to converge if

$$U_{k,m}^{(r+1)} = U_{k,m}^{(r)} = U_{k,m}$$

for some r and if the $U_{k,m}$ ($k, m = 1, 2, \ldots, M$) satisfy (2.9).

2.2.1 Formulation for the Dirichlet problem

For the Dirichlet problem, the solution vector **U** is computed by solving the matrix equation (2.11) given by

$$A\mathbf{U} = \mathbf{b}$$

where the matrix A is block tridiagonal, as in (2.12). The matrix B in (2.12) may be written as

$$B = -C - 2I, \qquad (2.61)$$

where

$$C = \begin{bmatrix} 2 & -1 & & & 0 \\ -1 & 2 & -1 & & \\ & \ddots & \ddots & \ddots & \\ & & -1 & 2 & -1 \\ 0 & & & -1 & 2 \end{bmatrix} \qquad (2.62)$$

is of of order M, and I is the identity matrix of order M.

It is now easy to show that

$$A = -D - E \qquad (2.63)$$

where

$$D = \begin{bmatrix} C & & & 0 \\ & C & & \\ & & \ddots & \\ 0 & & & C \end{bmatrix} \qquad (2.64)$$

is block diagonal and

$$E = \begin{bmatrix} 2I & -I & & & 0 \\ -I & 2I & -I & & \\ & \ddots & \ddots & \ddots & \\ & & -I & 2I & -I \\ 0 & & & -I & 2I \end{bmatrix} \qquad (2.65)$$

is block tridiagonal.

Using this splitting of the matrix A and the applying the pair of equations (2.59) and (2.60) to every mesh point in R (the interior of the square $0 < x, y < 1$), the resulting ADI method for the Dirichlet problem is given by

$$(2h^2 I + D)\mathbf{U}^{(r+\frac{1}{2})} = (2h^2 I - E)\mathbf{U}^{(r)} - \mathbf{b}$$

and $\qquad (2.66)$

$$(2h^2 I + E)\mathbf{U}^{(r+1)} = (2h^2 I - D)\mathbf{U}^{(r+\frac{1}{2})} - \mathbf{b}$$

where $\mathbf{U}^{(r)}$ denotes $\mathbf{U}$ at the rth iterate and $\mathbf{b}$ is given in (2.15).

Eliminating the intermediate vector $\mathbf{U}^{(r+\frac{1}{2})}$ from (2.66) gives an equation of the form

$$\mathbf{U}^{(r+1)} = F\mathbf{U}^{(r)} + \mathbf{c}, \tag{2.67}$$

where F is the matrix of order M^2 given by

$$F = (2h^2 I + E)^{-1}(2h^2 I + D)^{-1}(2h^2 I - D)(2h^2 I - E) \tag{2.68}$$

and $\mathbf{c}$ is the vector given by

$$\mathbf{c} = -(2h^2 I + E)^{-1}[(2h^2 I + D)^{-1}(2h^2 I - D) + I]\mathbf{b}. \tag{2.69}$$

It is easy to see that the matrices D and E, given by (2.64) and (2.65), commute; they therefore have a common system of orthonormal eigenvectors (Forsythe and Moler, 1967) and their eigenvalues λ_i, μ_j ($i, j = 1, 2, \ldots, M$) are given, respectively, by the equations

$$\lambda_i = 4\sin^2\{i\pi/2(M+1)\}, \quad \mu_j = 4\sin^2\{j\pi/2(M+1)\} \tag{2.70}$$

with $i, j = 1, 2, \ldots, M$.

The eigenvalues of the matrix F are therefore

$$\frac{(2h^2 - \lambda_i)(2h^2 - \mu_j)}{(2h^2 + \lambda_i)(2h^2 + \mu_j)}, \quad i, j = 1, 2, \ldots, M \tag{2.71}$$

and, since each λ_i, μ_j satisfies $0 < \lambda_i, \mu_j < 4$, it is easy to show that the spectral norm of F satisfies

$$\|F\| < 1. \tag{2.72}$$

It was seen in Chapter 1 that an iterative procedure of the form (2.67) is convergent when (2.72) is satisfied and so the ADI method considered in this section is convergent to $(D+E)\mathbf{U} = -\mathbf{b}$ which is equivalent to (2.11) since $A = -D-E$.

A more general treatment of ADI methods for the Dirichlet problem may be found in the text by Mitchell and Griffiths (1980).

2.2.2 Formulation for the Neumann problem

In the case of the Neumann problem, the solution vector $\mathbf{U}$ is computed by solving the matrix equation (2.18) given by

$$A\mathbf{U} = 2h\boldsymbol{\phi}$$

where, again, the matrix A is block tridiagonal as in (2.19). The matrix B in (2.19) may be written in the form

$$B = -G - 2I \tag{2.73}$$

where

$$G = \begin{bmatrix} 2 & -2 & & & & \\ -1 & 2 & -1 & & 0 & \\ & -1 & 2 & -1 & & \\ & & \ddots & \ddots & \ddots & \\ & & & -1 & 2 & -1 \\ & 0 & & & -2 & 2 \end{bmatrix} \quad (2.74)$$

is of order $M+2$ and I is now the identity matrix of order $M+2$. It may be shown that

$$A = -H - J, \quad (2.75)$$

where

$$H = \begin{bmatrix} G & & & \\ & G & & O \\ & & \ddots & \\ & O & & G \end{bmatrix} \quad (2.76)$$

is block diagonal and

$$J = \begin{bmatrix} 2I & -2I & & & & \\ -I & 2I & -I & & O & \\ & \ddots & \ddots & \ddots & & \\ & & & -I & 2I & -I \\ & O & & & -2I & 2I \end{bmatrix} \quad (2.77)$$

is block tridiagonal.

Using the matrices H and J in the application of (2.59) and (2.60) to all $(M+2)^2$ mesh point of $R \cup \partial R$, the resulting ADI method for the Neumann problem is given by

$$(2h^2 I + H)\mathbf{U}^{(r+\frac{1}{2})} = (2h^2 I - J)\mathbf{U}^{(r)} - 2h\phi$$

and $\quad (2.78)$

$$(2h^2 I + J)\mathbf{U}^{(r+1)} = (2h^2 I - H)\mathbf{U}^{(r+\frac{1}{2})} - 2h\phi,$$

where ϕ is given by (2.21).

Eliminating the intermediate vector $\mathbf{U}^{(r+\frac{1}{2})}$ from (2.78) gives

$$\mathbf{U}^{(r+1)} = K\mathbf{U}^{(r)} + \psi \quad (2.79)$$

where K is the matrix of order $(M+2)^2$ given by

$$K = (2h^2 I + J)^{-1}(2h^2 I + H)^{-1}(2h^2 I - H)(2h^2 I - J) \quad (2.80)$$

and ψ is the vector given by

$$\psi = -2h(2h^2 I + J)^{-1}[(2h^2 I + H)^{-1}(2h^2 I - H) + I]\phi. \quad (2.81)$$

The matrices H and J clearly commute and they therefore have a common system of orthonormal eigenvectors. Their eigenvalues λ_i and μ_j ($i, j = 0, 1, \ldots, M+1$) are given by

$$\lambda_i = 4\sin^2\{i\pi/2(M+1)\} \text{ and } \mu_j = 4\sin^2\{j\pi/2(M+1)\}$$

where $i, j = 0, 1, \ldots, M+1$. It follows that the eigenvalues of K are

$$\frac{(2h^2 - \lambda_i)(2h^2 - \mu_j)}{(2h^2 + \lambda_i)(2h^2 + \mu_j)}$$

and, since $\lambda_0 = \mu_0 = 0$, the spectral radius of K is unity; hence the sufficiency condition, $\|K\| < 1$, for the convergence of (2.79) to the solution vector $\mathbf{U}$, is violated.

However, the matrix $A = -H - J$ in (2.17) is negative semi-definite since the eigenvalues of H and J satisfy $\lambda_i, \mu_j \geq 0$ ($i, j = 0, 1, \ldots, M+1$) and using the results of Douglas and Pearcey (1963), the formulas given by (2.78) do converge to $(H+J)\mathbf{U} = -2h\boldsymbol{\phi}$ which is equivalent to the formulation of the Neumann problem given in (2.18).

2.2.3 Formulation for the Robbins problem

The solution vector $\mathbf{U}$ of the Robbins problem defined by (2.3), (2.23), (2.24), (2.25), (2.26) is computed by solving the matrix equation (2.55) given by

$$A\mathbf{U} = h\mathbf{g}$$

where, now, the matrix A is that outlined following (2.55).

It is not difficult to show that the matrix A may be written as

$$A = P - Q, \qquad (2.82)$$

where P and Q are of order $(M+2)^2$ and are given by

$$P = \begin{bmatrix} L & & & \\ & L & & O \\ & & \ddots & \\ O & & & L \end{bmatrix} \qquad (2.83)$$

and

$$Q = \begin{bmatrix} (hq_0+1)I & -I & & & \\ -\tfrac{1}{2}I & I & -\tfrac{1}{2}I & & O \\ & -\tfrac{1}{2}I & I & -\tfrac{1}{2}I & \\ & & \ddots & \ddots & \ddots \\ & O & & -\tfrac{1}{2}I & I & -\tfrac{1}{2}I \\ & & & & -I & (hq_1+1)I \end{bmatrix} \qquad (2.84)$$

In (2.83) L is a matrix of order $M+2$ given by

$$L = \begin{bmatrix} hp_0+1 & -1 & & & 0 \\ -\tfrac{1}{2} & 1 & -\tfrac{1}{2} & & \\ & \ddots & \ddots & \ddots & \\ & & -\tfrac{1}{2} & 1 & -\tfrac{1}{2} \\ 0 & & & -1 & hp_1+1 \end{bmatrix} \quad (2.85)$$

and in (2.84) I is the identity matrix of order $M+2$.

It follows that, for the Robbins problem, equations (2.59) and (2.60) may be written in the form

$$(2h^2 I + P)\mathbf{U}^{(r+\tfrac{1}{2})} = (2h^2 I - Q)\mathbf{U}^{(r)} - h\mathbf{g}$$

and
$$(2h^2 I + Q)\mathbf{U}^{(r+1)} = (2h^2 I - P)\mathbf{U}^{(r)} - h\mathbf{g}$$
(2.86)

where $\mathbf{g}$ is given by (2.56).

Eliminating the intermediate vector $\mathbf{U}^{(r+\tfrac{1}{2})}$ from (2.86) gives an equation of the form

$$\mathbf{U}^{(r+1)} = S\mathbf{U}^{(r)} + \mathbf{k} \quad (2.87)$$

where S is the matrix of order $(M+2)^2$ given in terms of P and Q by

$$S = (2h^2 I + Q)^{-1}(2h^2 I + P)^{-1}(2h^2 I - P)(2h^2 I - Q) \quad (2.88)$$

and $\mathbf{k}$ is the vector given in terms of $\mathbf{g}$ by

$$\mathbf{k} = -h(2h^2 I + Q)^{-1}[(2h^2 I + P)^{-1}(2h^2 I - P) + I]\mathbf{g}. \quad (2.89)$$

It is easy to see that the matrices P and Q commute and thus have a common system of orthonormal eigenvectors. Denoting the eigenvalues of P and Q by λ_i, μ_j $(i, j = 0, 1, \ldots, M+1)$, respectively, the eigenvalues of S are given by

$$\frac{(2h^2 - \lambda_i)(2h^2 - \mu_j)}{(2h^2 + \lambda_i)(2h^2 + \mu_j)} \quad (2.90)$$

where $i, j = 0, 1, \ldots, M+1$. The sufficient condition for (2.87) to converge to the solution vector $\mathbf{U}$ is $\|S\| < 1$; this means that the expression in (2.90) must be less than unity in modulus for all $i, j = 0, 1, \ldots, M+1$, a requirement which is satisfied provided

$$\lambda_i + \mu_j > 0 \text{ and } \lambda_i \mu_j > -4h^2 \quad (2.91)$$

for all $i, j = 0, 1, \ldots, M+1$.

The eigenvalues of $A = -P-Q$ are $-\lambda_i - \mu_j$ $(i, j = 0, 1, \ldots, M+1)$. The ADI method (2.86) is therefore convergent if and only if A is negative definite or

Sec. 2.3] **Irregular Boundaries** 61

negative semi-definite. Keast and Mitchell (1967) have shown that convergence follows if either

and
$$p_i p_j + p_i + p_j > 0 \text{ with } p_i + p_j > 0$$
$$q_i q_j + q_i + q_j > 0 \text{ with } q_i + q_j > 0$$
$(i, j) = (0, 1)$ or $(1, 0)$, (2.92)

or

and
$$p_i p_j + p_i + p_j = 0 \text{ with } p_i + p_j \geq 0$$
$$q_i q_j + q_i + q_j = 0 \text{ with } q_i + q_j \geq 0$$
$(i, j) = (0, 1)$ or $(1, 0)$. (2.93)

2.3 IRREGULAR BOUNDARIES

So far in this chapter, the boundary ∂R of the region, R, of integration has been rectangular so that it has been possible to choose a rectangular mesh with points lying on ∂R. This has made the application of a finite difference scheme, such as (2.9), for solving the elliptic partial differential equation, quite easy.

In the mathematical modelling of, for example, displacements in an elastic membrane, boundaries are not rectangular. Indeed, in the case of some membranes within the human body, rarely, if ever, are they even regular and it is very appropriate to consider curved boundaries.

In its steady state, the model membrane may be assumed to lie in a horizontal position. Rotation of the membrane about either coordinate axis causes gravitational forces to act on the membrane material, distorting the membrane and thus displacing the points of its surface. The magnitudes and directions of such displacements for various rotations of the membrane will be discussed in this section.

By way of illustration, the geometry of the membrane will be taken to be that discussed in the works by Hudetz (1973), Twizell and Curran (1977), Twizell (1980) and Finlayson (1982), where the shape of the boundary ∂R is shown in Fig. 2.1. The maximum length of this membrane, which is located in the human inner ear, is taken to be 0.256 cm, and the maximum breadth to be 0.192 cm.

The portion ACB in Fig. 2.1 is given by the ellipse

$$\frac{x^2}{0.009216} + \frac{y^2}{0.016384} = 1 \qquad (2.94)$$

(Twizell, 1980) and the portion ADB by the polynomial

$$x = -0.084 - 5.0y + 2.1701y^2 + 54.2535y^3 \qquad (2.95)$$

(Finlayson, 1982). The membrane is disretized by superimposing a square mesh of length $h = 0.032$ cm on R producing 30 interior nodes as in Fig. 2.1.

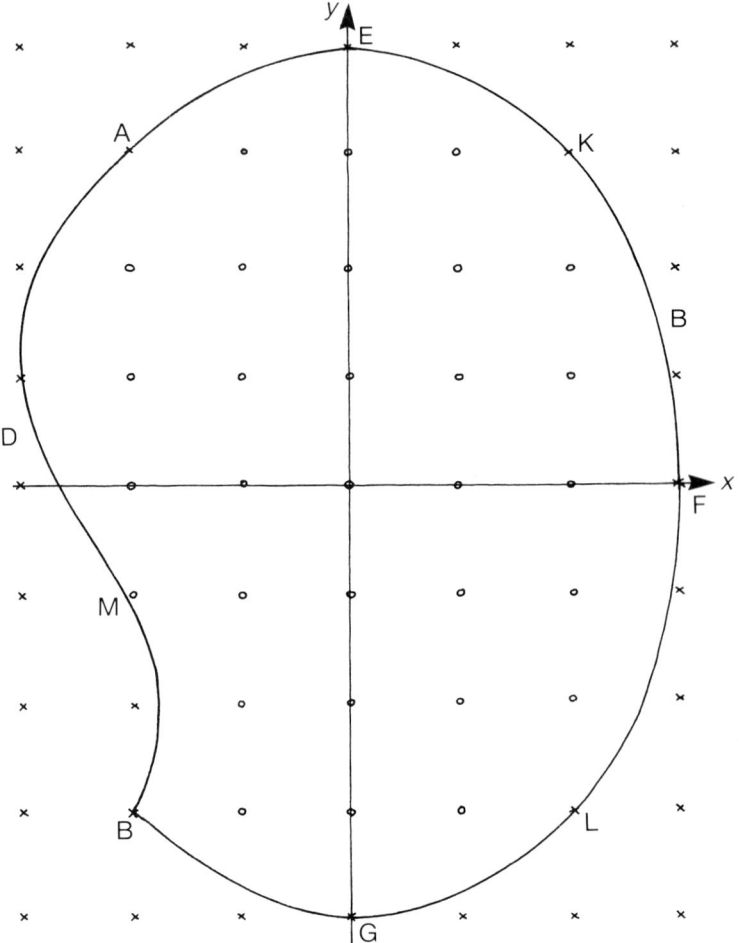

Fig. 2.1 — The model elastic membrane.

It is known from the small displacement theory of elasticity that the displacements u, v in the positive x, y directions of the points of the membrane, following a rotation of the membrane, are governed by a system of elliptic partial differential equations of the form

$$\frac{\partial^2 u}{\partial x^2} + \frac{1-\nu}{2}\frac{\partial^2 u}{\partial y^2} + \frac{1+\nu}{2}\frac{\partial^2 v}{\partial x \partial y} = -\frac{1-\nu^2}{E}F^{(x)} \qquad (2.96)$$

and

Sec. 2.3] Irregular Boundaries

$$\frac{\partial^2 v}{\partial y^2} + \frac{1-\nu}{2}\frac{\partial^2 v}{\partial x^2} + \frac{1+\nu}{2}\frac{\partial^2 u}{\partial x \partial y} = -\frac{1-\nu^2}{E}F^{(y)} \qquad (2.97)$$

in R together with the Dirichlet boundary conditions

$$u(x, y) = v(x, y) = 0 \qquad (2.98)$$

for all $(x, y) \in \partial R$. In (2.96), (2.97), ν represents the Poisson ratio and E the value of Young's modulus of the membrane material. The terms $F^{(x)}$, $F^{(y)}$ are the x, y components of the gravitational force acting on the membrane following a rotation about either one or both coordinate axes.

The coordinates of the 30 mesh points in Fig. 2.1 are given by (kh, mh) where $h = 0.032$, $k = -2, -1, 0, 1, 2$ and $m = -3, -2, \ldots, 2, 3$. It is noted that this discretization leads to three points of the mesh on ∂R (the points E, F, G in Fig. 2.1) and five other points very close to the boundary (the points A, K, L, B, M in Fig. 2.1).

The theoretical solution of the differential equations (2.96), (2.97) at the mesh point (kh, mh), given by $u(kh, mh)$ and $v(kh, mh)$, will be denoted by $u_{k,m}$ and $v_{k,m}$, and their values computed as a result of some finite difference replacements of the derivatives in (2.96), (2.97) by $U_{k,m}$ and $V_{k,m}$.

The derivatives in (2.96) and (2.97) are replaced at each interior mesh point by the central difference replacements

$$\frac{\partial^2 u}{\partial x^2} = (u_{k-1,m} - 2u_{k,m} + u_{k+1,m})/h^2 + O(h^2), \qquad (2.99)$$

$$\frac{\partial^2 v}{\partial x^2} = (v_{k-1,m} - 2v_{k,m} + v_{k+1,m})/h^2 + O(h^2), \qquad (2.100)$$

$$\frac{\partial^2 u}{\partial y^2} = (u_{k,m-1} - 2u_{k,m} + u_{k,m+1})/h^2 + O(h^2), \qquad (2.101)$$

$$\frac{\partial^2 v}{\partial y^2} = (v_{k,m-1} - 2v_{k,m} + v_{k,m+1})/h^2 + O(h^2), \qquad (2.102)$$

$$\frac{\partial^2 u}{\partial x \partial y} = (u_{k-1,m-1} - u_{k+1,m-1} - u_{k-1,m+1} + u_{k+1,m+1})/(4h^2) + O(h^2), \qquad (2.103)$$

$$\frac{\partial^2 v}{\partial x \partial y} = (v_{k-1,m-1} - v_{k+1,m-1} - v_{k-1,m+1} + v_{k+1,m+1})/(4h^2) + O(h^2). \qquad (2.104)$$

Substituting (2.99), (2.101), (2.104) in (2.96) gives the difference scheme

$$\frac{1-\nu}{2h^2} U_{k,m-1} + \frac{1}{h^2} U_{k-1,m} - \frac{2}{h^2}\left(1 + \frac{1-\nu}{2}\right) U_{k,m} + \frac{1}{h^2} U_{k+1,m}$$

$$+ \frac{1-\nu}{2h^2} U_{k,m+1} + \frac{1+\nu}{8h^2}\left(V_{k-1,m-1} - V_{k+1,m-1} - V_{k-1,m+1} + V_{k+1,m+1}\right)$$

$$= -\frac{1-\nu^2}{E} F^{(x)} \qquad (2.105)$$

and substituting (2.100), (2.102), (2.103) in (2.97) gives

$$\frac{1+\nu}{8h^2}\left(U_{k-1,m-1} - U_{k+1,m-1} - U_{k-1,m+1} + U_{k+1,m+1}\right)$$

$$+ \frac{1}{h^2} V_{k,m-1} + \frac{1-\nu}{2h^2} V_{k-1,m} - \frac{2}{h^2}\left(1 + \frac{1-\nu}{2}\right) V_{k,m} + \frac{1-\nu}{2h^2} V_{k+1,m}$$

$$+ \frac{1}{h^2} V_{k,m+1} = -\frac{1-\nu^2}{E} F^{(y)}. \qquad (2.106)$$

Expanding (2.105) with U, V replaced by u, v about $u_{k,m}$ and $v_{k,m}$ using Taylor's series, shows that (2.105) is consistent with (2.96) and has local truncation error with principal part

$$\frac{h^2}{12}\frac{\partial^4 u}{\partial x^4} + \frac{(1-\nu)h^2}{24}\frac{\partial^4 u}{\partial y^4} + \frac{(1+\nu)h^2}{12}\left(\frac{\partial^4 v}{\partial x^3 \partial y} + \frac{\partial^4 v}{\partial x \partial y^3}\right). \qquad (2.107)$$

It is then easy to see, by interchanging x with y and u with v, that (2.106) is consistent with (2.97) and has local truncation error with principle part

$$\frac{(1+\nu)h^2}{12}\left(\frac{\partial^4 u}{\partial x^3 \partial y} + \frac{\partial^4 u}{\partial x \partial y^3}\right) + \frac{(1-\nu)h^2}{24}\frac{\partial^4 v}{\partial x^4} + \frac{h^2}{12}\frac{\partial^4 v}{\partial y^4}. \qquad (2.108)$$

Equations (2.105) and (2.106) are each applied to the thirty interior mesh points in Fig. 2.1 giving a linear system of sixty equations in the sixty unknown values of U and V. Applying (2.105) to all thirty mesh points and then applying (2.106) ensures that the linear system may be written in block matrix form as

$$\begin{bmatrix} P & Q \\ Q & S \end{bmatrix} \begin{bmatrix} \mathbf{U} \\ \mathbf{V} \end{bmatrix} = \begin{bmatrix} \mathbf{F}^{(x)} \\ \mathbf{F}^{(y)} \end{bmatrix} \qquad (2.109)$$

where P, Q, S are square matrices of order 30, $\mathbf{U}$ and $\mathbf{V}$ are the vectors of values

Sec. 2.3] **Irregular Boundaries** 65

of U and V at the 30 interior grid points, and $\mathbf{F}^{(x)}$, $\mathbf{F}^{(y)}$ are the vectors which have all elements equal to $-(1-\nu^2)F^{(x)}/E$ and $-(1-\nu^2)F^{(y)}/E$ respectively.

When (2.105) and (2.106) are applied to mesh points adjacent to ∂R, mesh points on or outside ∂R must be used and function values outside R must be estimated in terms of function values at internal nodes. This is easily achieved by linear Lagrangian interpolation which was described in Chapter 1.

All interior mesh points adjacent to ∂R are in one or more of three types of position relative to the exterior nodal points adjacent to ∂R. These three types of position are depicted in Fig. 2.2 where a is the interior mesh point, b is the boundary point (where $u = v = 0$) and c is the exterior nodal point.

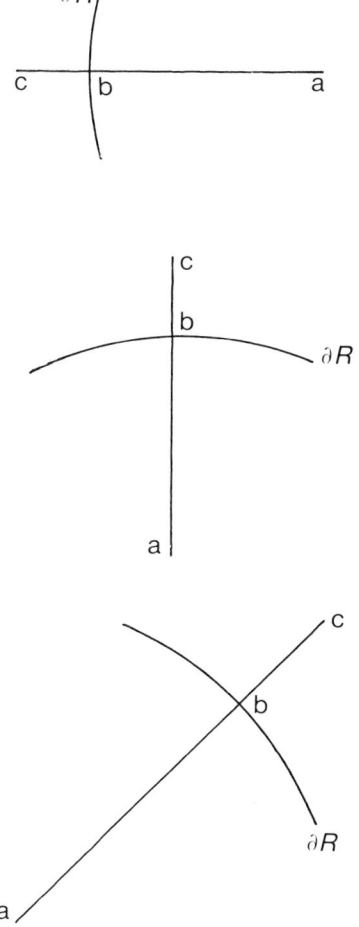

Fig. 2.2 – Three types of position of mesh points adjacent to ∂R.

Denoting, temporarily, the values of x, y, u, v at the mesh point a by x_a, y_a, u_a, v_a, with similar notations at the points b and c, linear Lagrangian interpolation in situation (i) of Fig. 2.2 gives

$$u_x = \frac{x-x_c}{x_a-x_c} u_a + \frac{x-x_a}{x_c-x_a} u_c$$

and

$$v_x = \frac{x-x_c}{x_a-x_c} v_a + \frac{x-x_a}{x_c-x_a} v_c.$$

Writing $x = x_b$, and noting that $u_b = v_b = 0$, leads to

$$u_c = \frac{x_b-x_c}{x_b-x_a} u_a \text{ and } v_c = \frac{x_b-x_c}{x_b-x_a} v_a \qquad (2.110)$$

where the coordinates x_a, x_c are known multiples of h. The y-coordinate of b in situation (i) of Fig. 2.2 is the same as the y-coordinate of a and c, and x_b can thus be found from (2.94) or (2.95), taking care to attach the appropriate sign after taking the square root in (2.94); u_c and v_c can then be written in terms of u_a and v_a using (2.110).

It is easy to see that in situation (ii) of Fig. 2.2 the corresponding values of u_c and v_c are given by

$$u_c = \frac{y_b-y_c}{y_b-y_a} u_a \text{ and } v_c = \frac{y_b-y_c}{y_b-y_a} v_a. \qquad (2.111)$$

The y-coordinate of the boundary point in situation (ii) is determined in the same way as the x-coordinate in situation (i) when a and c are on opposite sides of the ellipse (2.94). The mesh point numbered 19 in Fig. 2.1, however, is separated from the exterior point $(-2h, -2h)$ by the cubic polynomial (2.95). The x-coordinate of the corresponding boundary point is $x_b = -2h$; the value of y_b is thus found by solving the equation

$$f(y) \equiv 54.2535 y^3 + 2.1701 y^2 - 0.5 y - 0.084 + 2h = 0. \qquad (2.112)$$

Taking $y_b^{(0)} = -2h$, the Newton–Raphson method, which was described in Chapter 1 converges quadratically to y_b which is used in (2.111) to give u_c and v_c at the point $(-2h, -2h)$ in terms of the interior point numbered 19 which has coordinates $(-2h, -h)$.

Situation (iii) in Fig. 2.2 is much more difficult to deal with in that neither the x-coordinate nor the y-coordinate of the boundary point b is the same as the corresponding coordinate of a or c. It is known, however, that the boundary point lies at the intersection of a straight line of the form

$$y = x + rh \text{ or } y = -x + rh \qquad (2.113)$$

where r is some integer, positive or negative, which can easily be determined by inspection of Fig. 2.2, and a curve of one of the forms

$$x = \tfrac{3}{4}(0.016384 - y^2)^{1/2},\ x = -\tfrac{3}{4}(0.016384 - y^2),$$
$$x = -0.084 - 0.05y + 2.1701y^2 + 54.2535y^3. \tag{2.114}$$

The choice of line from (2.113) and curve from (2.114) clearly depends on the positions of a and c.

To determine the value of x_b, the appropriate form of y from (2.113) is substituted into the appropriate equation from (2.114). This results in some function of the form

$$x = \phi(x)$$

which may be solved iteratively to give the value of x_b by writing

$$x_b^{(s+1)} = \phi(x_b^{(s)}); s = 0, 1, 2, \ldots \tag{2.115}$$

Taking $x_b^{(0)} = \tfrac{1}{2}(x_a + x_c)$ as starting value, the one-point iteration formula (2.115) will give linear convergence to x_b (see Chapter 1) which is then used in (2.110) to give u_c and v_c.

All information required for the compilation of the linear system (2.109) is now known and this system may be solved using one of the algorithms described in Chapter 1 or a method developed especially for block sparse systems.

The overall displacement at the point (kh, mh) in the membrane is given by

$$(U_{k,m}^2 + V_{k,m}^2)^{1/2}$$

and is inclined at an angle

$$\tan^{-1}(V_{k,m}/U_{k,m})$$

to the positive x-axis.

2.4 NONRECTANGULAR COORDINATES

2.4.1 Transformation of rectangular coordinates

Occasionally, it is useful to have a finite-difference approximation to Laplace's equation (2.3) on a circular region. For a circular region R a finite difference approximation may be derived in terms of polar coordinates. This requires, first of all, the transformation of

$$\frac{\partial^2 u}{\partial x^2} + \frac{\partial^2 u}{\partial y^2} \tag{2.116}$$

into polar form by the formulas of plane polar transformation given by

$$x = r\cos\theta,\ y = r\sin\theta. \tag{2.117}$$

From (2.117) it is easy to derive the expressions

$$dx = \cos\theta \, dr - r\sin\theta \, d\theta$$
$$dy = \sin\theta \, dr + r\cos\theta \, d\theta$$

from which it follows that

$$dr = \cos\theta \, dx + \sin\theta \, dy$$

$$d\theta = -\frac{\sin\theta}{r} dx + \frac{\cos\theta}{r} dy$$

and, therefore, that

$$\frac{\partial r}{\partial x} = \cos\theta, \quad \frac{\partial r}{\partial y} = \sin\theta, \quad \frac{\partial \theta}{\partial x} = -\frac{\sin\theta}{r}, \quad \frac{\partial \theta}{\partial y} = \frac{\cos\theta}{r}. \tag{2.118}$$

Now,

$$\frac{\partial u}{\partial x} = \frac{\partial u}{\partial r}\frac{\partial r}{\partial x} + \frac{\partial u}{\partial \theta}\frac{\partial \theta}{\partial x} = \left(\cos\theta\frac{\partial}{\partial r} - \frac{\sin\theta}{r}\frac{\partial}{\partial \theta}\right)u \tag{2.119}$$

and

$$\frac{\partial u}{\partial y} = \frac{\partial u}{\partial r}\frac{\partial r}{\partial y} + \frac{\partial u}{\partial \theta}\frac{\partial \theta}{\partial y} = \left(\sin\theta\frac{\partial}{\partial r} + \frac{\cos\theta}{r}\frac{\partial}{\partial \theta}\right)u \tag{2.120}$$

so that

$$\frac{\partial^2 u}{\partial x^2} = \left(\cos\theta\frac{\partial}{\partial r} - \frac{\sin\theta}{r}\frac{\partial}{\partial \theta}\right)\left(\cos\theta\frac{\partial u}{\partial r} - \frac{\sin\theta}{r}\frac{\partial u}{\partial \theta}\right)$$

$$= \cos^2\theta\frac{\partial^2 u}{\partial r^2} + \frac{\cos\theta \sin\theta}{r^2}\frac{\partial u}{\partial \theta} - \frac{\cos\theta \sin\theta}{r}\frac{\partial^2 u}{\partial r \partial \theta}$$

$$-\frac{\sin\theta \cos\theta}{r}\frac{\partial^2 u}{\partial \theta \partial r} + \frac{\sin^2\theta}{r}\frac{\partial u}{\partial r} + \frac{\sin^2\theta}{r^2}\frac{\partial^2 u}{\partial \theta^2} + \frac{\sin\theta \cos\theta}{r^2}\frac{\partial u}{\partial \theta} \tag{2.121}$$

and

$$\frac{\partial^2 u}{\partial y^2} = \left(\sin\theta\frac{\partial}{\partial r} + \frac{\cos\theta}{r}\frac{\partial}{\partial \theta}\right)\left(\sin\theta\frac{\partial u}{\partial r} + \frac{\cos\theta}{r}\frac{\partial u}{\partial \theta}\right)$$

$$= \sin^2\theta\frac{\partial^2 u}{\partial r^2} + \frac{\sin\theta \cos\theta}{r}\frac{\partial^2 u}{\partial r \partial \theta} - \frac{\sin\theta \cos\theta}{r^2}\frac{\partial u}{\partial \theta}$$

$$+ \frac{\cos\theta \sin\theta}{r}\frac{\partial^2 u}{\partial \theta \partial r} + \frac{\cos^2\theta}{r}\frac{\partial u}{\partial r} + \frac{\cos^2\theta}{r^2}\frac{\partial^2 u}{\partial \theta^2} - \frac{\sin\theta \cos\theta}{r^2}\frac{\partial u}{\partial \theta}$$

$$\tag{2.122}$$

Sec. 2.4]   **Nonrectangular Coordinates**   69

Adding (2.121) and (2.122) leads to

$$\frac{\partial^2 u}{\partial x^2} + \frac{\partial^2 u}{\partial y^2} = \frac{\partial^2 u}{\partial r^2} + \frac{1}{r}\frac{\partial u}{\partial r} + \frac{1}{r^2}\frac{\partial^2 u}{\partial \theta^2} \tag{2.123}$$

which gives the polar form of the Laplacian operator (2.116).

In three space dimensions, in which the region is a circular cylinder, the Laplacian is

$$\frac{\partial^2 u}{\partial x^2} + \frac{\partial^2 u}{\partial y^2} + \frac{\partial^2 u}{\partial z^2} = \frac{\partial^2 u}{\partial r^2} + \frac{1}{r}\frac{\partial u}{\partial r} + \frac{1}{r^2}\frac{\partial^2 u}{\partial \theta^2} + \frac{\partial^2 u}{\partial z^2}. \tag{2.124}$$

In three space dimensions, in which the region is a sphere, the Laplacian

$$\frac{\partial^2 u}{\partial x^2} + \frac{\partial^2 u}{\partial y^2} + \frac{\partial^2 u}{\partial z^2} \tag{2.125}$$

is transformed using

$$\begin{aligned}x &= w \cos \phi, \quad y = w \sin \phi \\ z &= r \cos \theta, \quad w = r \sin \theta\end{aligned}, \tag{2.126}$$

so that, from (2.123),

$$\frac{\partial^2 u}{\partial x^2} + \frac{\partial^2 u}{\partial y^2} = \frac{\partial^2 u}{\partial w^2} + \frac{1}{w}\frac{\partial u}{\partial w} + \frac{1}{u^2}\frac{\partial^2 u}{\partial \phi^2} \tag{2.127}$$

and

$$\frac{\partial^2 u}{\partial z^2} + \frac{\partial^2 u}{\partial w^2} = \frac{\partial^2 u}{\partial r^2} + \frac{1}{r}\frac{\partial u}{\partial r} + \frac{1}{r^2}\frac{\partial^2 u}{\partial \theta^2}. \tag{2.128}$$

From equation (2.120),

$$\frac{\partial u}{\partial w} = \sin \theta \frac{\partial u}{\partial r} + \frac{\cos \theta}{r}\frac{\partial u}{\partial \theta}.$$

Hence

$$\begin{aligned}\frac{\partial^2 u}{\partial x^2} + \frac{\partial^2 u}{\partial y^2} + \frac{\partial^2 u}{\partial z^2} &= \frac{\partial^2 u}{\partial r^2} + \frac{1}{r^2}\frac{\partial^2 u}{\partial \theta^2} + \frac{1}{r^2 \sin^2 \theta}\frac{\partial^2 u}{\partial \phi^2} \\ &\quad + \frac{1}{r}\frac{\partial u}{\partial r} + \frac{1}{r \sin \theta}\left(\sin \theta \frac{\partial u}{\partial r} + \frac{\cos \theta}{r}\frac{\partial u}{\partial \theta}\right) \\ &= \frac{\partial^2 u}{\partial r^2} + \frac{1}{r^2}\frac{\partial^2 u}{\partial \theta^2} + \frac{1}{r^2 \sin^2 \theta}\frac{\partial^2 u}{\partial \phi^2} + \frac{2}{r}\frac{\partial u}{\partial r} + \frac{\cot \theta}{r^2}\frac{\partial u}{\partial r},\end{aligned}$$

that is,

$$\frac{\partial^2 u}{\partial x^2} + \frac{\partial^2 u}{\partial y^2} + \frac{\partial^2 u}{\partial z^2} = \frac{1}{r^2}\left[\frac{\partial}{\partial r}\left(r^2 \frac{\partial u}{\partial r}\right) + \frac{1}{\sin\theta}\frac{\partial}{\partial \theta}\left(\sin\theta \frac{\partial}{\partial \theta}\right) + \frac{1}{\sin^2\theta}\frac{\partial^2 u}{\partial \phi^2}\right].$$
(2.129)

2.4.2 The Dirichlet problem with a circular region
In subsection 2.1.1 the Dirichlet problem was discussed for Laplace's equation (2.3) with functional boundary values given around the unit square. When the region R is a circle of radius a, centre at the origin, Laplace's equation is expressed in the form

$$\frac{\partial^2 u}{\partial r^2} + \frac{1}{r}\frac{\partial u}{\partial r} + \frac{1}{r^2}\frac{\partial^2 u}{\partial \theta^2} = 0; (r, \theta) \in R \qquad (2.130)$$

and the Dirichlet boundary conditions are expressed in the form

$$u(a, \theta) = f(\theta); \ (a, \theta) \in \partial R. \qquad (2.131)$$

In equation (2.130) $0 \leq r < a$ and in (2.130) and (2.131) $0 \leq \theta < 2\pi$.
The region R is discretized at $KM+1$ points, including the centre $(0, 0)$ by drawing K concentric circles within ∂R and M radii in such a way that one radius is coincident with $\theta = 0$, the circles (including ∂R) are the same distance apart, and adjacent radii make an angle $2\pi/M$ with each other as in Fig. 2.3. The Dirichlet problem requires the solution at each of the $KM+1$ points of the mesh. Defining $\delta r = a/K$ and $\delta\theta = 2\pi/M$, the theoretical value of the solution at the mesh point $(k\delta r, m\delta\theta)$ is $u(k\delta r, m\delta\theta)$ for $k = 0, 1, \ldots, K+1$ and $m = 1, 2, \ldots, M$. The computed value of the solution at this point is $U(k\delta r, m\delta\theta)$ which will be denoted by $U_{k,m}$. It is clear, using this notation, that $k = 0$ refers to the centre of R and $k = K+1$ refers to points on ∂R.

Laplace's equation in the form (2.130) cannot be applied at the origin where $r = 0$. The value of U at the origin will, however, be required in the application of (2.130) to the M mesh points $(\delta r, m\delta\theta), m = 1, 2, \ldots, M$.

An estimate of U at the origin can be determined by recourse to the cartesian form of (2.130) given by (2.3). Replacing (2.3) by the five point formula (2.9) gives

$$U(\delta r, 0) + U(\delta r, \pi/2) + U(\delta r, \pi) + U(\delta r, 3\pi/2) - 4U(0, m) + 0((\delta r)^4) = 0.$$
(2.132)

Rotation of the axes through an angle $\delta\theta$ leads to a similar equation; repetition through $K-2$ further rotations of $\delta\theta$ and addition of the K equations of the form (2.132) leads to

$$U_{0,m} = \sum_{k=1}^{K} U_{1,k}/K. \qquad (2.133)$$

Sec. 2.4] Nonrectangular Coordinates

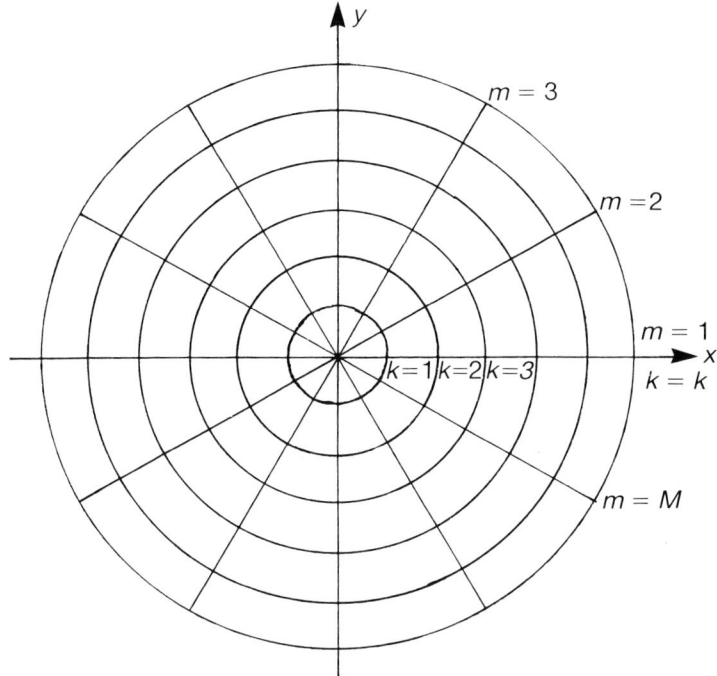

Fig. 2.3 — Discretization of circular region with K concentric circles and M radial lines.

The three derivatives in (2.130) may now be approximated by the finite difference replacements

$$\frac{\partial^2 u}{\partial r^2} = (U_{k-1,m} - 2U_{k,m} + U_{k+1,m})/(\delta r)^2 + 0 \ (\delta r)^2 \ , \quad (2.134)$$

$$\frac{\partial u}{\partial r} = (U_{k+1,m} - U_{k-1,m})/2\delta r + 0 \ ((\delta r)^2), \quad (2.135)$$

$$\frac{\partial^2 u}{\partial \theta^2} = (U_{k,m-1} - 2U_{k,m} + U_{k,m+1})/(\delta \theta)^2 + 0 \ ((\delta \theta)^2). \quad (2.136)$$

Substitution of (2.134), (2.135), (2.136) into (2.130) gives, when applied to the mesh point $(k\delta r, m\delta\theta)$, a difference equation which simplifies to

$$\frac{M^2}{4\pi^2 k^2} U_{k,m-1} + \left(1 - \frac{1}{2k}\right) U_{k-1,m} - 2\left(1 + \frac{M^2}{4\pi^2 k^2}\right) U_{k,m-1}$$
$$+ \left(1 + \frac{1}{2k}\right) U_{k+1,m} + \frac{M^2}{4\pi^2 k^2} U_{k,m+1} = 0. \quad (2.137)$$

72   Elliptic Equations: Finite Difference Methods   [Ch. 2

In problems where Poisson's equation

$$\frac{\partial^2 u}{\partial x^2} + \frac{\partial^2 u}{\partial y^2} = F(x, y) \qquad (2.138)$$

is to be solved in the same circular region R, equation (2.130) becomes

$$\frac{\partial^2 u}{\partial r^2} + \frac{1}{r}\frac{\partial u}{\partial r} + \frac{1}{r^2}\frac{\partial^2 u}{\partial \theta^2} = F(r, \theta) \qquad (2.139)$$

and the difference scheme (2.137) becomes

$$\frac{M^2}{4\pi^2 k^2} U_{k,m-1} + \left(1 - \frac{1}{2k}\right) U_{k-1,m} - 2\left(1 + \frac{M^2}{4\pi^2 k^2}\right) U_{k,m}$$
$$+ \left(1 + \frac{1}{2k}\right) U_{k+1,m} + \frac{M^2}{4\pi^2 k^2} U_{k,m+1} = \frac{a^2}{K^2} F_{k,m}, \qquad (2.140)$$

where $F_{k,m} \equiv F(k\delta r, m\delta\theta)$. The local truncation errors of (2.137) and (2.140) are both

$$0\left((\delta r)^2 + (\delta\theta)^2\right).$$

In compiling the linear system which gives the computed solution of the differential equation, the Poisson problem will be considered (for the solution of Laplace's equation can then be obtained by writing $F \equiv 0$). Equation (2.140) is applied to the KM mesh points $(k\delta r, m\delta\theta)$, $k = 1,\ldots,K$ for each $m = 1,\ldots,M$. This leads to a linear system of the form (2.11), namely

$$A\mathbf{U} = \mathbf{b}, \qquad (2.141)$$

where, now, A is a KM-square matrix, $\mathbf{U}$ is the vector of computed values of u, and $\mathbf{b}$ is the vector of values of F and f; $\mathbf{U}$ and $\mathbf{b}$ each have KM elements.

The ordering of the rows of A and the elements of $\mathbf{U}$ and $\mathbf{b}$ are determined by the order in which (2.140) is applied to the grid points, as described in the preceding paragraph. The elements of the vector $\mathbf{U}$ are thus ordered as follows:

$$\mathbf{U} = [U_{1,1}, U_{2,1},\ldots,U_{K,1}; U_{1,2}, U_{2,2},\ldots,U_{K,2};\ldots;U_{1,M}, U_{2,M},\ldots,U_{K,M}]^T, \qquad (2.142)$$

where T denotes transpose.

The matrix A is sparse and consists of M^2 square submatrices each of order K. The rows of A are ordered in the same way as the elements of $\mathbf{U}$ given by (2.142). The non-zero elements of A are as follows, where it must be remembered that radius M is used for radius 0 (zero) and radius 1 is used for radius $M+1$:

(i) Row 1 corresponds to the application of (2.140) to point 1 on radius 1. In this application the term $U_{0,1}$ is replaced by its estimate given in (2.133). The non-zero elements of row 1 thus have the following locations and values:
column 1, value is

$$-2\left(1 + \frac{M^2}{4\pi^2} - \frac{1}{4K}\right) ;$$

column 2, value is

$$\frac{3}{2} ;$$

column $K+1$, value is

$$\frac{1}{2K} + \frac{M^2}{4\pi^2} ;$$

columns $(m-1)K+1$, for $m = 3, \ldots, M-1$, values are

$$\frac{1}{2K} ;$$

column $(M-1)K+1$, value is

$$\frac{1}{2K} + \frac{M^2}{4\pi^2} .$$

(ii) Row k, for $k = 2, 3, \ldots, K-1$, corresponds to the application of (2.140) to point k on radius 1. The non-zero elements of row k have the following locations and values ($k = 2, 3, \ldots, K-1$):
column $k-1$, value is

$$1 - \frac{1}{2k} ;$$

column k, value is

$$-2\left(1 + \frac{M^2}{4\pi^2 k^2}\right) ;$$

column $k+1$, value is

$$1 + \frac{1}{2k} ;$$

column $K+k$, value is

$$\frac{M^2}{4\pi^2 k^2};$$

column $(M-1)K+k$, value is

$$\frac{M^2}{4\pi^2 k^2}.$$

(iii) Row K corresponds to the application of (2.140) to the grid point $(K\delta r, \delta\theta)$. It is noted that the point $K+1$ on radius 1 is a boundary point so that $U_{K+1,1} = f_1$, where $f_m \equiv f(m\delta\theta)$. The non-zero elements of row K have the following locations and values:
column $K-1$, value is

$$1 - \frac{1}{2K};$$

column K, value is

$$-2\left(1 + \frac{M^2}{4\pi^2 K^2}\right);$$

column $2K$, value is

$$\frac{M^2}{4\pi^2 K^2};$$

column KM, value is

$$\frac{M^2}{4\pi^2 K^2}.$$

(iv) Row $(m-1)K+1$, for $m = 2, 3, \ldots, M-1$, corresponds to the application of (2.140) to the grid point $(\delta r, m\delta\theta)$. In this application the term $U_{0,m}$ is replaced by its estimate given in (2.133). The non-zero elements of row $(m-1)K+1$ have the following locations and values for each $m = 2, 3, \ldots, M-1$:
columns $(j-1)K+1$ for $j = 1, 2, \ldots, M$ except $j = m-1, m, m+1$, values are

$$\frac{1}{2K};$$

column $(m-2)K+1$, value is

$$\frac{1}{2K} + \frac{M^2}{4\pi^2};$$

column $(m-1)K+1$, value is

$$-2\left(1+\frac{M^2}{4\pi^2}-\frac{1}{4K}\right);$$

column $(m-1)K+2$, value is

$$\frac{3}{2};$$

column $mK+1$, value is

$$\frac{M^2}{4\pi^2}.$$

(v) Row $(m-1)K+k$, for $m = 2, 3, \ldots, M-1$ and $k = 2, 3, \ldots, K-1$, corresponds to the application of (2.140) to point k on radial line m. The non-zero elements of this row have the following locations and values:
column $(m-2)K+k$, value is

$$\frac{M^2}{4\pi^2 k^2};$$

column $(m-1)K+k-1$, value is

$$1-\frac{1}{2k};$$

column $(m-1)K+k$, value is

$$-2\left(1+\frac{M^2}{4\pi^2 k^2}\right);$$

column $(m-1)K+k+1$, value is

$$1+\frac{1}{2k};$$

column $mK+k$, value is

$$\frac{M^2}{4\pi^2 k^2}.$$

(vi) Row mK, for $m = 2, 3, \ldots, M-1$, corresponds to the application of (2.140) to point K on radial line m; point $K+1$ on radial line m is a boundary point so that $U_{K+1,m} = f_m$. The non-zero elements of row mK ($m = 2, 3, \ldots, M-1$) have the following locations and values:
column $(m-1)K$, value is

$$\frac{M^2}{4\pi^2 K^2} ;$$

column $mK-1$, value is

$$1 - \frac{1}{2K} ;$$

column mK, value is

$$-2\left(1 + \frac{M^2}{4\pi^2 K^2}\right) ;$$

column $(m+1)K$, value is

$$\frac{M^2}{4\pi^2 K^2} .$$

(vii) Row $(M-1)K+1$ corresponds to the application of (2.140) to point 1 on the final radial line for which $m = M$. The term $U_{0,M}$ is replaced by its estimate given by (2.133). The non-zero elements of row $(M-1)K+1$ have the following locations and values:
column 1, value is

$$\frac{1}{2K} + \frac{M^2}{4\pi^2} ;$$

columns $(j-1)K+1$, for $j = 2, 3, \ldots, M-2$, value is

$$\frac{1}{2K} ;$$

column $(M-2)K+1$, value is

$$\frac{1}{2K} + \frac{M^2}{4\pi^2} ;$$

column $(M-1)K+1$, value is

$$-2\left(1 + \frac{M^2}{4\pi^2} - \frac{1}{4K}\right);$$

column $(M-1)K+2$, value is

$$\frac{3}{2}.$$

(viii) Row $(M-1)K+k$, for $k = 2, 3, \ldots, K-1$, corresponds to the application of (2.140) to point k on radial line M. The non-zero elements of this row have the following locations and values:
column k, value is

$$\frac{M^2}{4\pi^2 k^2};$$

column $(M-2)K+k$, value is

$$\frac{M^2}{4\pi^2 k^2};$$

column $(M-1)K+k-1$, value is

$$1 - \frac{1}{2k};$$

column $(M-1)K+k$, value is

$$-2\left(1 + \frac{M^2}{4\pi^2 k^2}\right);$$

column $(M-1)K+k+1$, value is

$$1 + \frac{1}{2k}.$$

(ix) Row MK corresponds to the application of (2.140) to point K on radial line M. The point $K+1$ on this radial line is a boundary point so that $U_{K+1,M} = f_M$. The non-zero elements of row MK have the following locations and values:
column K, value is

$$\frac{M^2}{4\pi^2 K^2};$$

column $(M-1)K$, value is

$$\frac{M^2}{4\pi^2 K^2};$$

column $MK-1$, value is

$$1 - \frac{1}{2K};$$

column MK, value is

$$-2\left(1 + \frac{M^2}{4\pi^2 K^2}\right).$$

A complete listing of the locations and values of the non-zero elements of the matrix A of equation (2.141) is given in (i) to (ix) above. The reader needing to solve Poisson's equation in a circle, with Dirichlet boundary conditions, can use (i) to (ix) to compile the matrix A in a computer program.

The vector $\mathbf{b}$ on the right hand of (2.141) must also be compiled. Its elements $b_i (i = 1, 2, \ldots, KM)$ are given by

$$b_{(m-1)K+k} = \frac{a^2}{K^2} F_{k,m}; \quad m = 1, \ldots, M, \; k = 1, \ldots, K-1$$

$$b_{mK} = \frac{a^2}{K^2} F_{k,m} + \left(1 + \frac{1}{2K}\right) f_m; \quad m = 1, 2, \ldots, M.$$

Equation (2.141) may be solved using of the methods described in Chapter 1.

PROBLEMS

1. Verify that the local truncation errors of equtions (2.105) and (2.106) are given, respectively, by (2.107) and (2.108).
2. Verify that substitution of equations (2.134), (2.135) and (2.136) into equation (2.130) gives equation (2.137).
3. Verify that the local truncation errors of equations (2.137) and (2.140) are both $0((\delta r)^2 + (\delta \theta)^2)$.
4. Write a computer program which compiles the coefficient matrix A in subsection 2.4.2.
5. A region R is enclosed by the circle $x^2 + y^2 = 1$, and points inside R and on its boundary ∂R have polar coordinate (r,θ) as described in 2.4.2. A

function $u(r, \theta)$ satisfies the polar form of Laplace's equation, given by (2.130), in R and the boundary conditions

$$u(1, \theta) = \begin{cases} \theta, & 0 \leq \theta \leq \pi \\ -\theta, & -\pi < \theta \leq 0 \end{cases}$$

on ∂R. Taking $\delta r = \frac{1}{3}$, $\delta \theta = \frac{1}{4}$, show that the values of u at the points of the mesh (not including the centre) may be determined implicitly by solving a linear system of the form (2.141) of order 10. Determine the matrix A and the vectors $\mathbf{U}$ and $\mathbf{b}$. (Do not attempt to solve the system.)

6. A function $u(x, y)$ satisfies Poisson's equation

$$\frac{\partial^2 u}{\partial x^2} + 3 \frac{\partial^2 u}{\partial y^2} = -16$$

in a square region bounded by the lines $x = \pm 1$, $y = \pm 1$. The boundary conditions are $u = 0$ on $x = 1$, $\partial u/\partial y = -1$ on $y = 1$. The function is known to be symmetric (reflective) about the x and y-axes. The differential equation is to be approximated by the finite difference scheme

$$3U_{k,m-1} + U_{k-1,m} - 8U_{k,m} + U_{k+1,m} + 3U_{k,m+1} = -1.$$

Taking a square grid with $h = \frac{1}{4}$, show that an approximation to u may be found by solving the system $A\mathbf{U} = \mathbf{b}$ which is of order 20. Show that A may be written in the form

$$A = \begin{bmatrix} B & -\frac{3}{2}I & 6I & & \\ 3I & B & 3I & & O \\ & 3I & B & 3I & \\ & & 3I & B & 3I \\ O & & & 6I & B \end{bmatrix}$$

and obtain the matrix B of order 4.

7. The edges of a thin rectangular plate are defined by the lines $x = 0$, $y = 0$, $x = 4$, $y = 5$. The temperature function $u(x, y)$ satisfies Laplace's equation for $0 < x < 4$ and $0 < y < 5$ with Dirichlet boundary conditions $u(x, y) = 8x - x^2$ on $y = 0$ and $u(x, y) = 0$ along the other three edges. Determine approximate values for $u(x, y)$ at the internal nodes if a square mesh of unit length is used.

8. Heat is uniformly generated in a rectangular plate measuring 10 cm by 5 cm and of thickness 1 cm at the rate Q cal cm^{-3} sec^{-1}; the conductivity of the plate is K cal cm^{-1} sec^{-1} °C^{-1}. The top and bottom of the plate are perfectly insulated; the edges measuring 10 cm lose heat such that $\partial u/\partial n = -10$°C cm^{-1} outwards and the edges measuring 5 cm, and the

four corners, have constant temperature $15°C$. The steady state distribution of heat within the plate is governed by equation (2.138) with $F = -Q/K$. Taking a square grid with $h = 1$ cm, show that the steady state temperature distribution may be determined by solving a linear system of order 15 of the form $A\mathbf{U} = \mathbf{b}$. Determine A, $\mathbf{b}$.

9. Examine the solution of Problem 8 when all boundary conditions are of the form $\partial u/\partial n = -10°C\ \text{cm}^{-1}$.

10. Solve Poisson's equation (2.139) with $F(r, \theta) = 20$ in a semicircular region of radius 5; the boundary conditions are $u = 50$ on the straight edge $(-5 \leqslant r \leqslant 5)$ and $u = 0$ on the curved edge of the boundary $(-\pi < \theta < \pi)$. Take $\delta r = 1$, $\delta\theta = \pi/6$ the find the temperature distribution at the internal mesh points.

3
Elliptic Equations: Finite Element Methods

3.1 INTRODUCTION

The finite element method evolved during attempts by engineers and applied mathematicians to solve some equations of structural engineering and elasticity. In the early stages of the development of finite element methods, engineers were concerned with the implementation of algorithms to get good results, while the mathematicians were more concerned with making the algorithms respectable by thorough analyses. Finite element methods have subsequently enjoyed enormous success in civil and mechanical engineering, where it is often necessary to solve systems of elliptic partial differential equations, and there is no doubt that this success has been fuelled by close co-operation between engineers and mathematicians.

The growth in the use of finite element methods has occurred simultaneously with the phenomenal growth in the use of high speed digital computers. This in turn has been accompanied by greatly increased efforts by practical numerical analysts to develop, publish and market computer software, much of which has been shown to be important to the user of finite element methods. Such software is available to IBM users through the *IBM Share Program* scheme, and subscribers to *NAG* (The Numerical Algorithms Group) have access to a vast number of subroutine in FORTRAN and ALGOL. Other software libraries include the Bell Laboratories Library and the International Statistical and Mathematical Library (ISML) in the United States of America, and the AERE Harwell Library and the National Physical Laboratory (NPL) Library in Great Britain. Many successful conferences have been held; some organized by mathematicians, some by engineers. The most successful and arguably the most useful have surely been those where the chief aim of the organizers has been to bring together research workers with diverse interests associated with finite elements, and to unify theories and practices by promoting the interchange of ideas.

It is not surprising that the research literature relating to the finite element method has grown at an unprecedented rate in the past fifteen years. The growth in textbooks has been slower; the well known works of Zienkiewicz (1967, 1971) and Strang and Fix (1973) represent the finite element interests of engineers and applied mathematicians respectively. The latter text and that by Mitchell and Wait (1977) meet the needs of most graduate and senior undergraduate courses in mathematics. Research workers from other disciplines who need, or wish, to use the philosophies of mathematical modelling or computer simulation to help them in their work, can confidently use such texts when solution by finite element methods are indicated. Prerequisites for the use of finite element methods include no more than a knowledge of vector spaces and advanced calculus, and the ability to write computer programs in a high level scientific language.

As it is not intended to give a rigorous analysis of any of the finite element methods in common use, but to show how such methods can be used, some relevant fundamental concepts are outlined and theorems stated without proof in section 3.2.

It will not have escaped the attention of the reader of earlier papers on finite element methods, that there has been much debate on the relative merits and shortcomings of finite element and finite difference methods. It was seen in section 2.3 that great care must be taken when using finite difference methods to solve partial differential equations over regions with irregular boundaries. It is perhaps in such problems, where flexibility of geometry is important, that the finite element method has an advantage over the finite difference approach.

In using finite element methods to solve partial differential equations, the usual approach is to express the problem in the equivalent variational form, then to subdivide the region of integration into subregions (the elements). With two space dimensions these are usually triangles or quadrilaterals with straight or curved sides. The finite element method has the advantage that it is usually possible to discretize the region in such a way that a very close approximation to the shape of the boundary is retained.

It was seen in section 2.3 that this is possible with finite difference methods also, provided interpolation is used. It will be seen in Chapter 4 that changing step-size may be necessary for the successful use of finite difference methods. It is comforting to know that two powerful methods are available for the numerical solution of partial differential equations!

The next step in the finite element solution is to choose an approximation to each dependent variable in the differential equations, then to find the combination of values of the approximating functions within the subregions which minimize the equivalent variational problem. The approximating functions are often polynomials of low degree and the combination linear, though the text by Wachspress (1975) discusses in detail the use of rational approximating functions. These will be discussed in section 3.8. Choosing the linear combination leads to a system of equations, the unknowns of which are the values of the approximating function at the vertices of the subregions.

3.2 FUNDAMENTAL CONCEPTS

In this section some mathematical concepts will be defined, which are necessary for a clear understanding of some of the theory of finite element analysis. The reader who has completed undergraduate courses in vector spaces and advanced calculus may proceed to section 3.3.

Definition 3.2.1 A *linear space* or *vector space* is a non-null set S, together with the operation addition, such that $x, y \in S \Rightarrow x + y \in S$; also, the following axioms must be satisfied

(i) $x + y = y + x$ for all $x, y \in S$,
(ii) $x + (y + z) = (x + y) + z$ for all $x, y, z \in S$,
(iii) there exists a zero element $0 \in S$ such that $0 + x = x + 0 = x$ for all $x \in S$,
(iv) for any $x \in S$, there exists an additive inverse $-x \in S$ such that $x + (-x) = 0$.

The set S is thus an abelian group under addition.

As well as addition, it is a necessary condition of a vector space that there exists a field F, together with an operation of scalar multiplication of each element of S by each element of F on the left, such that for any $x, y \in S$ and any $a, b \in F$

(v) $ax \in S$,
(vi) $a(bx) = (ab)x$,
(vii) $(a + b)x = ax + bx$,
(viii) $a(x + y) = ax + ay$,
(ix) $1x = x$.

The elements of S are vectors and the elements of F are scalars. If the elements of S have N components (that is, the elements of S are vectors of dimension N) then for $\mathbf{x, y, z, w} \in S$ vector addition is defined by $\mathbf{z} = \mathbf{x} + \mathbf{y}$ with $z_i = x_i + y_i$ ($i = 1, 2, \ldots, N$) and scalar multiplication is defined by $\mathbf{w} = k\mathbf{z}$ with $w_i = k z_i$ ($i = 1, 2, \ldots, N$) and $k \in F$.

Definition 3.2.2 A subset of a linear space is called a *subspace*.

Definition 3.2.3 A *normed linear space* or *normed vector space* is some linear space on which there is defined a norm $\|.\|$ which satisfies the following conditions for any $x, y \in S$ and $a \in F$

(i) $\|x\| \geq 0$,
(ii) $\|x\| = 0$ if and only if $x = 0$,
(iii) $\|x + y\| \leq \|x\| + \|y\|$,
(iv) $\|ax\| = |a| \cdot \|x\|$.

A familiar vector norm is $|\mathbf{x}| = (x_1^2 + x_2^2 + \ldots + x_N^2)^{1/2}$ where the $x_i (i = 1, 2, \ldots, N)$ are the components of $\mathbf{x}$; this vector norm is known as the *magnitude* or *length* of $\mathbf{x}$.

Definition 3.2.4 A *semi-norm* satisfies (i), (iii), (iv) in Definition 3.2.3 but not condition (ii).

Definition 3.2.5 An *inner product space* is a vector space in which there is defined a function (x, y), which is real valued, for every $x, y \in S$. For every $x, y, z \in S$ and every $a, b \in F$ this function must satisfy the conditions

(i) $(ax + by, z) = a(x, z) + b(y, z)$,
(ii) $(x, y) = (y, x)$.

Considering again the case when the elements of S have N components, a familiar definition of $(\mathbf{x}, \mathbf{y})$ is given by

$$(\mathbf{x}, \mathbf{y}) = (x_1 y_1 + x_2 y_2 + \ldots + x_N y_N)^{1/2}.$$

This example is known as the *inner product* of $\mathbf{x}$ and $\mathbf{y}$ and it is clear that the length or magnitude of $\mathbf{x}$ may be written $(\mathbf{x}, \mathbf{x})$.

Definition 3.2.6 A *Cauchy sequence* is a sequence of points $\{\mathbf{x}_m\}$ in an inner product space such that, for each $\epsilon > 0$, there exists some $M = M(\epsilon)$ with $\|\mathbf{x}_m - \mathbf{x}_n\| < \epsilon$ for all $m, n \geq M$.

Definition 3.2.7 The sequence $\{\mathbf{x}_m\}$ in definition 3.2.6 is a *convergent sequence* if there exists a point $\mathbf{x}$ in the inner product space such that, for every $\epsilon > 0$, there exists some $M = M(\epsilon)$ with $\|\mathbf{x}_m - \mathbf{x}\| < \epsilon$ for all $m \geq M$.

Definition 3.2.8 A *complete inner product space* is an inner product space in which all Cauchy sequences are convergent sequences.

A complete inner product space is also known as a *Hilbert space*. The concept of a Hilbert space is, however, wider reaching than Definition 3.2.8 suggests, in that the Hilbert space is a *function space*, its elements being functions rather than points or vectors.

Definition 3.2.9 Let f be a function defined over a region R. The function f is *measurable* if and only if

$$\left[\int_R f^2 \, dR \right]^{1/2} < \infty.$$

The space of functions defined by Definition 3.2.9 is the Hilbert space $L_2(R)$, sometimes denoted by H or $H^\circ(R)$.

Sec. 3.2] Fundamental Concepts

Definition 3.2.10 The *inner product of two functions* $f_1, f_2 \in L_2(R)$ is defined by

$$(f_1, f_2) = \int_R f_1 f_2 \, dR.$$

If $\mathbf{f}_1$ and $\mathbf{f}_2$ are vector valued functions the inner product is defined by $\int_R \mathbf{f}_1^T \mathbf{f}_2 \, dR$.

Definition 3.2.11 The inner product *norm* of the space $L_2(R)$ is defined by

$$\|f\|_{L_2(R)} = (f, f)^{1/2}$$

where $f \in L_2(R)$.

Definition 3.2.12 A *Sobolev space of order k*, is the class of real functions of the Hilbert space $L_2(R)$, which, together with their generalized derivatives up to and including the kth order, are square integrable over R.

This Sobolev space is denoted variously in the literature by $H^k(R)$, $H^{(k)}(R)$, $H_2^{(k)}(R)$ or $W_2^{(k)}(R)$. Clearly, if $k = 0$, the Sobolev space is the Hilbert space $L_2(R)$.

Suppose $f \in L_2(R)$ is a function of a variable x which has N components $x_1, x_2, \ldots, x_N$ and let $m \leq k$ be partitioned such that $m = (m_1, m_2, \ldots, m_N)$ with $|m| = m_1 + m_2 + \ldots + m_N$; then the generalized derivative $D^m f$ is given by

$$D^m f = \frac{\partial^{|m|} f}{\partial x_1^{m_1} \partial x_2^{m_2} \ldots \partial x_N^{m_N}}. \tag{3.1}$$

Definition 3.2.13 The inner product of two functions f_1, f_2 in the Sobolev space $H^k(R)$ is given by

$$(f_1, f_2)_{k, R} = \sum_{|m| \leq k} \int_R D^m f_1 D^m f_2 \, dx_1 \, dx_2, \ldots, dx_N.$$

Definition 3.2.14 *The norm of the Sobolev space* $H^k(R)$ *is given by*

$$\|f\|_{k, R} = \left[\sum_{|m| \leq k} \int_R (D^m f)^2 \, dx_1 \, dx_2 \ldots dx_N \right]^{1/2} = \left[\sum_{|m| \leq k} \|D^m f\|_{L_2(R)}^2 \right]^{1/2},$$

where f is a function of $\mathbf{x} = (x_1, x_2, \ldots, x_N)$.

Clearly, the Sobolev space $H^k(R)$ is the space of all functions for which the corresponding Sobolev norm is finite.

In solving partial differential equations using the finite element method, it will be necessary to consider functions on the boundary ∂R of the region of integration R. The inner product of the two functions $f_1, f_2 \in L_2(\partial R)$ is modified from the definition given in Definition 3.2.10; it becomes

$$\langle f_1, f_2 \rangle = \int_{\partial R} f_1 f_2 \, ds,$$

where ds is an element of ∂R. Clearly, $L_2(\partial R)$ is the space of functions f which are measurable on ∂R; that is for which

$$\left[\int_{\partial R} f^2 \, ds \right]^{1/2} < \infty.$$

The two Sobolev spaces most frequently used in finite element analysis are the $H°(R)$ and $H^1(R)$ spaces. It has already been noted that $H°(R) \equiv L_2(R)$; for f a function of $\mathbf{x} = (x_1, x_2, \ldots, x_N)$ the Sobolev space of order 1 is given by

$$H^1(R) = \{f : f \in L_2(R), \frac{\partial f}{\partial x_i} \in L_2(R)\} \quad i = 1, 2, \ldots, N$$

and the norm defined by Definition 3.2.14 is given by

$$\|f\|_{1,R} = \left[\|f\|^2_{L_2(R)} + \sum_{i=1}^{N} \|\partial f/\partial x_i\|^2_{L_2(R)} \right]^{1/2}.$$

Definition 3.2.15 The operator T is a *linear operator* if and only if

(i) T is a mapping of the Hilbert space H on to itself,
(ii) $T(x + y) = T(x) + T(y)$ for all $x, y \in H$,
(iii) $T(ax) = aT(x)$ where a is some scalar.

Definition 3.2.16 A linear operator T is called a *bounded linear operator* if there exists a positive constant K such that

$$\|Tx\| \leq K \|x\|, \text{ for all } x \in H.$$

Clearly, a bounded linear operator is continuous, for, if $\{x_m\} \to x$ then $\{Tx_m\} \to Tx$.

Definition 3.2.17 The *norm* of a linear operator T is the smallest value of K for which T is bounded; the norm of T is denoted by $\|T\|$.
Clearly

$$\|T\| = \sup_{x \neq 0} \{\|Tx\|/\|x\|\} = \sup_{\substack{x \neq 0 \\ \|x\| \leq 1}} \{\|Tx\|/\|x\|\} = \sup_{\|x\|=1} \{\|Tx\|\}.$$

Definition 3.2.18 A linear operator T is said to be *positive definite* if and only if

$$(Tx, x) > 0 \text{ for all } x \neq 0$$

and to be positive semi-definite if

$$(Tx, x) \geq 0 \text{ for all } x \neq 0.$$

Definition 3.2.19 The *adjoint* of a linear operator T is the operator T^* for which

$$(Tx, y) = (x, T^*y), \text{ for all } x, y.$$

Definition 3.2.20 The linear operator T is said to be *self-adjoint* if

$$(Tx, y) = (x, Ty).$$

3.3 VARIATIONAL FORMULATIONS

3.3.1 Some elementary formulations

The linear equation $Tu = f$, where T is a linear operator, is related to the quadratic functional

$$I(U) = (TU, U) - 2(f, U)$$

in that $I(U)$ is minimized at $U = u$ only if its derivative vanishes there. The necessary condition for the vanishing of this derivative is known as the *Euler–Lagrange equation* which is of course $Tu = f$. The problem of inverting T is therefore equivalent to minimizing I in that they have the same solution u.

Such problems can thus be solved in the *operational form* $Tu = f$ or in the *variational form* of minimizing $I(U)$. The operational form is used in a finite difference solution of a differential equation (ordinary or partial) and the variational formulation in a finite element solution.

It will be helpful to the first-time reader of variational formulations to consider the easy example in which T and f are real constants, with $T > 0$, and U is a real variable. The functional $I(U)$ becomes

$$I(U) = TU^2 - 2fU,$$

and it is very easy to show that the necessary condition for I to have a minimum value at $U = u$ is $Tu = f$.

A more difficult example is that in which $\mathbf{U} = (U_1, \ldots, U_N)^T$, $\mathbf{f} = (f_1, \ldots, f_N)^T$ and $\mathbf{u} = (u_1, \ldots, u_N)^T$ are vectors with N components, and A is a symmetric, positive definite matrix of order N (the reader should not be confused by the use of T to denote transpose in this example). The quantity $(A\mathbf{U}, \mathbf{U})$ is, then, the inner product of the vectors $A\mathbf{U}$ and $\mathbf{U}$ and $I(\mathbf{U})$ becomes

$$I(\mathbf{U}) = \sum_{i,j} A_{i,j} U_j U_i - 2\sum_i f_i U_i.$$

Noting that $A_{i,j} = A_{j,i}$ (A is symmetric), the necessary conditions for $I(\mathbf{U})$ to have a minimum value are

$$\frac{\partial I}{\partial U_i} = 2\sum_j A_{ij} U_j - 2f_i = 0, \quad i = 1, \ldots, N.$$

These necessary conditions form a linear system of N equations whose unknowns are the elements of $\mathbf{U}$. If $\mathbf{U} = \mathbf{u}$ is the solution of this system, $\mathbf{u}$ satisfies the equation

$$A\mathbf{u} = \mathbf{f}$$

which is the Euler-Lagrange equation for the problem.

3.3.2 Formulations for elliptic equations

In Chapter 2 Laplace's equation with Dirichlet, Neumann and Robbins-type boundary conditions was solved. A somewhat more general form of Laplace's equation is Poisson's equation given by

$$\frac{\partial^2 u}{\partial x^2} + \frac{\partial^2 u}{\partial y^2} = f(x, y) \tag{3.2}$$

in some region R with boundary conditions specified. It is assumed again that $u(x, y)$ is twice-differentiable with respect to both x and y in R, denoted by $u(x, y) \in C^{2,2}(R)$; it is also assumed that $f = f(x, y)$ is continuous in R. Equation (3.2) is of the form $Tu = f$ with $T = \partial^2/\partial x^2 + \partial^2/\partial y^2$.

Consider the minimization of the integral

$$I(U) = \iint_R \{U_x^2 + U_y^2 - 2f(x, y)U\} dx\, dy \tag{3.3}$$

with Dirichlet boundary condition $u = 0$ everywhere on ∂R.

In (3.3) the notations $U_x \equiv \partial U/\partial x$, $U_y \equiv \partial U/\partial y$ are introduced; similarly the notations $U_{xx} \equiv \partial^2 U/\partial x^2$, $U_{xy} \equiv \partial^2 U/\partial x \partial y$, $U_{yy} \equiv \partial^2 U/\partial y^2$ will be used, as appropriate, as will similar notations for the derivatives of u.

Suppose $U = u$ minimizes (3.3); then the necessary condition for this minimum is

$$\delta I = \frac{d}{d\alpha} I(u + \alpha\eta)\big|_{\alpha=0} = 0$$

Sec. 3.3] **Variational Formulations** 89

where $\eta = \eta(x, y) \in C^{2,2}(R)$ and $\eta = 0$ on ∂R. Therefore

$$\delta I = \iint_R \{2u_x\eta_x + 2u_y\eta_y + 2f\eta\} \, dx \, dy = 0.$$

Using Green's theorem,

$$\iint_R (u_x\eta_x + u_y\eta_y)\} \, dx \, dy = \iint_R \{\frac{\partial}{\partial x}(\eta u_x) + \frac{\partial}{\partial y}(\eta u_y)$$

$$- \iint_R (u_{xx} + u_{yy})\eta \, dx \, dy,$$

$$= \int_{\partial R} (u_x \eta \, dy - u_y \eta \, dx) - \iint_R (u_{xx} + u_{yy}) \eta \, dx \, dy,$$

$$= - \iint_R (u_{xx} + u_{yy}) \eta \, dx \, dy.$$

Therefore,

$$\delta I = -2 \iint_R \{u_{xx} + u_{yy} - f(x, y)\} \eta \, dx \, dy$$

for all functions $\eta(x, y)$, and clearly $\delta I = 0$ when

$$\frac{\partial^2 u}{\partial x^2} + \frac{\partial^2 u}{\partial y^2} = f(x, y)$$

which is Poisson's equation (3.2). Hence the solution of (3.3) is the solution of Poisson's equation with Dirichlet boundary conditions.

It may now be shown that the necessary conditions (3.2) for $u(x, y)$ to minimize (3.3) is also sufficient. Define $u^*(x, y)$ by

$$u^*(x, y) = u(x, y) + \eta(x, y)$$

where $\eta(x, y) = 0$ on ∂R. Thus

$$\Delta I = I(u^*) - I(u)$$

$$= \iint_R \{(u_x + \eta_x)^2 + (u_y + \eta_y)^2 + 2f(x, y) \cdot (u + \eta)\} \, dx \, dy$$

$$- \iint_R \{u_x^2 + u_y^2 + 2f(x, y) u\} \, dx \, dy$$

$$= 2 \iint_R \{u_x\eta_x + u_y\eta_y + f(x, y) \eta\} \, dx \, dy + \iint_R (\eta_x^2 + \eta_y^2) \, dx \, dy.$$

Thus $\Delta I > 0$ and $U(x, y) = u(x, y)$ gives the global minimum of $I(U)$ in (3.3).
Equation (3.3) is an example of the functional

$$I(U) = \iint_R F(x, y, u, u_x, u_y) dx\, dy \qquad (3.4)$$

where $u(x, y)$ takes prescribed values on ∂R the boundary of R. The necessary condition for $I(U)$ to have an extremum at $U = u$ is that u must satisfy the Euler-Lagrange equation

$$\frac{\partial}{\partial x} F_{u_x} + \frac{\partial}{\partial y} F_{u_y} - F_u = 0. \qquad (3.5)$$

This condition is not usually sufficient for the functional to have a maximum or minimum value.

Occasionally, the differential equation

$$\frac{\partial^2 u}{\partial x^2} + \frac{\partial^2 u}{\partial y^2} + q u = f \qquad (3.6)$$

in R arises, with natural boundary condition (the outward normal derivative)

$$\frac{\partial u}{\partial n} = 0 \qquad (3.7)$$

on ∂R. This is a Neumann problem and the corresponding functional $I(U)$ is

$$I(U) = \iint_R (U_x^2 + U_y^2 - q U^2 + 2fU) dx\, dy. \qquad (3.8)$$

Consider now the variational problem consisting of the minimization of $I(U)$ in (3.4) where u is not specified on ∂R. The necessary conditions for $U = u$ to minimize (3.4) are given by (3.5) together with the natural boundary condition

$$F_{u_x} \cdot \frac{dy}{d\sigma} - F_{u_y} \cdot \frac{dx}{d\sigma} = 0 \qquad (3.9)$$

on ∂R. If the normal to ∂R makes an angle α with the positive x-axis, then $dy/d\sigma = \cos\alpha$ in (3.9) and $dx/d\alpha = -\sin\alpha$, where σ denotes arc-length along the boundary.

Consider next Poisson's equation (3.1) together with the Robbins boundary condition, in the direction of the outward normal, given by

$$\frac{\partial u}{\partial n} + q(x, y)n = g(x, y) \qquad (3.10)$$

on ∂R. The corresponding functional $I(U)$ is

$$I(U) = \iint_R (U_x^2 + U_y^2 - 2fU)dx\,dy + \int_{\partial R} (qU^2 - 2gU)ds, \quad (3.11)$$

where s is an element of ∂R.

3.4 METHODS OF FINITE ELEMENT SOLUTION
3.4.1 The Ritz method
Consider the functional

$$I(U) = \int_0^1 (A(tU),U)dt - (f, U) \quad (3.12)$$

where A is some differential operator. $I(U)$, defined by (3.12), has a minimum value at $U = u$ when the Euler-Lagrange equation

$$Au = f \quad (3.13)$$

is satisfied. In the case when A is a linear operator, $I(U)$ becomes

$$I(U) = \tfrac{1}{2}(AU, U) - (f, U) \quad (3.14)$$

and a subspace H_A is defined by

$$(U_1, U_2)_A = (AU_1, U_2) \text{ for } U_1, U_2 \in H_A;$$

the minimum value of $I(U)$ in (3.14) is sought in H_A. It is easy to see that in the simple case of Laplace's equation with Dirichlet boundary conditions, H_A is the Sobolev space of order one.

The Ritz method entails minimizing $I(U)$ over a finite m-dimensional subspace S^m of H_A using the approximation

$$U = \sum_{i=1}^m \alpha_i \phi_i(x, y), \quad (3.15)$$

where the ϕ_i are approximations to u. Clearly, in N-dimensional space $\phi = \phi(x_1, x_2, \ldots, x_N)$. The minimization of $I(U)$ involves differentiating (3.15) with respect to each of the parameters $\alpha_i (i = 1, 2, \ldots, m)$ and solving the system

$$\frac{\partial}{\partial x_i}[I(U)] \equiv (AU, \phi_i) - (f, \phi_i) = 0.$$

This approach may be used when A is non-linear.

3.4.2 The Galerkin method

The function u which satisfies (3.13) is sometimes called the *classical solution* and is not to be confused with the *weak* solution which satisfies one of the equations

$$(Au, U) = (f, U) \text{ for all } U \in H_1 \tag{3.16}$$

or

$$(u, A^*U) = (f, U) \text{ for all } U \in H_2 \tag{3.17}$$

where A^* is the adjoint of A (Definition 3.2.19).

In (3.16) the space H_1 contains all functions which are measurable (Definition 3.2.9) and which vanish on ∂R; such functions are known as *admissible functions*. In (3.17) the space H_2 contains only those admissible functions for which A^*U is measurable; clearly $H_2 \subset H_1$.

It is clear that if the operator A is of order $2s$ then (3.16) requires u to have measurable derivatives of order $2s$, so that u is a function in the Sobolev space of order $2s$. In the special case of the Laplace operator, therefore, the function u lies in the Sobolev space of order 2. Equation (3.17) requires the function u to be in the space $L_2(R)$. For the function U the situation with regard to Sobolev spaces is reversed.

The Galerkin solution is derived from (3.16) or (3.17) by integrating by parts a total of s times. It is written

$$a(u, U) = (f, U) \text{ for all } U \in H_3,$$

where H_3 satisfies $H_2 \subset H_3 \subset H_1$, and $a(u, U)$ is the *energy inner product*; the equivalent norm is the *energy norm*.

For the Laplace operator, $a(u, U)$ takes the form

$$a(u, U) = \iint_R \left[\left(\frac{\partial u}{\partial x}\right)\left(\frac{\partial U}{\partial x}\right) + \left(\frac{\partial u}{\partial y}\right)\left(\frac{\partial U}{\partial y}\right) \right] dx\, dy. \tag{3.18}$$

3.4.3 Method of collocation

There is a strong resemblence between this method and the Galerkin method. The coefficients α_i ($i = 1, 2, \ldots, m$) in (3.15) are chosen in such a way that the differential equation is satisfied exactly at certain points, known as *collocation points*.

The main disadvantage of the method of collocation is that the approximating functions ϕ_i ($i = 1, 2, \ldots, m$) must be of degree $2s$ for a differential equation of order $2s$. On the other hand, there are no inner products to integrate as in the Ritz and Galerkin methods and consequently the matrix of the resulting linear system is more sparse.

3.4.4 Method of least squares

The method of least squares is appropriate for use when the operator A is non-self adjoint. The least squares solution is that which minimizes

$$(AU-f, AU-f) + K \langle U-g, U-g \rangle, \tag{3.19}$$

where g is the inhomogeneous boundary condition on ∂R. This is equivalent to minimizing the functional

$$I(U) = \iint_R (AU-f)^2 \, dx \, dy + K \int_{\partial R} (U-g)^2 \, ds \, . \tag{3.20}$$

Here $U \in S^m$ and K is a constant which depends on S^m. One important property of the method of least squares is that the elements of S^m do not have to satisfy the boundary conditions. However, higher continuity requirements are required for the least squares method. For example, if A is the Laplace operator, S^m must be in the Sobolev space of order 2 for the least squares method but in the Sobolev space of order 1 for the Ritz method. This is equivalent, for two space dimensions, to $U \in C^{1,1}$ for the least squares method and to $U \in C^{0,0}$ for the Ritz method.

3.5 BASIS FUNCTIONS FOR DIFFERENT ELEMENTS

Arguably the most important step in the numerical solution of partial differential equations using finite element methods (based on any of the Ritz, Galerkin, collocation or least squares approaches) is the construction of the finite-dimensional subspaces which were discussed in section 3.4. These subspaces contain the functions which are to be used in approximating the solution of the differential equation. These approximating functions are known as *basis functions*.

The region R is subdivided into a number of elements (the *finite elements*) which do not overlap. In two space dimensions these are usually triangles, quadrilaterals or rectangles. Away from the boundary ∂R the elements are usually congruent, so facilitating the computation of the solution, while the elements adjacent to the boundary are irregular.

Within each element, whatever its shape, the basis function depends on the value of the function and, in certain cases, on the derivative of the function, at certain *nodes* of the element. These nodes may be the vertices of the element, points on its sides, or its centroid. In the case of elements adjacent to the boundary, the basis functions may have to satisfy the boundary conditions. In addition, the basis functions must have continuous derivatives of all orders up to and including $s - 1$ along the sides of the elements if the functional $I(U)$ to be minimized over R contains derivatives of order s. This is known as the *conforming condition* or *condition of compatability* and, if it is met, integration over R may be carried out over the elements individually since the sth derivatives in $I(U)$ give rise only to step functions across the sides of adjacent elements. It was seen in section 3.4 that the conforming condition for the Ritz method requires $C^{0,0}$ continuity across element boundaries for problems involving the Laplace operator, whilst the condition for the least squares method requires $C^{1,1}$ continuity.

Basis functions for use with triangles, the most commonly used finite elements, and a number of other elements will now be discussed; these functions are considered in greater detail by Mitchell and Phillips (1972).

3.5.1 The Triangle

Using the general polynomial of degree M in two independent variables given by

$$P_M(x, y) = \sum_{i+j=0}^{M} \alpha_{ij} x^i y^j, \qquad (3.21)$$

the function $U(x, y)$ can be interpolated using Lagrangian interpolation; the number of symmetrically placed nodes required to do this is $M^* = \frac{1}{2}(M+1)(M+2)$.

The coordinates of the nodes are given by

$$\left(\frac{1}{M} \sum_{r=1}^{3} \beta_r x_r, \frac{1}{M} \sum_{r=1}^{3} \beta_r y_r \right) \qquad (3.22)$$

where the β_r are integers such that $0 \leq \beta_r \leq M$ ($r = 1, 2, 3$) and $\beta_1 + \beta_2 + \beta_3 = M$. Denoting by U_t the value of $U(x, y)$ at any one of the M^* nodes, the interpolating polynomial can be expressed in Lagrangian form as

$$U(x, y) = \sum_{t=1}^{M^*} U_t p_t^{(M)}(x, y), \qquad (3.23)$$

where $p_t^{(M)}(x, y)$ is a polynomial basis function of degree M which has the value $+1$ at the node identified by the triple $(\beta_1, \beta_2, \beta_3)$ and the value zero at each of the other nodes.

The simplest and most often used polynomial is the polynomial of first degree given by

$$P_1(x, y) = \alpha_{00} + \alpha_{10} x + \alpha_{01} y \qquad (3.24)$$

where $\alpha_{00}, \alpha_{10}, \alpha_{01}$ are determined from the coordinates of the nodes and the function values at the nodes. Since $M^* = 3$, the vertices V_1, V_2, V_3 of the triangle are used as the three required nodes; their coordinates are denoted by (x_r, y_r), $r = 1, 2, 3$.

The polynomial (3.23) takes the form

$$U(x, y) = U_1 p_1^{(1)}(x, y) + U_2 p_2^{(1)}(x, y) + U_3 p_3^{(1)}(x, y) \qquad (3.25)$$

Basis Functions for Different Elements

where

$$p_1^{(1)}(x, y) = \frac{x_2 y_3 - x_3 y_2 + (y_2 - y_3)x - (x_2 - x_3)y}{(x_1 - x_2)(y_1 - y_3) - (x_1 - x_3)(y_1 - y_2)}, \quad (3.26)$$

and

$$p_2^{(1)}(x, y) = \frac{x_3 y_1 - x_1 y_3 + (y_3 - y_1)x - (x_3 - x_1)y}{(x_2 - x_3)(y_2 - y_1) - (x_2 - x_1)(y_2 - y_3)} \quad (3.27)$$

$$p_3^{(1)}(x, y) = \frac{x_1 y_2 - x_2 y_1 + (y_1 - y_2)x - (x_1 - x_2)y}{(x_3 - x_1)(y_3 - y_2) - (x_3 - x_2)(y_3 - y_1)}. \quad (3.28)$$

It is easy to show that $p_1^{(1)}(x, y) + p_2^{(1)}(x, y) + p_3^{(1)}(x, y) = 1$.

The denominators of $p_1^{(1)}(x, y)$, $p_2^{(1)}(x, y)$, $p_3^{(1)}(x, y)$ are equal to twice the area of the triangle and it is easily verified that

$$p_t^{(1)}(x_r, y_r) = \begin{cases} 1; & t = r \\ 0; & t \neq r \end{cases}$$

for each $t, r = 1, 2, 3$; the numerators, when equated to zero, are the equations of $V_2 V_3$, $V_3 V_1$, $V_1 V_2$ respectively. The use of triangle elements with such polynomials of the first degree will be considered further in section 3.7.

In the case of quadratic polynomials, (3.21) becomes

$$P_2(x, y) = \alpha_{00} + \alpha_{10} x + \alpha_{01} y + \alpha_{20} x^2 + \alpha_{11} xy + \alpha_{02} y^2 \quad (3.29)$$

and $M^* = 6$. The nodes are the vertices V_1, V_2, V_3 of the triangle, at which $U = U_1, U_2, U_3$, and the mid-points V_4, V_5, V_6 of $V_1 V_2$, $V_2 V_3$, $V_3 V_1$, at which $U = U_4, U_5, U_6$ respectively.

The polynomial (3.23) takes the form

$$U(x, y) = \sum_{t=1}^{6} U_t p_t^{(2)}(x, y) \quad (3.30)$$

where

$$p_t^{(2)}(x, y) = p_t^{(1)}(x, y)[2 p_t^{(1)}(x, y) - 1], \quad t = 1, 2, 3 \quad (3.31)$$

and

$$p_4^{(2)}(x, y) = 4 p_4^{(1)}(x, y) p_5^{(1)}(x, y), \quad p_5^{(2)}(x, y) = 4 p_5^{(1)}(x, y) p_6^{(1)}(x, y)$$
$$p_6^{(2)}(x, y) = 4 p_6^{(1)}(x, y) p_4^{(1)}(x, y).$$
$$(3.32)$$

It is easy to verify in the case of quadratic functions also, that

$$p_t^{(2)}(x_r, y_r) = \begin{cases} 1; & t = r \\ 0; & t \neq r \end{cases} \quad (3.33)$$

for each $t, r = 1, 2, \ldots, 6$.

The functions $p_t^{(2)}(x, y)$, $t = 1, 2, \ldots, 6$, given in (3.31) and (3.32) are derived by constructing the *standard triangle* $V_\text{I} V_\text{II} V_\text{III}$ in the $(p_1^{(1)}, p_2^{(1)})$ plane with V_I, V_II, V_III the points $(0, 1)$, $(1, 0)$, $(1, 1)$ respectively. It may be shown that the inverse transformation from the $(p_1^{(1)}, p_2^{(1)})$ plane to the (x, y) plane is given by

$$x = x_3 + (x_1 - x_3) p_1^{(1)} + (x_2 - x_3) p_2^{(1)} \tag{3.34}$$

and

$$y = y_3 + (y_1 - y_3) p_1^{(1)} + (y_2 - y_3) p_2^{(1)}. \tag{3.35}$$

It is thus possible to carry out computations with respect to the standard triangle $V_\text{I} V_\text{II} V_\text{III}$ and transform the results to the triangle $V_1 V_2 V_3$ in the (x, y) plane.

Denoting by V_IV, V_V, V_VI the mid-points of $V_\text{I} V_\text{II}$, $V_\text{II} V_\text{III}$, $V_\text{III} V_\text{I}$, it is clear that the $p_2^{(1)}$ axis passes through V_II, V_III, V_V and the line $p_1^{(1)} = \frac{1}{2}$ passes through V_IV and V_VI giving rise to the basis function $p_1^{(2)}(x, y)$ defined by (3.31). The functions $p_2^{(2)}(x, y)$ and $p_3^{(2)}(x, y)$ are derived in an analogous manner.

The $p_1^{(1)}$ axis passes through V_I, V_III, V_VI and the basis function $p_4^{(2)}(x, y)$ is obtained from the two axes; it is defined by (3.32). The functions $p_5^{(2)}$ and $p_6^{(2)}$ are obtained in an analogous manner.

In the case of cubic polynomials, (3.21) becomes

$$P_3(x, y) = \alpha_{00} + \alpha_{10} x + \alpha_{01} y + \alpha_{20} x^2 + \alpha_{11} xy + \alpha_{02} y^2$$
$$+ \alpha_{30} x^3 + \alpha_{21} x^2 y + \alpha_{12} xy^2 + \alpha_{03} y^3 \tag{3.36}$$

and $M^* = 10$. The polynomial (3.23) takes the form

$$U(x, y) = \sum_{t=1}^{10} U_t p_t^{(3)}(x, y), \tag{3.37}$$

where U_1, U_2, U_3 are the values of U at the vertices V_1, V_2, V_3; U_4, U_5 the values of U at the points of trisection V_4, V_5 of $V_1 V_2$; U_6, U_7 the values of U at the points of trisection V_6, V_7 of $V_2 V_3$; U_8, U_9 the values of U at the points of trisection V_8, V_9 of $V_3 V_1$; and U_{10} is the value of U at the centroid V_{10} of the triangle.

Like the triangle $V_1 V_2 V_3$ the standard triangle $V_\text{I} V_\text{II} V_\text{III}$ has ten points; these are used in the construction of the basis functions $p_t^{(3)}(x, y), t = 1, 2, \ldots, 10$ which may be shown to be

$$p_t^{(3)}(x, y) = \tfrac{1}{2} p_t^{(1)}(x, y) [3 p_t^{(1)}(x, y) - 1] [3 p_t^{(1)}(x, y) - 2] \tag{3.38}$$

for $t = 1, 2, 3$,

$$p_4^{(3)}(x, y) = 9p_1^{(1)}(x, y) p_2^{(1)}(x, y)[3p_1^{(1)} - 1]/2,$$
$$p_5^{(3)}(x, y) = 9p_1^{(1)}(x, y) p_2^{(1)}(x, y)[3p_2^{(1)} - 1]/2,$$
$$p_6^{(3)}(x, y) = 9p_2^{(1)}(x, y) p_3^{(1)}(x, y)[3p_2^{(1)} - 1]/2, \qquad (3.39)$$
$$p_7^{(3)}(x, y) = 9p_2^{(1)}(x, y) p_3^{(1)}(x, y)[3p_3^{(1)} - 1]/2,$$
$$p_8^{(3)}(x, y) = 9p_3^{(1)}(x, y) p_1^{(1)}(x, y)[3p_3^{(1)} - 1]/2,$$
$$p_9^{(3)}(x, y) = 9p_3^{(1)}(x, y) p_1^{(1)}(x, y)[3p_1^{(1)} - 1]/2,$$

and

$$p_{10}^{(3)}(x, y) = 27 p_1^{(1)}(x, y) p_2^{(1)}(x, y) p_3^{(1)}(x, y). \qquad (3.40)$$

In the general case the function $U(x, y)$ is interpolated at M^* points on the triangle. This interpolant degenerates to a polynomial of degree M on a side of the triangle; there are $M + 1$ points on the side of the triangle at which $U(x, y)$ is interpolated, and so the interpolating polynomial is unique.

In a finite element triangular discretization, each side belongs to two triangles and, if the function (3.23) interpolates at M^* symmetrically placed points in the triangle, it will degenerate to the unique polynomial of degree M on the common side.

Over the complete triangular discretization, therefore, the interpolating function is continuous along and across the sides of internal triangles and thus has $C^{0,0}$ continuity over the region R.

To obtain $C^{1,1}$ continuity over R, the function $U(x, y)$ and its derivative U_x and U_y are interpolated. In fact, $C^{0,0}$ continuity can be obtained by interpolating $U(x, y)$, U_x and U_y at a smaller number of points. This is achieved using those polynomials of (3.21) which are of odd degree, and the generalized derivatives defined by (3.1) with only two independent variables x and y.

The first odd polynomial which can involve the use of derivatives is $P_3(x, y)$ which has the form given in (3.36). An alternative way of determining the ten coefficients of $P_3(x, y)$ is to interpolate $U(x, y)$, U_x, U_y at each vertex of the triangle together with the function value $U(x, y)$ at the centroid. Eliminating the function value at the centroid reduces the cubic interpolating polynomial $P_3(x, y)$ to one which interpolates quadratic polynomials exactly.

It is known that this new polynomial, which has $C^{0,0}$ continuity, is of the form

$$P_3^*(x, y) = \sum_{t=1}^{3} [U_t q_t(x, y) + (U_x)_t r_t(x, y) + (U_y)_t s_t(x, y)], \qquad (3.41)$$

where

$$q_1(x, y) = p_1^{(1)}(x, y)[p_1^{(1)}(x, y)\{3 - 2p_1^{(1)}(x, y)\} + 2p_2^{(1)}(x, y)p_3^{(1)}(x, y)],$$
$$q_2(x, y) = p_2^{(1)}(x, y)[p_2^{(1)}(x, y)\{3 - 2p_2^{(1)}(x, y)\} + 2p_3^{(1)}(x, y)p_1^{(1)}(x, y)],$$
$$q_3(x, y) = p_3^{(1)}(x, y)[p_3^{(1)}(x, y)\{3 - 2p_3^{(1)}(x, y)\} + 2p_1^{(1)}(x, y)p_2^{(1)}(x, y)],$$

$$r_1(x, y) = [p_1^{(1)}(x, y)]^2 [(x_2 - x_1)p_2^{(1)}(x, y) + (x_3 - x_1)p_3^{(1)}(x, y)]$$
$$+ \tfrac{1}{2}(-2x_1 + x_2 + x_3)p_1^{(1)}(x, y)p_2^{(1)}(x, y)p_3^{(1)}(x, y),$$
$$r_2(x, y) = [p_2^{(1)}(x, y)]^2 [(x_1 - x_2)p_1^{(1)}(x, y) + (x_3 - x_2)p_3^{(1)}(x, y)]$$
$$+ \tfrac{1}{2}(x_1 - 2x_2 + x_3)p_1^{(1)}(x, y)p_2^{(1)}(x, y)p_3^{(1)}(x, y),$$
$$r_3(x, y) = [p_3^{(1)}(x, y)]^2 [(x_1 - x_3)p_1^{(1)}(x, y) + (x_2 - x_3)p_2^{(1)}(x, y)]$$
$$+ \tfrac{1}{2}(x_1 + x_2 - 2x_3)p_1^{(1)}(x, y)p_2^{(1)}(x, y)p_3^{(1)}(x, y),$$
$$s_1(x, y) = [p_1^{(1)}(x, y)]^2 [(y_2 - y_1)p_2^{(1)}(x, y) + (y_3 - y_1)p_3^{(1)}(x, y)]$$
$$+ \tfrac{1}{2}(-2y_1 + y_2 + y_3)p_1^{(1)}(x, y)p_2^{(1)}(x, y)p_3^{(1)}(x, y),$$
$$s_2(x, y) = [p_2^{(1)}(x, y)]^2 [(y_1 - y_2)p_1^{(1)}(x, y) + (y_3 - y_2)p_3^{(1)}(x, y)]$$
$$+ \tfrac{1}{2}(y_1 - 2y_2 + y_3)p_1^{(1)}(x, y)p_2^{(1)}(x, y)p_3^{(1)}(x, y),$$
$$s_3(x, y) = [p_3^{(1)}(x, y)]^2 [(y_1 - y_3)p_1^{(1)}(x, y) + (y_2 - y_3)p_2^{(1)}(x, y)]$$
$$+ \tfrac{1}{2}(y_1 + y_2 - 2y_3)p_1^{(1)}(x, y)p_2^{(1)}(x, y)p_3^{(1)}(x, y). \tag{3.42}$$

The continuity of $P_3^*(x, y)$ may be extended to $C^{1,1}$ by the addition of the *corrective terms* of Zienkiewicz (1967, p. 117) given by

$$A_1(x, y) = p_1^{(1)}(x, y)[p_2^{(1)}(x, y)]^2 [p_3^{(1)}(x, y)]^2 / \{[1 - p_2^{(1)}(x, y)][1 - p_3^{(1)}(x, y)]\},$$
$$A_2(x, y) = p_2^{(1)}(x, y)[p_3^{(1)}(x, y)]^2 [p_1^{(1)}(x, y)]^2 / \{[1 - p_3^{(1)}(x, y)][1 - p_1^{(1)}(x, y)]\},$$
$$A_3(x, y) = p_3^{(1)}(x, y)[p_1^{(1)}(x, y)]^2 [p_2^{(1)}(x, y)]^2 / \{[1 - p_1^{(1)}(x, y)][1 - p_2^{(1)}(x, y)]\}. \tag{3.43}$$

The function $Q(x, y)$, defined by

$$Q(x, y) = \sum_{t=1}^{3} [U_t q_t(x, y) + (U_x)_t r_t(x, y) + (U_y)_t s_t(x, y) + A_t(x, y)], \tag{3.44}$$

therefore has $C^{1,1}$ continuity across the sides of the finite element mesh of triangles in R.

Other corrective functions, all of which vanish on the sides of the triangle and which have linear normal derivatives of the function $U(x, y)$ across the sides of the triangle, are given by Clough and Tocher (1965), Irons (1969) and Dupuis and Göel (1970). $C^{1,1}$ continuity is also achieved by the interpolating polynomial $P_5(x, y)$ over a finite element mesh of triangles covering R.

3.5.2 The quadrilateral

It is not possible to construct a polynomial in x and y which interpolates $U(x, y)$ at the four vertices of a quadrilateral. In order to achieve this, rational basis functions are used. The form of each term of these interpolating functions is seen to depend solely on the geometry of the quadrilateral which is shown in Fig. 3.1.

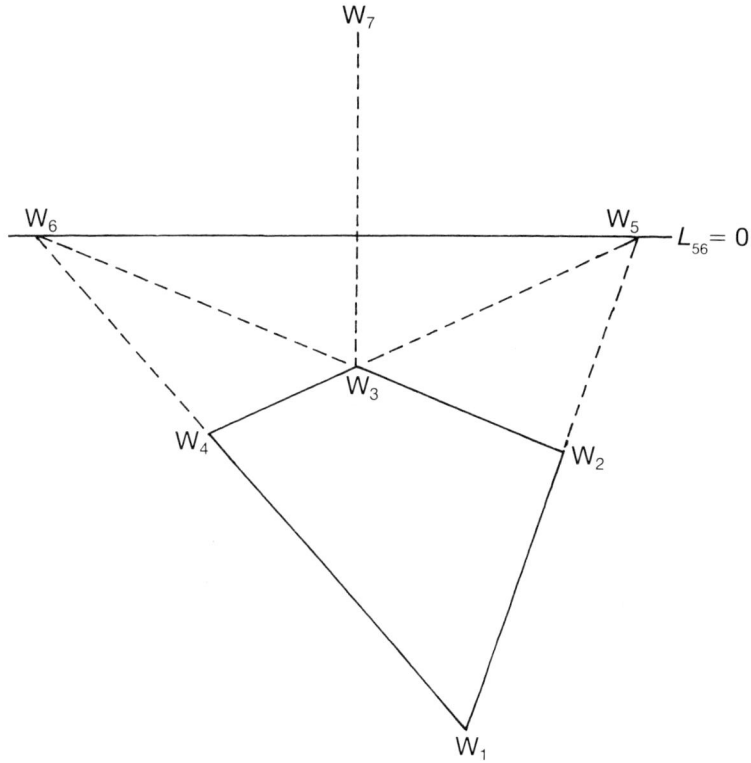

Fig. 3.1 – Quadrilateral element.

Suppose W_1, W_2, W_3, W_4 are the vertices of the quadrilateral which lies in the (x, y) plane and let $L_{ab} = 0$ be the equation of the edge joining the vertices W_a, W_b. The point of intersection of the lines $L_{12} = 0$ and $L_{43} = 0$ is the point W_5, and W_6 is the point of intersection of the lines $L_{23} = 0$ and $L_{14} = 0$. The point W_7 in Fig. 3.1 is vertically above W_3 and is a distance z from it.

The equation of the quadric surface containing the lines $L_{12} = 0$, $L_{14} = 0$, $L_{27} = 0$ and $L_{47} = 0$ is given by

$$\Pi_{412}\Pi_{472} + \alpha\Pi_{147}\Pi_{127} = 0, \qquad (3.45)$$

where Π_{412}, for example, is the plane in (x, y, z) space containing the lines $L_{12} = 0$ and $L_{14} = 0$, and α is some parameter. Equation (3.45) is a quadratic in z and by giving α an appropriate value the z^2 term is made to vanish and the term

$$z = \frac{L_{41}L_{12}}{L_{56}} \left(\frac{L_{56}}{L_{41}L_{12}}\right)_3 \tag{3.46}$$

remains, where the subscript 3 indicates normalization at the vertex W_3. It is easy to see that, by extending (3.46) and replacing z by U, the function which interpolates $U(x, y)$ at the four vertices of the quadrilateral is

$$U(x, y) = \frac{L_{23}L_{34}}{L_{56}} \left(\frac{L_{56}}{L_{23}L_{34}}\right)_1 U_1 + \frac{L_{34}L_{41}}{L_{56}} \left(\frac{L_{56}}{L_{34}L_{41}}\right)_2 U_2$$

$$+ \frac{L_{41}L_{12}}{L_{56}} \left(\frac{L_{56}}{L_{41}L_{12}}\right)_3 U_3 + \frac{L_{12}L_{23}}{L_{56}} \left(\frac{L_{56}}{L_{12}L_{23}}\right)_4 U_4. \tag{3.47}$$

It is easy to verify that (3.47) is linear along each of the four sides of the quadrilateral and it follows that $C^{0,0}$ continuity is achieved by using a finite element discretization of non-overlapping quadrilaterals.

3.5.3 The parallelogram, rectangle and trapezium

In the case of the parallelogram, rectangle and trapezium, the line $L_{56} = 0$ is not involved and (3.47) becomes

$$U(x, y) = \frac{L_{23}L_{34}}{(L_{23}L_{34})_1} U_1 + \frac{L_{34}L_{41}}{(L_{34}L_{41})_2} U_2 + \frac{L_{41}L_{12}}{(L_{41}L_{12})_3} U_3 + \frac{L_{12}L_{23}}{(L_{12}L_{23})_4} U_4. \tag{3.48}$$

The function $U(x, y)$ defined by (3.48) is clearly linear along each side of the parallelogram or rectangle and $C^{0,0}$ continuity is maintained over a patchwork of parallelograms or rectangles.

In the case of a network of rectangles, $C^{1,1}$ continuity is achieved using a function of the form

$$P_4(x, y) = \sum_{i=0}^{3} \sum_{j=0}^{3} \alpha_{ij} x^i y^j \tag{3.49}$$

in each rectangle. The sixteen coefficients α_{ij} ($i, j = 0, 1, 2, 3$) are determined in terms of P_4, $\partial P_4/\partial x$, $\partial P_4/\partial y$ and $\partial^2 P_4/\partial x \partial y$ at all four vertices of the rectangle.

3.5.4 Curved elements

All elements considered so far in this section have had straight sides and, unless the boundary ∂R of the region R is polygonal, any finite element discretization of R into a non-overlapping patchwork of triangles, quadrilaterals, parallelograms, or trapezia will leave a residue or skin adjacent to the boundary.

The skin will not be present if elements with one or more curved sides are used. An example in which curved elements are used will be discussed in section 3.8.

3.6 CONVERGENCE AND ERROR ANALYSIS

Whereas engineers have been concerned largely with the implementation of finite element methods, mathematicians, in particular numerical analysts, have devoted their attention to analyses of the techniques. The questions of convergence and rate of convergence of approximations, and bounds on the errors in computed solutions, are important in any finite element method.

The mathematical analyses which lead to convergence rates and error bounds require more than a knowledge of vector spaces and advanced calculus and, as the aim of this chapter is to show the reader how finite element methods are used in the solution of elliptic equations, such analyses will not be included. Instead the reader is referred to the texts by Strang and Fix (1973) and Mitchell and Wait (1977), and the many research papers referenced therein.

The use of the finite element method in solving the Dirichlet problem of subsection 2.1.1 will be considered in the following section.

3.7 THE DIRICHLET PROBLEM

As with finite difference methods, the easiest elliptic problem to solve using finite element methods is the integration of Laplace's equation

$$\frac{\partial^2 u}{\partial x^2} + \frac{\partial^2 u}{\partial y^2} = 0 \tag{3.50}$$

over some region R, with Dirichlet boundary conditions

$$u = f(x, y) \tag{3.51}$$

on the boundary ∂R of R.

It was shown in section 3.3.2 that the solution of (3.50) is equivalent to minimizing the functional

$$I(U) = \iint_R (U_x^2 + U_y^2) \mathrm{d}x \, \mathrm{d}y \tag{3.52}$$

over all functions $U(x, y)$ satisfying (3.51) on ∂R.

The next step is to subdivide the region R into finite elements. Choosing triangles results in the approximation of a general region R by a polygonal region. A general region is partially triangulated in Fig. 3.2.

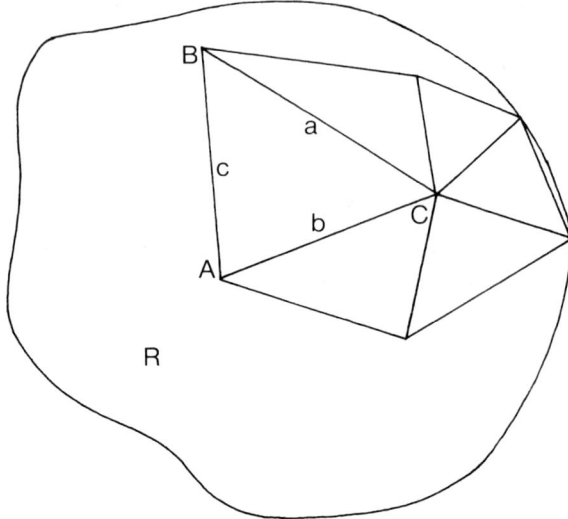

Fig. 3.2 – Partial triangulation of a general region R.

On each triangle T_i of the polygonal region, the solution of (3.50) is approximated by

$$U(x, y) = p + qx + ry; \tag{3.53}$$

as shown in section 3.5.1 this leads to $C^{0,0}$ continuity over the polygonal region. The terms a, b, c are constants and are unique to T_i.

Suppose triangle T_i has vertices (x_j, y_j), $j = 1, 2, 3$; its area A_i is then given by

$$A_i = \tfrac{1}{2} \begin{vmatrix} 1 & x_1 & y_1 \\ 1 & x_2 & y_2 \\ 1 & x_3 & y_3 \end{vmatrix}$$

$$= \tfrac{1}{2}[(x_2 - x_1)(y_3 - y_1) - (x_3 - x_1)(y_2 - y_1)] \ . \tag{3.54}$$

At each nodal vertex the function U is given by

$$U_j = p + qx_j + ry_j \ (j = 1, 2, 3). \tag{3.55}$$

This leads to a system of equations written in matrix form as

$$\begin{bmatrix} U_1 \\ U_2 \\ U_3 \end{bmatrix} = \begin{bmatrix} 1 & x_1 & y_1 \\ 1 & x_2 & y_2 \\ 1 & x_3 & y_3 \end{bmatrix} \begin{bmatrix} p \\ q \\ r \end{bmatrix} \ . \tag{3.56}$$

Equation (3.56) is easily inverted to give

$$\begin{bmatrix} p \\ q \\ r \end{bmatrix} = \frac{1}{2A_i} \begin{bmatrix} x_2 y_3 - x_3 y_2 & x_3 y_1 - x_1 y_3 & x_1 y_2 - x_2 y_1 \\ y_2 - y_3 & y_3 - y_1 & y_1 - y_2 \\ x_3 - x_2 & x_1 - x_3 & x_2 - x_1 \end{bmatrix} \begin{bmatrix} U_1 \\ U_2 \\ U_3 \end{bmatrix}. \quad (3.57)$$

Consider now equation (3.52) with U given by (3.55) so that $U_x = q$ and $U_y = r$. Then

$$I(U) = \iint_R (q^2 + r^2) dx\, dy$$

$$= (q^2 + r^2) \iint_R dx\, dy$$

$$= (q^2 + r^2) A_i,$$

so that

$$\frac{\partial}{\partial U} I(U) = 2A_i \left[q \frac{\partial q}{\partial U_j} + r \frac{\partial r}{\partial U_j} \right]$$

$$= F_j, \quad (3.58)$$

say, for $j = 1, 2, 3$. The necessary conditions for $I(U)$ to have a minimum value are that every

$$F_j = 0 \ (j = 1, 2, 3) \quad (3.59)$$

simultaneously.

It is easy to show, using q and r from (3.57), that

$$2 \left[q \frac{\partial q}{\partial U_j} + r \frac{\partial r}{\partial U_j} \right] = [\{(x_3 - x_2)^2 + (y^2 - y_3)^2\} U_1$$

$$+ \{(x_3 - x_2)(x_1 - x_3) + (y_2 - y_3)(y_3 - y_1)\} U_2$$

$$+ \{(x_3 - x_2)(x_2 - x_1) + (y_2 - y_3)(y_1 - y_3)(y_1 - y_2)\} U_3]/(4A_i^2). \quad (3.60)$$

In any triangle with sides a, b, c, opposite angles A, B, C, respectively (see Fig. 3.2), it is well known that

$$\mathbf{a} = (x_2 - x_3)\mathbf{i} + (y_2 - y_3)\mathbf{j},$$
$$\mathbf{b} = (x_3 - x_1)\mathbf{i} + (y_3 - y_1)\mathbf{j},$$
$$\mathbf{c} = (x_1 - x_2)\mathbf{i} + (y_1 - y_2)\mathbf{j},$$

where **i, j** are unit vectors in the positive x, y directions, respectively, and that

$$\mathbf{a.a} = a^2 = (x_2 - x_3)^2 + (y_2 - y_3)^2,$$
$$\mathbf{a.b} = -a\,b\,\cos C = (x_2 - x_3)(x_3 - x_1) + (y_2 - y_3)(y_3 - y_1),$$
$$\mathbf{a.c} = -a\,c\,\cos B = (x_2 - x_3)(x_1 - x_2) + (y_2 - y_3)(y_1 - y_2).$$

It follows, then, that

$$q\frac{\partial q}{\partial U_1} + r\frac{\partial r}{\partial U_1} = [a^2 U_1 - (a\,b\,\cos C)U_2 - (a\,c\,\cos B)U_3]/(4A_i^2) \qquad (3.61)$$

and similarly that

$$q\frac{\partial q}{\partial U_2} + r\frac{\partial r}{\partial U_2} = [-(b\,a\,\cos C)U_1 + b^2 U_2 - (b\,c\,\cos A)U_3]/(4A_i^2), \qquad (3.62)$$

$$q\frac{\partial q}{\partial U_3} + r\frac{\partial r}{\partial U_3} = [-(c\,a\,\cos B)U_1 - (b\,c\,\cos A)U_2 + c^2 U_3]/(4A_i^2). \qquad (3.63)$$

It is also known that for the triangle ABC of Fig. 3.2

$$A_i = \tfrac{1}{2} b\,c \sin A = \tfrac{1}{2} a\,c \sin B = \tfrac{1}{2} a\,b \sin C. \qquad (3.64)$$

Equations (3.64) give values of ab, ac, bc for use in (3.61), (3.62), (3.63). Using (3.64) a^2, b^2, c^2 can also be found in terms of the area A_i and the angles A, B, C as follows:

$$a^2 = \frac{abac}{bc} = \frac{2A_i}{\sin C} \cdot \frac{2A_i}{\sin B} \cdot \frac{\sin A}{2A_i}$$

$$= 2A_i \frac{\sin A}{\sin B \sin C} = 2A_i \frac{\sin(B+C)}{\sin B \sin C}$$

$$= 2A_i \frac{\sin B \cos C + \cos B \sin C}{\sin B \sin C}$$

so that

$$a^2 = 2A_i(\cot B + \cot C). \qquad (3.65)$$

Similarly

$$b^2 = 2A_i(\cot C + \cot A). \qquad (3.66)$$

and

$$c^2 = 2A_i(\cot A + \cot B). \qquad (3.67)$$

Substituting for a^2, b^2, c^2, ab, bc, ca in (3.61), (3.62) and (3.63) gives

$$q\frac{\partial q}{\partial U_1} + r\frac{\partial r}{\partial U_1} = [(\cot B + \cot C)U_1 - (\cot C)U_2 - (\cot B)U_3]/(2A_i),$$

$$q\frac{\partial q}{\partial U_2} + r\frac{\partial r}{\partial U_2} = [-(\cot C)U_1 + (\cot C + \cot A)U_2 - (\cot A)U_3]/(2A_i),$$

$$q\frac{\partial q}{\partial U_3} + r\frac{\partial r}{\partial U_3} = [-(\cot B)U_1 - (\cot A)U_2 + (\cot A + \cot B)U_3]/(2A_i),$$

so that

$$\begin{bmatrix} F_1 \\ F_2 \\ F_3 \end{bmatrix} = \begin{bmatrix} \cot B + \cot C & -\cot C & -\cot B \\ -\cot C & \cot C + \cot A & -\cot A \\ -\cot B & -\cot A & \cot A + \cot B \end{bmatrix} \begin{bmatrix} U_1 \\ U_2 \\ U_3 \end{bmatrix}. \quad (3.68)$$

Clearly, the terms F_1, F_2, F_3 are dependent on the geometry of the triangle T_i. It was noted earlier that $I(U)$ is minimized when each $F_j = 0$ simultaneously ($j = 1, 2, 3$), and the values of U_1, U_2, U_3 which satisfy (3.59) are easily determined for triangle T_i from (3.68).

As an example, the region R will be taken to be the unit square bounded by the lines $x = 0, x = 1, y = 0, y = 1$ and the boundary conditions to be

$$f(x, y) = 1 \text{ on } \partial R. \quad (3.69)$$

The region R will be triangulated as in Fig. 3.3(a), where each triangle is one of two isosceles forms (b) or (c) in which $A = 90°C$, $B = C = 45°$. For this triangulation, (3.68) becomes

$$\begin{bmatrix} F_1 \\ F_2 \\ F_3 \end{bmatrix} = \begin{bmatrix} 2 & -1 & -1 \\ -1 & 1 & 0 \\ -1 & 0 & 0 \end{bmatrix} \begin{bmatrix} U_1 \\ U_2 \\ U_3 \end{bmatrix}. \quad (3.70)$$

Each node of Fig. 3.3(a) has six common triangles; for the node with coordinates (kh, mh), $h = \frac{1}{4}$, the six triangles are shown in Fig. 3.3(d).

Applying $F_1 = 0$ to triangles 1 and 4 in Fig. 3.3(d), $F_2 = 0$ to triangles 2 and 5, and $F_3 = 0$ to triangles 3 and 6, and adding, gives

$$(2U_{k,m} - U_{k+1,m} - U_{k,m-1}) + (-U_{k+1,m} + U_{k,m})$$
$$+ (-U_{k,m+1} + U_{k,m}) + (2U_{k,m} - U_{k,m+1} - U_{k-1,m})$$
$$+ (-U_{k-1,m} + U_{k,m}) + (-U_{k,m-1} + U_{k,m}) = 0.$$

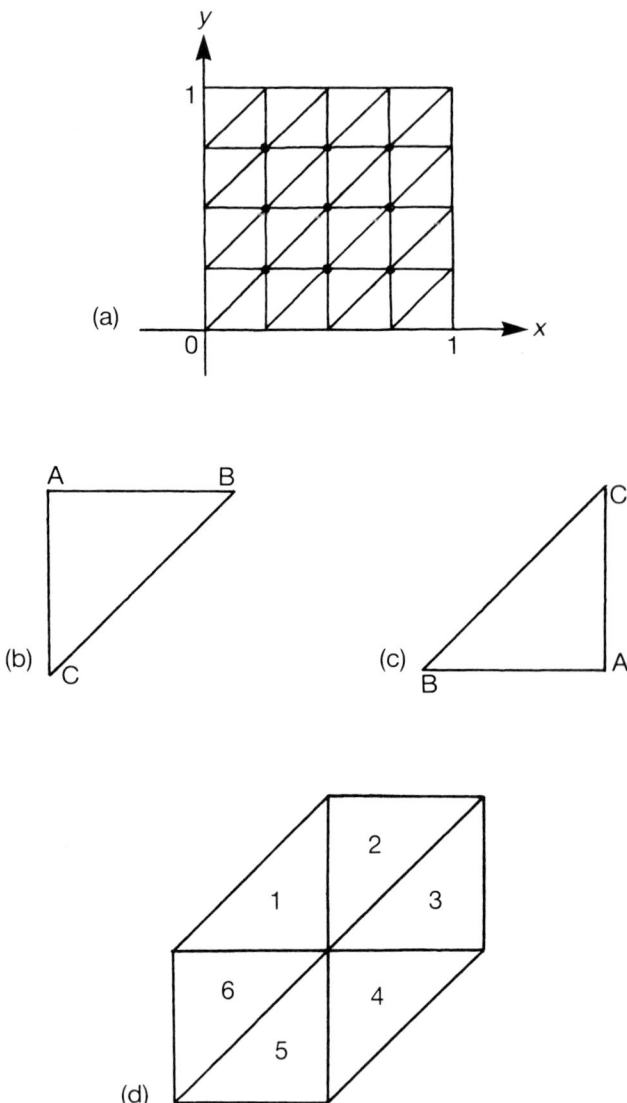

Fig. 3.3 – Triangulation of unit square for model problem: (a) triangulation of the square, (b) Type I isosceles triangle, (c) Type II isosceles triangle, (d) the six common triangles of any interior node of the square.

Rearranging, and dividing by -2, gives

$$U_{k+1,m} + U_{k,m+1} + U_{k-1,m} + U_{k,m-1} - 4U_{k,m} = 0 \qquad (3.71)$$

which is the five-point difference scheme (2.9) derived using finite difference methods in Chapter 2; the finite difference and finite element approaches have produced the same formula for the solutions of Laplace's equation in a square region with Dirichlet boundary conditions.

Applying (3.71) to each of the nine nodes of Fig. 3.3(a) gives

$$\begin{bmatrix} -4 & 1 & 0 & 1 & 0 & 0 & 0 & 0 & 0 \\ 1 & -4 & 0 & 0 & 1 & 0 & 0 & 0 & 0 \\ 0 & 1 & -4 & 1 & 0 & 1 & 0 & 0 & 0 \\ 1 & 0 & 0 & -4 & 1 & 0 & 1 & 0 & 0 \\ 0 & 1 & 0 & 1 & -4 & 1 & 0 & 1 & 0 \\ 0 & 0 & 1 & 0 & 1 & -4 & 0 & 0 & 1 \\ 0 & 0 & 0 & 1 & 0 & 0 & -4 & 1 & 0 \\ 0 & 0 & 0 & 0 & 1 & 0 & 1 & -4 & 1 \\ 0 & 0 & 0 & 0 & 0 & 1 & 0 & 1 & -4 \end{bmatrix} \begin{bmatrix} U_{1,1} \\ U_{2,1} \\ U_{3,1} \\ U_{1,2} \\ U_{2,2} \\ U_{3,3} \\ U_{1,3} \\ U_{2,3} \\ U_{3,3} \end{bmatrix} = \begin{bmatrix} -2 \\ -1 \\ -2 \\ -1 \\ 0 \\ -1 \\ -2 \\ -1 \\ -2 \end{bmatrix} \qquad (3.72)$$

which is of the form of equation (2.11) of Chapter 2 and its associated equations (2.12) and (2.13). Equation (3.72) is easily solved using the conjugate gradient method outlined in subsection 1.2.2 of Chapter 1.

The solution of (3.72) minimizes the functional $I(U)$, given by (3.52), over the unit square and satisfies the Dirichlet boundary condition (3.69); it is therefore the solution of Laplace's equation (3.50) in the unit square, together with (3.69) on the boundary of the unit square.

3.8 RATIONAL BASIS FUNCTIONS

In this section, a circular region R is discretized into triangles. Elements having two straight sides and one curved side are used, enabling the circular boundary to be retained. Linear approximating functions are not suitable and rational approximating functions are developed, which are dependent on the geometry of the elements.

The particular formulation to be considered is that developed for the doctoral thesis of Smith (1981) whose finite element discretization uses 193 nodes. The discretization is reproduced in Fig. 3.4.

3.8.1 The rational functions

The reader can see from Fig. 3.4 that there are three types of element and each will be considered using the terminology of Wachspress (1975).

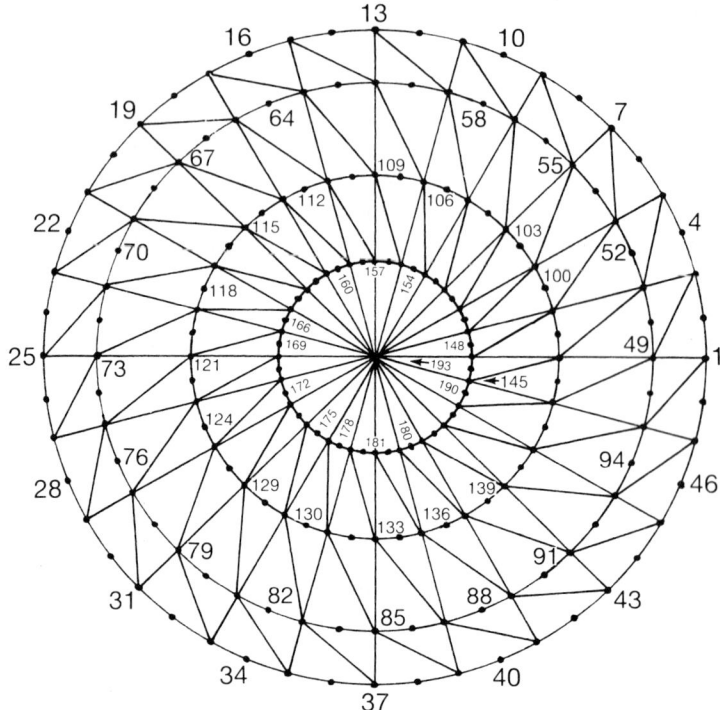

Fig. 3.4 – Discretization of the circular region for rational basis functions.

The first element to be considered is that shown in Fig. 3.5. It will be called a *3-con I* and occurs only in the innermost circle of Fig. 3.5.

The approximation to U within this element is taken to be

$$U(x, y) = W_1 k_1 U_1 + W_2 k_2 U_2 + W_3 k_3 U_3 + W_4 k_4 U_4, \qquad (3.73)$$

where the W_i ($i = 1, 2, 3, 4$) are wedge functions or basis functions, the k_i ($i = 1, 2, 3, 4$) are normalizing factors which reduce the coefficients of U_i to unity when evaluated at node i, and the U_i ($i = 1, 2, 3, 4$) are the values of the function at the nodes of the element. Each 3-con I is of order 4 having two straight edges and one curved edge.

The four wedge functions in (3.73) are given by

$$\begin{aligned}
W_1 &= (2, 3)_2/Q_\mathrm{I}, \\
W_2 &= (3, 1)\,(4, M)/Q_\mathrm{I}, \\
W_3 &= (1, 2)\,(4, L)/Q_\mathrm{I}, \\
W_4 &= (1, 2)\,(3, 1)/Q_\mathrm{I},
\end{aligned} \qquad (3.74)$$

Sec. 3.8] Rational Basis Functions

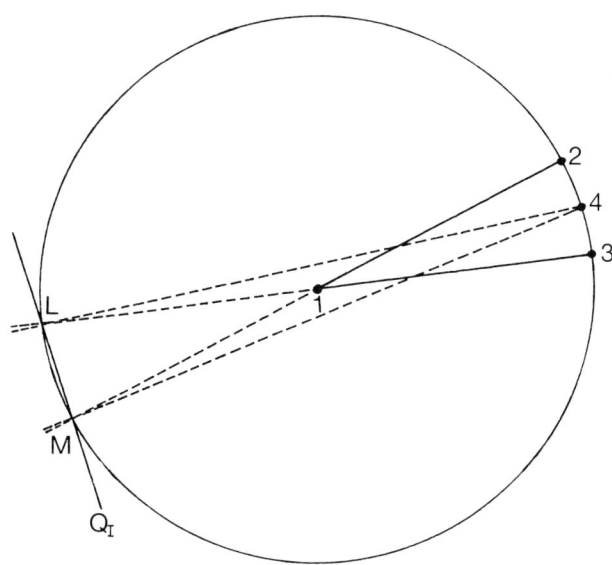

Fig. 3.5 – 3-con I.

where, in the notation of Wachspress (1975), (3, 1) is that function which vanishes along the line joining nodes 3 and 1, (4, M) is that function which vanishes along the line joining node 4 (the mid-point of the arc joining nodes 2 and 3) with point M, and so on. The function $(2, 3)_2$ is that which vanishes along the conic joining nodes 2 and 3; the subscript 2 denotes degree 2, the absence of a subscript denotes first degree. The function Q_I is that which vanishes along the line joining the external intersection points of the sides of the 3-con, namely L and M in Fig. 3.5.

The normalizing factors are given by

$$
\begin{aligned}
k_1 &= [Q_I/(2, 3)_2]_1, \\
k_2 &= [Q_I/\{(3, 1)(4, M)\}]_2, \\
k_3 &= [Q_I/\{(1, 2)(4, L)\}]_3, \\
k_4 &= [Q_I/\{(1, 2)(3, 1)\}]_4,
\end{aligned}
\qquad (3.75)
$$

where the subscript i outside the square brackets indicate that the contents of the square brackets are evaluated at node i ($i = 1, 2, 3, 4$).

In Fig. 3.4 there are 24 positions of the 3-con I; the first has the side joining nodes 1 and 3 coincident with the x-axis, node 1 being at the origin (position $r = 1$), the other positions of the element occurring after $15°$ rotation ($r = 2, \ldots, 24$). Denoting by A_1 the inclination of the line joining node 3 and the origin to the x-axis, by A_2 the inclination of the line joining node 2 and the origin to the

x-axis, and by A_3 the inclination of the line joining node 4 and the origin to the x-axis, it is easy to see that, for $r = 1, 2, \ldots, 24$

$$A_1 = (r-1)\pi/12,$$
$$A_2 = r\pi/12, \qquad (3.76)$$
$$A_3 = (2r-1)\pi/24.$$

Denoting, next, the radius of the innermost circle of Fig. 3.4 by a, the coordinates of the nodes of 3-con I are

Node 1: $(0, 0)$,
Node 2: $(a \cos A_2, a \sin A_2)$,
Node 3: $(a \cos A_1, a \sin A_1)$,
Node 4: $(a \cos A_3, a \sin A_3)$,

The point L has coordinates $(-a \cos A_1, -a \sin A_1)$ and the point M has coordainates $(-a \cos A_2, -a \sin A_2)$.

The functions used in (3.74) are given by

$(2, 3)_2 = x^2 + y^2 - a^2$
$(3, 1) \ = y \cos A_1 - x \sin A_1,$
$(1, 2) \ = y \cos A_2 - x \sin A_2,$
$(4, M) = (a \cos A_2 + a \cos A_3)y - (a \sin A_2 + a \sin A_3)x + a^2 \sin(A_2 - A_3),$
$(4, C) \ = (a \cos A_1 + a \cos A_3)y - (a \sin A_1 + a \sin A_3)x + a^2 \sin(A_1 - A_3),$
$Q_{\mathrm{I}} = (L, M) = (a \sin A_1 - a \sin A_2)x - (a \cos A_1 - a \cos A_2)y$
$\qquad + a^2 \sin(A_1 - A_2).$

The second type of element in Fig. 3.4 is that shown in Fig. 3.6. This element is called a *3-con II* and occurs in the second, third and fourth concentric circles; like the 3-con I, the 3-con II is of order 4 having two straight sides and one curved side.

The approximation to U within each 3-con II is taken to be of the form (3.73). The basis functions are given by

$$W_1 = (2, 3)_2/Q_{\mathrm{II}},$$
$$W_2 = (3, 1)(4, P)/Q_{\mathrm{II}},$$
$$W_3 = (1, 2)(4, N)/Q_{\mathrm{II}}, \qquad (3.77)$$
$$W_4 = (1, 2)(3, 1)/Q_{\mathrm{II}}$$

and the normalizing factors by

$$k_1 = [Q_{\mathrm{II}}/(2, 3)_2]_1,$$
$$k_2 = [Q_{\mathrm{II}}/\{(3, 1)(4, P)\}]_2,$$
$$k_3 = [Q_{\mathrm{II}}/\{(1, 2)(4, N)\}]_3, \qquad (3.78)$$
$$k_4 = [Q_{\mathrm{II}}/\{(1, 2)(3, 1)\}]_4.$$

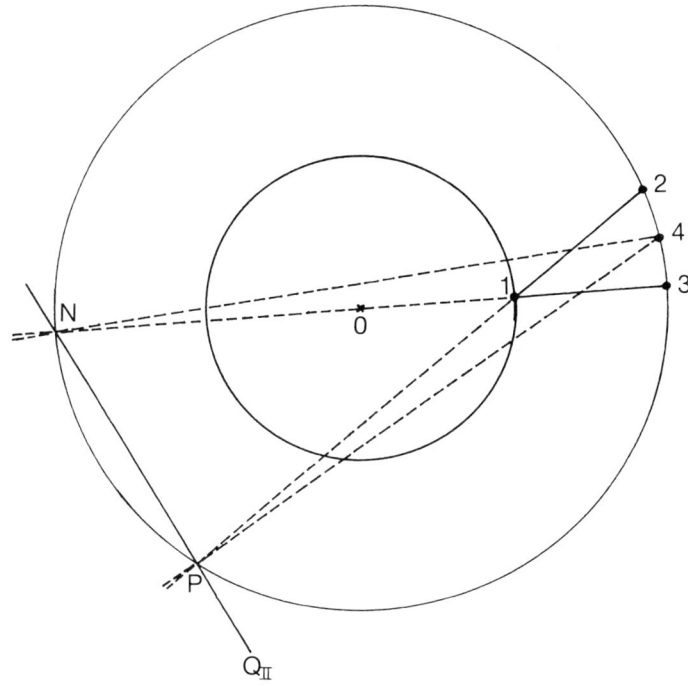

Fig. 3.6 – 3-con II.

The radii of the second, third and fourth concentric circles of Fig. 3.4 are ja, where, now, a is the length of the side of the element joining nodes 1 and 3, and j is some scalar. The coordinates of the nodes of 3-con II are thus

Node 1: $((j-1)a \cos A_1, (j-1)a \sin A_1)$,
Node 2: $(ja \cos A_2, ja \sin A_2)$,
Node 3: $(ja \cos A_1, ja \sin A_1)$,
Node 4: $(ja \cos A_3, ja \sin A_3)$,

where the angles A_1, A_2, A_3 are determined as for elements of the type 3-con I. The point N has coordinates $(-ja \cos A_1, -ja \sin A_1)$ and the point P has coordinates (A_8, A_7) where A_7, A_8 are given below.

The functions which vanish along the straight and conic sides of 3-con II are used in (3.77) and are given by

$(2, 3)_2 = x^2 + y^2 - j^2 a^2$,
$(3, 1) = y \cos A_1 - x \sin A_1$,
$(1, 2) = A_6 x - A_4 y - A_5$,
$(4, P) = (A_7 - ja \sin A_3)x + (ja \cos A_3 - A_8)y$
$\qquad + A_8 ja \sin A_3 - A_7 ja \cos A_3$,

$$(4, N) = (ja \cos A_1 + ja \cos A_3)y - (ja \sin A_1 + ja \sin A_3)x$$
$$+ j^2 a^2 \sin(A_1 - A_3),$$
$$Q_{II} = (N, P) = (A_7 + ja \sin A_1)x - (A_8 + ja \cos A_1)y$$
$$+ A_7 ja \cos A_1 - A_8 ja \sin A_1,$$

where

$$A_4 = ja \cos A_2 - (j-1)a \cos A_1,$$
$$A_5 = -j(j-1)a^2 \sin(A_1 - A_2),$$
$$A_6 = ja \sin A_2 - (j-1)a \sin A_1,$$
$$A_7 = [-A_4 A_5 - ja^2 A_6 \{j - (j-1) \cos(A_1 - A_2)\}]/(A_4^2 + A_6^2),$$
$$A_8 = (A_4 A_7 + A_5)/A_6,$$

The third and last type of element in Fig. 3.4 is called a *3-con III* and is shown in Fig. 3.7. It occurs in the second, third and fourth concentric circles and, like the others, is of order 4 having two straight sides and one conic side.

The approximation to the function U within each 3-con III is also to be taken of the form (3.73). The wedge functions, or basis functions, are given by

$$W_1 = (2, 3)(4, T)/Q_{III},$$
$$W_2 = (3, 1)(4, S)/Q_{III},$$
$$W_3 = (1, 2)_2/Q_{III},$$
$$W_4 = (2, 3)(3, 1)/Q_{III}$$

(3.79)

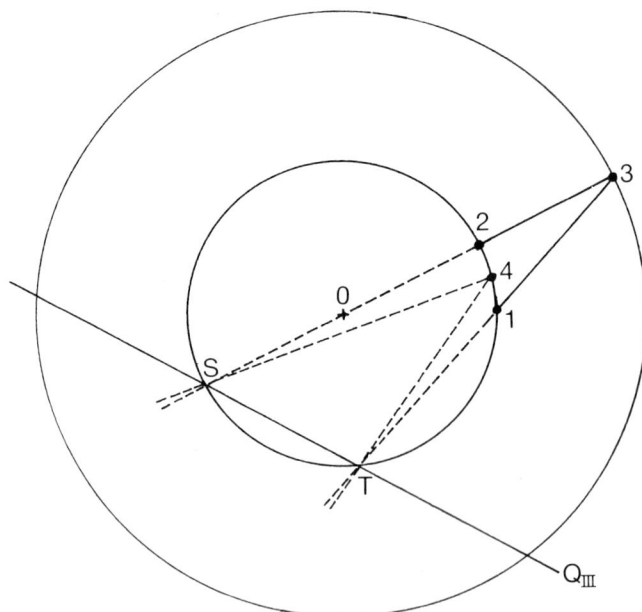

Fig. 3.7 – 3-con III.

Sec. 3.8] **Rational Basis Functions** 113

and the normalizing factors by

$$\begin{align}
k_1 &= [Q_{\text{III}}/\{(2, 3)(4, T)\}]_1, \\
k_2 &= [Q_{\text{III}}/\{(3, 1)(4, S)\}]_2, \\
k_3 &= [Q_{\text{III}}/(1, 2)_2]_3, \\
k_4 &= [Q_{\text{III}}/\{(2, 3)(3, 1)\}]_4.
\end{align} \tag{3.80}$$

The coordinates of the nodes of 3-con III are

Node 1: $((j-1)a \cos A_1, (j-1)a \sin A_1)$,
Node 2: $((j-1)a \cos A_2, (j-1)a \sin A_2)$,
Node 3: $(ja \cos A_2, ja \sin A_2)$,
Node 4: $((j-1)a \cos A_3, (j-1)a \sin A_3)$,

where the angles A_1, A_2, A_3 are determined as for the 3-con I. The point S has coordinates $(-(j-1)a \cos A_2, -(j-1)a \sin A_2)$ and the point T has coordinates (A_{10}, A_9) where A_9, A_{10} are given below.

The functions used in (3.79) which vanish along the sides of the elements of the type 3-con III are

$$\begin{align}
(1, 2)_2 &= x^2 + y^2 - (j-1)^2 a^2, \\
(3, 1) &= -A_6 x + A_4 y + A_5, \\
(2, 3) &= y \cos A_2 - x \sin A_2, \\
(4, T) &= (A_9 - (j-1)a \sin A_3)x + ((j-1)a \cos A_3 - A_{10})y \\
&\quad - A_9(j-1)a \cos A_3 + A_{10}(j-1)a \sin A_3, \\
(4, S) &= ((j-1)a \cos A_2 + (j-1)a \cos A_3)y - ((j-1)a \sin A_2 \\
&\quad + (j-1)a \sin A_3)x + (j-1)^2 a^2 \sin(A_2 - A_3), \\
Q_{\text{III}} &= (S, T) = (A_9 + (j-1)a \sin A_2)x - (A_{10} + (j-1)a \cos A_2)y \\
&\quad + A_9(j-1)a \cos A_2 - A_{10}(j-1)a \sin A_2,
\end{align}$$

where

$$\begin{align}
A_9 &= [-A_4 A_5 - (j-1)a^2 A_6\{j \cos(A_1 - A_2) - (j-1)\}]/(A_4^2 + A_6^2), \\
A_{10} &= (A_4 A_9 + A_5)/A_6.
\end{align}$$

3.8.2 Computational aspects

The variational formulation is developed by substituting equation (3.73) into the appropriate expression for $I(U)$. The forms of W_i, k_i ($i = 1, 2, 3, 4$) used in (3.73) are given by the pairs of systems of equations (3.74) and (3.75), (3.77) and (3.78), or (3.79) and (3.80), depending on the type of element over which integration is taking place. The process is very similar to that described in section 3.7 and will not be described further. The reader is referred to the thesis by Smith (1981) or the paper by Smith and Twizell (1980) for a detailed description of the variational formulation.

It is easy to see that, because of symmetry, it is necessary to evaluate the integrals for each type of element only once within the bands between concentric circles. This reduces the computation time required for setting up the matrix to 1/24 of the maximum time. The reader will find that the evaluation of the integrals is difficult because of the complexity of the functions W_i ($i = 1$, 2, 3, 4) used in (3.73). This may be overcome by integrating analytically with respect to one space variable and then by integrating numerically with respect to the other space variable. Suitable numerical techniques were outlined in Chapter 1 and are described in detail in most numerical texts, see, for instance, the book by Burden *et al.* (1981), Gerald and Wheatley (1984) or Ralston (1965). Some of the functions give terms involving natural logarithms when integrated analytically and this poses a problem when the arguments of the logarithms (which are, in fact, the factors of Q_I, Q_{II}, Q_{III}) are negative. Further investigation by the reader will show that Q_I, Q_{II}, Q_{III} are always negative and hence may be made always positive, the change of sign being taken care of by the normalizing factors.

The elements 3-con I, 3-con II and 3-con III are said to be *well set* if and only if the boundary curves intersect transversally at the vertices and the extensions of the boundary segments do not enter the element. This requirement is always satisfied by 3-con I and 3-con II. A constraint must be placed on the construction of 3-con III, however, to ensure that it does not become *ill set*, otherwise the rational basis functions cannot then be constructed.

The element 3-con III becomes ill set when the line Q_{III} enters the element. This happens unless the ratio of the radius of the inner circle to the radius of the outer circle of the band enclosing the 3-con III, is less than the cosine of the angle upon which the finite element discretization is based. In the formulation developed by Smith (1981) an angle of $\pi/12$ is used throughout and so the condition becomes

$$\frac{(j-1)a}{ja} < \cos\frac{\pi}{12} \cong 0.9659258.$$

It is pointed out by Smith that this restriction is not very severe.

A number of numerical experiments, in which the rational basis functions described in this section are used, are described in the thesis by Smith (1981) and in the associated paper by Smith and Twizell (1980).

PROBLEMS

1. The function $u(x, y)$ satisfies Laplace's equation (3.50) and minimizes the functional in (3.52). Use the finite element method to obtain u at the three interior points of the region bounded by the x-axis and the lines $y = \sqrt{3}x/2$ and $y = 8\sqrt{3}h - \sqrt{3}x/2$. The boundary conditions on the three sides are $u = 2$, $u = 0$ and $u = 1$, respectively, and the region is to be divided into equilateral triangular elements of side $2h$.

2. Use the Euler-Lagrange equation

$$\frac{\partial f}{\partial y} - \frac{d}{dx}\frac{\partial f}{\partial y'} = 0$$

to show that minimizing the functional

$$I(y) = \int_a^b [(y')^2 + 2fy]\,dx \;;\; y(a) = \alpha, y(b) = \beta$$

is equivalent to solving the boundary valve problem

$$y'' = f(x, y) \;;\; y(a) = \alpha, y(b) = \beta.$$

Divide the interval $[a, b]$ into $n + 1$ equal intervals of width h by the points $a = x_0 < x_1 < x_2 < \ldots < x_n < x_{n+1} = b$ and approximate y in each subinterval by $y = ax + b$.

Use the finite element method to determine a linear system of order n for the values of y at the n points $x_1, x_2, \ldots, x_n$ in the case $f(x, y) = 1$.

3. The function $u(x, y)$ satisfies Poisson's equation (3.2) in a polygonal region R with boundary condition $u = 0$ everywhere on ∂R. The problem is to be solved using the appropriate functional I and a finite element method based on piecewise linear basis functions. The region R is to be triangulated into elements having vertices (x_i, y_i) $i = 1, 2, \ldots, n$. The transformation $x = x(\xi, \eta), y = y(\xi, \eta)$ where

$$x(\xi, \eta) \equiv x_i + (x_j - x_i)\xi + (x_k - x_i)\eta \equiv x_i + X_j\xi + X_k\eta,$$
$$y(\xi, \eta) \equiv y_i + (y_j - y_i)\xi + (y_k - y_i)\eta \equiv y_i + Y_j\xi + Y_k\eta$$

maps the standard triangle T in the (ξ, η) plane, with vertices $(0, 0), (1, 0), (0, 1)$ on to the triangle T^e in the (x, y) plane with vertices $(x_i, y_i), (x_j, y_j), (x_k, y_k)$ and has inverse $\xi = \xi(x, y), \eta = \eta(x, y)$. Under this transformation, the functional becomes

$$I(v)\bigg|_{T^e} = \int_T \left\{ a\left(\frac{\partial V}{\partial \xi}\right)^2 - 2b\frac{\partial V}{\partial \xi}\frac{\partial V}{\partial \eta} + c\left(\frac{\partial V}{\partial \eta}\right)^2 + 2JFV \right\} d\xi d\eta,$$

where $v(x, y) = V(\xi, \eta), f(x, y) = F(\xi, \eta), J = X_jY_k - X_kY_j, a = (X_k^2 + Y_k^2)/J,$
$b = (X_jX_k + Y_jY_k)/J$ and $c = (X_j^2 + Y_j^2)/J$.

If, in the triangle T^e, the solution is approximated by

$$w(x, y) = w_i\{1 - \xi(x, y) - \eta(x, y)\} + w_j\,\xi(x, y) + w_k\,\eta(x, y),$$

where w_i, w_j, w_k are to be determined, derive the solution matrix (the stiffness matrix) for T^e. Hence derive the matrix for the case in which T^e has vertices $(x_i, y_i) = (4, 4), (x_j, y_j) = (5, 4)$ and $(x_k, y_k) = (5, 3)$.

4

Hyperbolic Equations

4.1 THE METHOD OF CHARACTERISTICS FOR SECOND ORDER EQUATIONS

Hyperbolic partial differential equations usually arise when waves or vibrations occur in a physical system, and mathematical modelling of such a system invariably involves the solution of a hyperbolic equation or of a hyperbolic system.

In this chapter, hyperbolic equations in one and two independent space variables will be considered; time will also be an independent variable.

The *Method of Characteristics* is universally regarded as the most effective method for solving hyperbolic problems in two independent variables (one space variable together with time). Indeed, a study of other methods of solution, such as those based on finite difference or finite element techniques, should not be undertaken until the reader has become thoroughly familiar with the important role played by characteristics in the solution of hyperbolic partial differential equations.

However, it is important to study other methods of solution of hyperbolic equations in one space variable and time too, for these often generalize in a natural way to equations with two or three space variables and time, where characteristic methods are less satisfactory.

It was seen in Chapter 1 that the general quasi-linear partial differential equation of the second order may be written in the form

$$a \frac{\partial^2 u}{\partial x^2} + b \frac{\partial^2 u}{\partial x \partial t} + c \frac{\partial^2 u}{\partial t^2} = e, \tag{4.1}$$

where a, b, c, e are functions of u, x, t, $\partial u/\partial x$, $\partial u/\partial t$, but not of the second order derivatives. The abbreviations p, q, r, s, y will be used for $\partial u/\partial x$, $\partial u/\partial t$, $\partial^2 u/\partial x^2$, $\partial^2 u/\partial x \partial t$, $\partial^2 u/\partial t^2$, respectively.

Sec. 4.1] The Method of Characteristics for Second Order Equations

It was seen in Chapter 1 that u, p, q satisfy the total differential formulas

$$du = p \, dx + q \, dt, \tag{4.2}$$

$$dp = r \, dx + s \, dt, \tag{4.3}$$

$$dq = s \, dx + y \, dt, \tag{4.4}$$

respectively, and that a, b, c satisfy

$$a \left(\frac{dt}{dx} \right)^2 - b \frac{dt}{dx} + c = 0. \tag{4.5}$$

Whenever $b^2 - 4ac > 0$ in (4.5), equation (4.1) is said to be *hyperbolic*, for then dt/dx has two distinct values at each point in the (x, t) plane. This means that two curves, called *characteristic curves*, may be drawn through each point in the plane.

The simplest second order hyperbolic equation is the wave equation

$$\frac{\partial^2 u}{\partial x^2} - \frac{\partial^2 u}{\partial t^2} = 0$$

which has general solution

$$u(x, t) = F(x + t) + G(x - t)$$

where F and G are arbitrary differentiable functions. If the initial conditions $u(x, 0) = f(x)$ and $\partial u/\partial t \, (x, 0) = g(x)$, for $-\infty < x < +\infty$ are specified, the wave equation together with these initial conditions form the *Cauchy initial value problem* which has solution

$$u(x, t) = \tfrac{1}{2} \left[f(x + t) + f(x - t) + \int_{x-t}^{x+t} g(\theta) \, d\theta \right]. \tag{4.6}$$

For the wave equation, $a = 1, b = 0, c = -1, e = 0$ so that $dt/dx = \pm 1$ and the characteristic curves are the straight lines $x \pm t =$ constant. The solution of the Cauchy initial value problem can be interpreted geometrically as in Fig. 4.1, where the value of u at the point T depends on the initial data on the part of the x-axis between A and B. The points A and B are the points where the characteristics through T cut the x-axis and it is these points which give the limits of the proper integral in (4.6). The triangle TAB in Fig. 4.1 is called the *domain of dependence* of the point T and the characteristics drawn in the triangle form the *characteristic mesh*.

Returning now to the general hyperbolic equation, it is easy to see from (4.5) that dt/dx does not have constant values when a, b, c are not constants and so the resulting characteristic mesh may not be uniform.

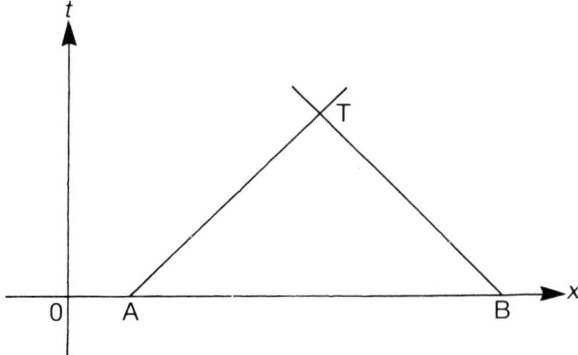

Fig. 4.1 — Domain of dependence of a point T for the simple wave equation.

Multiplying equation (4.3) by $a \, dt$, equation (4.4) by $c \, dx$, and adding gives

$$\begin{aligned} a \, dt \, dp + c \, dx \, dq &= (ar + cy) \, dx \, dt + as(dt)^2 + cs(dx)^2, \\ &= (-bs + e) \, dx \, dt + as(dt)^2 + cs(dx)^2, \text{ (from (4.1))} \\ &= e \, dx \, dt + s\{a(dt)^2 - b \, dx \, dt + c(dx)^2\}, \end{aligned} \quad (4.7)$$

so that on curves for which (4.5) holds

$$a \, dt \, dp + c \, dx \, dq = e \, dx \, dt. \tag{4.8}$$

Curves in the (x, t) plane for which (4.5) and (4.8) hold are thus the characteristic curves introduced earlier.

Suppose, then, that (4.1) is hyperbolic and that values of u, p, q are given at M points $T_m^{(1)}$ ($m=1, \ldots, M$), which are not necessarily equally spaced, along some initial curve L in the (x, t) plane which is *not* a characteristic. The exclusion of characteristic curves as initial curves is necessary because the solution would then be unique along the initial curve but nowhere else. Furthermore, it is impossible in such cases to use the method of characteristics to advance the solution from the M points on the initial curve to points off the initial curve.

Consider, for example, the hyperbolic equation

$$\frac{\partial^2 u}{\partial x^2} + 2 \frac{\partial^2 u}{\partial x \partial t} - 3 \frac{\partial^2 u}{\partial t^2} = 0,$$

for which (4.5) becomes

$$\left(\frac{dt}{dx}\right)^2 - 2 \frac{dt}{dx} - 3 = 0,$$

giving $dt/dx = 3$ or -1. The characteristics of the hyperbolic equation are therefore $t - 3x =$ constant and $t + x =$ constant. Suppose, then, that the

Sec. 4.1] The Method of Characteristics for Second Order Equations

initial curve L is the characteristic curve $t - 3x = 0$. Equation (4.8), divided by dx, becomes $3\,dp - 3\,dq = 0$ which is of the form $p - q = $ constant. The differential relationship (4.8) is therefore satisfied by, for example, the initial conditions $u = -1$, $p = -3$, $q = 1$ and it is easy to show that the family of solutions

$$u(x, t) = -1 + (t - 3x) + A(t - 3x)^2$$

satisfies these initial conditions also, where A is an arbitrary constant. The fact that A is an arbitrary constant, shows that the solution of the hyperbolic equation is unique along the characteristic $t - 3x = 0$ but nowhere else in the domain of dependence.

Returning to the M initial points lying on the initial curve L for the general hyperbolic equation (4.1), there will pass two distinct characteristics with slopes obtained by solving (4.5) for dt/dx. If these slopes are f and g they are given by

$$f = \frac{b + (b^2 - 4ac)^{1/2}}{2a}, \tag{4.9}$$

and

$$g = \frac{b - (b^2 - 4ac)^{1/2}}{2a} \tag{4.10}$$

and the two characteristics will be referred to as *f-type* and *g-type characteristics*. The two types of characteristics are thus families of curves, the equations of which are the solution of the ordinary differential equations

$$\frac{dt}{dx} = f \text{ and } \frac{dt}{dx} = g, \tag{4.11}$$

with initial conditions specified by the data points $T_m^{(1)}(m = 1, 2, \ldots, M)$.

The domain of dependence is a curvilinear triangle. The f-type characteristic through the initial data point $T_m^{(1)}$ meets the g-type characteristic through the data point $T_{m+1}^{(1)}$ at the point $T_m^{(2)}(m = 1, \ldots, M-1)$; the f-type characteristic through the mesh point $T_m^{(2)}$ meets the g-type characteristic through the mesh point $T_{m+1}^{(2)}$ at the point $T_m^{(3)}(m = 1, \ldots, M-2)$; in general, the f-type characteristic through the mesh point $T_m^{(n)}$ meets the g-type characteristic through the mesh point $T_{m+1}^{(n)}$ at the mesh point $T_m^{(n+1)}(n = 1, \ldots, M-1; m = 1, \ldots, M-n)$, giving rise to the curvilinear characteristic grid of Fig. 4.2.

The values of u, p, q at the mesh points of this grid are dependent only on those values of u, p, q at the initial data points $T_m^{(1)}(m = 1, \ldots, M)$ on the curve L. Should u, p, q be required at points not on the mesh, inside or outside the domain of dependence, the values of u, p, q at further points on L will be required. At the general mesh point $T_m^{(n)}$ the value of u will be denoted by $u_m^{(n)}$ with similar notations for $x, t, p, q, a, b, c, e, f, g$ at that point.

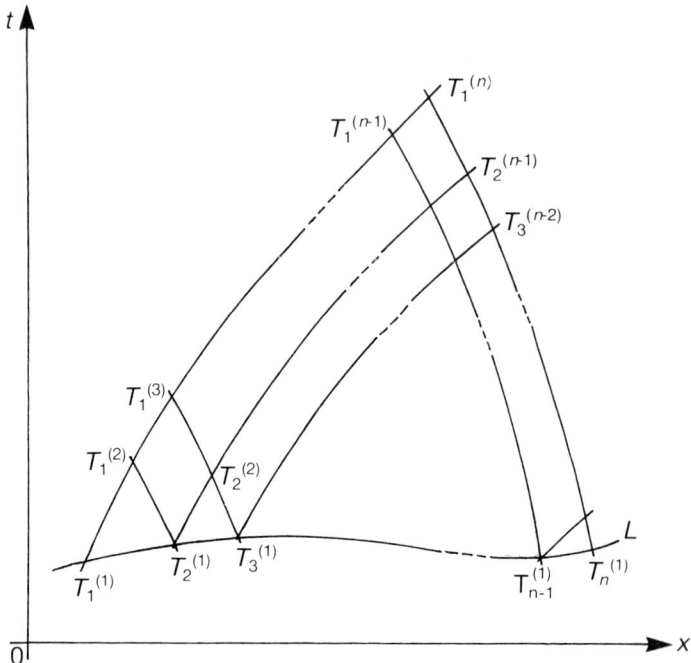

Fig. 4.2 – General curvilinear mesh.

The method of characteristics takes the curvilinear triangle $T_m^{(n)} T_{m+1}^{(n)} T_m^{(n+1)}$ ($n = 1, \ldots, M-1; m = 1, \ldots, M-n$) and approximates the curves $T_m^{(n)} T_m^{(n+1)}$ and $T_{m+1}^{(n)} T_m^{(n+1)}$ with the straight lines

$$t - t_m^{(n)} = f_m^{(n)}(x - x_m^{(n)}) \tag{4.12}$$

and

$$t - t_{m+1}^{(n)} = g_{m+1}^{(n)}(x - x_{m+1}^{(n)}). \tag{4.13}$$

The point of intersection of these lines gives a first approximation to the grid point $T_m^{(n+1)}$ whose coordinates are $(x_m^{(n+1)}, t_m^{(n+1)})$. Now equation (4.8) is divided by dx and applied along $T_m^{(n)} T_m^{(n+1)}$ to give

$$a_m^{(n)}(p_m^{(n+1)} - p_m^{(n)}) f_m^{(n)} + c_m^{(n)}(q_m^{(n+1)} - q_m^{(n)})$$
$$= e_m^{(n)}(t_m^{(n+1)} - t_m^{(n)}) \tag{4.14}$$

and along $T_{m+1}^{(n)} T_m^{(n+1)}$ to give

$$a_{m+1}^{(n)}(p_m^{(n+1)} - p_{m+1}^{(n)}) g_{m+1}^{(n)} + c_{m+1}^{(n)}(q_m^{(n+1)} - q_{m+1}^{(n)})$$
$$= e_{m+1}^{(n)}(t_m^{(n+1)} - t_{m+1}^{(n)}). \tag{4.15}$$

Sec. 4.1] The Method of Characteristics for Second Order Equations

Equations (4.14) and (4.15) are solved simultaneously to give first approximations to their unknowns $p_m^{(n+1)}$ and $q_m^{(n+1)}$.

A first approximation to $u_m^{(n+1)}$ is now determined from the total differential equation (4.2) giving

$$U_m^{(n+1)} = U_m^{(n)} + \frac{1}{2}(x_m^{(n+1)} - x_m^{(n)})(p_m^{(n+1)} + p_m^{(n)})$$

$$+ \frac{1}{2}(t_m^{(n+1)} - t_m^{(n)})(q_m^{(n+1)} + q_m^{(n)}), \qquad (4.16)$$

where, as in earlier chapters, upper case U is used to distinguish a computed solution from the theoretical solution.

The coefficients a, b, c, e of the hyperbolic equation (4.1) can now be evaluated at the mesh point $T_m^{(n+1)}$ and substitution of these values of a, b, c, into the right-hand sides of (4.9) and (4.10) gives first approximations to the slopes of the two characteristics advancing from this point.

In the special case of constant coefficients a, b, c, e in (4.1), the values of x, t, p, q, u, a, b, c, e, f, g which have so far been determined at the mesh point $T_m^{(m+1)}$ cannot be improved, but in the general case of variable coefficients a, b, c, e these values are all predicted values and can be corrected, as follows.

Equations (4.12) and (4.13) become

$$t - t_m^{(n)} = \frac{1}{2}(f_m^{(n+1)} + f_m^{(n)})(x - x_m^{(n)}) \qquad (4.17)$$

and

$$t - t_{m+1}^{(n)} = \frac{1}{2}(g_m^{(n+1)} + g_{m+1}^{(n)})(x - x_{m+1}^{(n)}) \qquad (4.18)$$

which are solved simultaneously to give corrected values of $x_m^{(n+1)}$ and $t_m^{(n+1)}$. The reader who has studied elementary methods for the numerical solution of ordinary differential equations will have realized that (4.12) and (4.13) are applications of the Euler predictor formula for solving the equations of (4.11), and that (4.17) and (4.18) are applications of the Euler corrector formula. This corrector is sometimes called the modified Euler formula and is, of course, the trapezoidal rule for integration. Error estimates for the coordinates of the mesh point $T_m^{(n)}$ are provided by the theory relating to the Euler predictor-corrector algorithm for solving ordinary differential equations. The errors in $x_m^{(n+1)}$ and $t_m^{(n+1)}$ yielded by (4.12) and (4.13) are respectively

$$0\left(h_{m,n}^2\right) \text{ and } 0\left(l_{m,n}^2\right),$$

where $h_{m,n} = x_m^{(n+1)} - x_m^{(n)}$ and $l_{m,n} = t_m^{(n+1)} - t_m^{(n)}$, and the errors yielded by (4.17) and (4.18) are, respectively,

$$0\left(h_{m,n}^3\right) \text{ and } 0\left(l_{m,n}^3\right).$$

Using the corrected coordinates of the mesh point $T_m^{(n+1)}$, equations (4.14) and (4.15) are written in the forms

$$\frac{1}{2}\left(a_m^{(n+1)} + a_m^{(n)}\right)\left(p_m^{(n+1)} - p_m^{(n)}\right) \cdot \frac{1}{2}\left(f_m^{(n+1)} + f_m^{(n)}\right)$$

$$+ \frac{1}{2}\left(c_m^{(n+1)} + c_m^{(n)}\right)\left(q_m^{(n+1)} - q_m^{(n)}\right)$$

$$= \frac{1}{2}\left(e_m^{(n+1)} + e_m^{(n)}\right)\left(t_m^{(n+1)} - t_m^{(n)}\right) \qquad (4.19)$$

and

$$\frac{1}{2}\left(a_m^{(n+1)} + a_{m+1}^{(n)}\right)\left(p_m^{(n+1)} - p_{m+1}^{(n)}\right) \cdot \frac{1}{2}\left(g_m^{(n+1)} + g_{m+1}^{(n)}\right)$$

$$+ \frac{1}{2}\left(c_m^{(n+1)} + c_{m+1}^{(n)}\right)\left(q_m^{(n+1)} - q_{m+1}^{(n)}\right)$$

$$= \frac{1}{2}\left(e_m^{(n+1)} + e_{m+1}^{(n)}\right)\left(t_m^{(n+1)} - t_{m+1}^{(n)}\right), \qquad (4.20)$$

which are solved simultaneously to yield corrected values of $p_m^{(n+1)}$ and $q_m^{(n+1)}$. The corrected values of x, t, p, q, at $T_m^{(n+1)}$ are then used to compute a corrected value of $u_m^{(n+1)}$ from equation (4.16).

Corrected values of a, b, c, e are now computed using their forms given in (4.1) and used in (4.9) and (4.10) to compute corrected values of the slopes of the two characteristics through the mesh point $T_m^{(n+1)}$. An iterative procedure using equations (4.17), (4.18), (4.19), (4.20), (4.16), (4.1), (4.9) and (4.10) is thus established and is carried out until all of x, t, p, q, u at $T_m^{(n+1)}$ have converged.

4.2 A SPECIMEN PROBLEM

The method of characteristics will be used to solve the equation

$$(x^2 + 1)\frac{\partial^2 u}{\partial x^2} - \lambda^2 (t^2 + 1)\frac{\partial^2 u}{\partial t^2} = 6(x + x^3 - \lambda^2 t - \lambda^2 t^3) \qquad (4.21)$$

for which a solution is $u = x^3 + t^3$, and which is hyperbolic for all real values of x, t and all real, non-zero λ. This specimen problem was used in Twizell (1976) and is chosen because of its difficult geometric properties. Depending on the value of λ the characteristic mesh generated, which, as in the previous section, is in the shape of a curvilinear triangle, may range from being very flat to being very long and thin; λ thus has great effect on the vertical elongation of the triangle.

Table 4.1

m	x	t	p	q	u
1	0.15	0.30	0.0675	0.2700	0.030375
2	0.20	0.28	0.1200	0.2352	0.029952
3	0.27	0.27	0.2187	0.2187	0.039366
4	0.33	0.26	0.3267	0.2028	0.053513
5	0.42	0.28	0.5292	0.2352	0.096040
6	0.48	0.29	0.6912	0.2523	0.134981
7	0.54	0.30	0.8748	0.2700	0.184464
8	0.60	0.31	1.0800	0.2883	0.245791
9	0.65	0.32	1.2675	0.3072	0.307393

A set of nine initial points, each satisfying $u = x^3 + t^3$, will be used; the values of x, t, p, q, u at these points are given in Table 4.1. In this example, the initial curve is not known explicitly; it can be seen, however, that it is probably a polynomial of low degree (or can be closely approximated by a polynomial of low degree in the interval $x_1^{(1)} \leqslant x \leqslant x_9^{(1)}$).

Using equation (4.9) and (4.10) the slopes of the characteristics are given by

$$f = -g = \frac{\lambda^2(t^2+1)}{(x^2+1)},$$

so that for large λ^2 the characteristics have steep slopes with large increments in t between successive mesh points, and for small values of λ^2 the characteristics have shallow slopes with very small increments in t between successive mesh points. For both these extremes, some ill-conditioning is evident in systems (4.12), (4.13) and (4.17), (4.18) leading to difficult geometric properties of the curvilinear triangle $T_1^{(1)}T_9^{(1)}T_1^{(9)}$, and to some large errors in the solution U. For λ^2 very large, the curvilinear triangle becomes very long and it is here that the largest errors are encountered. For λ^2 very small, the curvilinear triangle is very flat: indeed, $b^2 - 4ac$ is close to zero and the specimen equation (4.21) is therefor almost of the parabolic type (of course, for $\lambda^2 = 0$ the specimen equation becomes a second order ordinary differential equation).

For the data set given in Table 4.1, the relative error in u, defined by $|(u - U)/u|$, where U is the computed solution and $u = x^3 + t^3$ is the theoretical solution, was computed in a numerical experiment with $\lambda^2 = 1000, 100, 10, 1, 0.1,$

0.01, 0.001. The domains of dependence, in the forms of curvilinear triangles, are shown in Fig. 4.3 for $\lambda^2 = 1000, 100, 10, 1$ and the complete mesh generated for $\lambda^2 = 1$ is shown in Fig. 4.4.

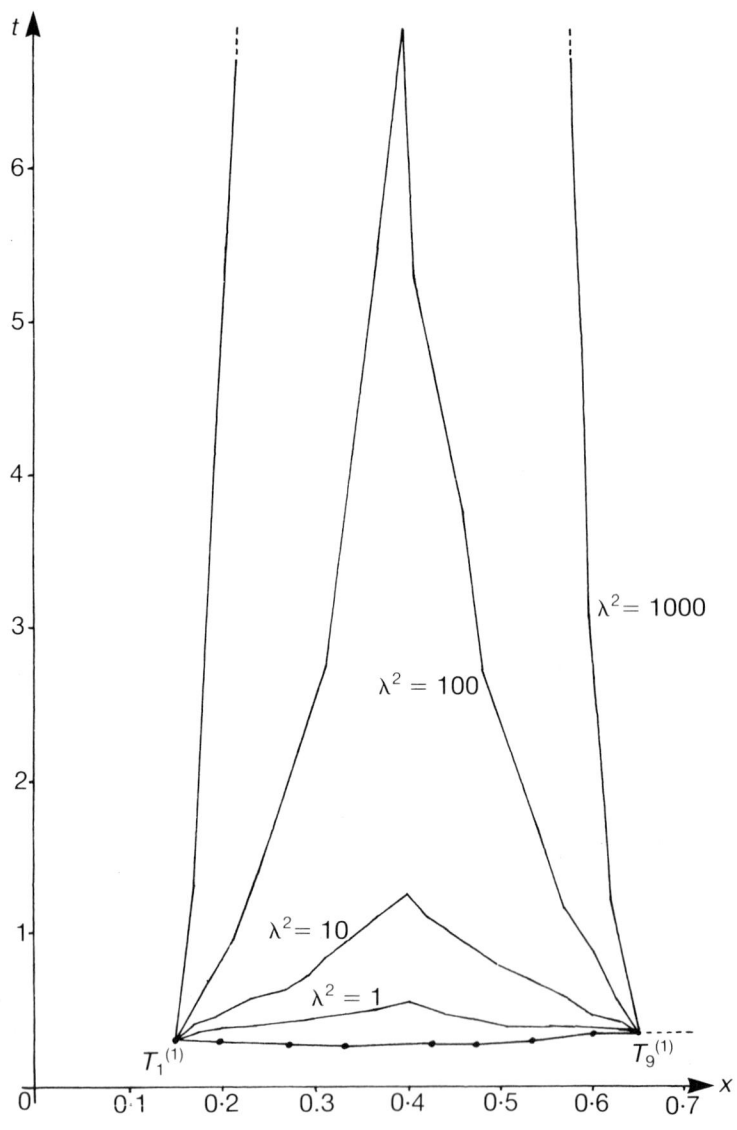

Fig. 4.3 – Curvilinear triangles for the model problem with $\lambda^2 = 1000, 100, 10, 1$.

Sec. 4.2] A Specimen Problem

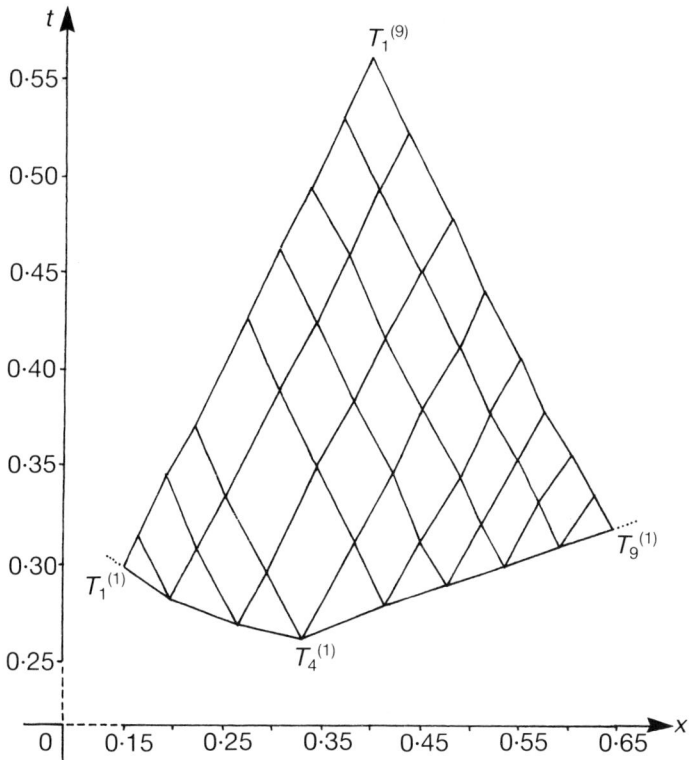

Fig. 4.4 – Curvilinear triangle for the model problem with $\lambda^2 = 1$.

The position of the vertex $T^{(9)}$ is given in Table 4.2 for each value of λ^2 together with the relative error in the solution at that point. The reader will recall that the notation 3.7(−1), for instance, means 3.7×10^{-1}.

Table 4.2

λ^2	$x_1^{(9)}$	$t_1^{(9)}$	Error
1000	0.39	15150	3.7(−1)
100	0.39	6.92	5.6(−2)
10	0.39	1.23	8.2(−3)
1	0.40	0.56	1.5(−3)
0.1	0.42	0.39	1.9(−3)
0.01	0.49	0.33	3.9(−3)
0.001	0.47	0.32	2.2(−2)

It is easy to see, in this example, that the smallest relative errors are obtained for $\lambda^2 = 1$ when the characteristics intersect at an angle which is reasonably close to 90°, and that the errors begin to increase as λ^2 becomes very large or very small. It is also easy to see that the vertical elongation is related to the magnitude of λ^2.

Increasing the number of initial points within $x_1^{(1)}$ and $x_9^{(1)}$ will have the effects of reducing the vertical elongation of the curvilinear triangle for each λ^2 and of decreasing the magnitude of the relative error in the solution.

The method of characteristics is analogous to the family of one-step methods, such as the Euler-modified Euler predictor-corrector method, used in the numerical solution of first order ordinary differential equation. Improved accuracy may be obtained using methods analogous to multi-step methods in ordinary differential equations. Such methods have been developed in Twizell (1975, 1976) but are beyond the scope of this text.

4.3 PROPAGATION OF DISCONTINUITIES IN INITIAL CONDITIONS

One important feature of hyperbolic equations is that discontinuities in initial conditions are propagated into the domain of dependence along characteristics.

Suppose, for instance, that in Fig. 4.5 the initial conditions along the curve L between $T_{m-j}^{(1)}$ and $T_m^{(1)}$ satisfy $u = \phi(x, t)$ and along L between $T_m^{(1)}$ and $T_{m+k}^{(1)}$ by $u = \psi(x, t)$, with $\phi\left(x_m^{(1)}, t_m^{(1)}\right) \neq \psi\left(x_m^{(1)}, t_m^{(1)}\right)$. There is thus a discontinuity in u at the initial point $T_m^{(1)}$. In the arguments which follow, discontinuities in the first derivatives $p_m^{(1)}$ and $q_m^{(1)}$ may also be introduced but they do not serve to strengthen the arguments and will not be considered.

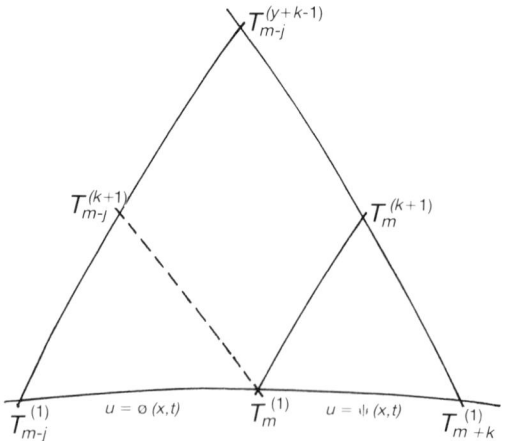

Fig. 4.5 — The propagation of discontinuities in initial conditions.

It was seen earlier in the chapter that the solution at the point $T_m^{(k+1)}$ is calculated in terms of the initial conditions at $T_m^{(1)}$ and $T_{m+k}^{(1)}$. Hence the solution at every point within the curvilinear triangle $T_m^{(1)} T_{m+k}^{(1)} T_m^{(k+1)}$ is a function of ψ. Similarly, the solution at every point within the curvilinear strip $T_{m-j}^{(1)} T_m^{(1)} T_m^{(k+1)} T_{m-j}^{(j+k+1)}$ of Fig. 4.5, is a function of ϕ propagated along an f-type characteristic and a function of ψ propagated along a g-type characteristic.

Suppose now that the initial point $T_{m-j}^{(1)}$ tends to the initial point $T_m^{(1)}$; as this happens the curvilinear strip of Fig. 4.5 tends to the f-type characteristic through $T_m^{(1)}$. Remembering that $\phi \neq \psi$ at this initial point shows that the discontinuity is propagated along the f-type characteristic through $T_m^{(1)}$.

It is clear, therefore, that a characteristic can separate two different solutions. One other important feature of hyperbolic equations of the second order is that two solutions, together with their first order derivatives may be continuous across the common characteristic, but their second order derivatives (and higher order derivatives, as it happens) may be discontinuous across the common characteristic.

This is illustrated by considering the simple wave equation

$$\frac{\partial^2 u}{\partial x^2} - \frac{\partial^2 u}{\partial t^2} = 0$$

which has initial conditions given along the line $t = 0$. It is easy to show that $f = 1$, $dp = dq$ and $g = -1$, $dp = -dq$ for this problem, so that $p-q$ is constant along each f-type characteristic $t-x =$ constant and $p+q$ is constant along each g-type characteristic $t+x =$ constant.

Considering, in particular, the f-type characteristic $t-x = 0$ it is easy to verify that one solution along this characteristic is $p = 2$, $q = -2$, $u = 1$ and that these values satisfy the alternative solutions

$$u_1(x, t) = 2 \sin(x-t) + \cos(x-t),$$
$$u_2(x, t) = 1 + 2(x-t) + (x-t)^2.$$

However, it is seen that $\partial^2 u_1/\partial x^2 \neq \partial^2 u_2/\partial x^2$ and $\partial^2 u_1/\partial t^2 \neq \partial^2 u_2/\partial t^2$ so that second order derivatives are discontinuous across the f-type characteristic $t - x = 0$.

It is easy to demonstrate this discontinuity in second order derivatives for the general hyperbolic equation of the form (4.1). Looking again at equation (4.7), which may be rearranged to give

$$s = \left\{ a \left(\frac{dt}{dx}\right)^2 - b\frac{dt}{dx} + c \right\} \bigg/ \left\{ a\frac{dt}{dx}\frac{dp}{dx} + c\frac{dq}{dx} - e\frac{dt}{dx} \right\}, \qquad (4.22)$$

it is clear from (4.5) and (4.8) that s is indeterminate along every characteristic and can therefore be given an arbitrary value along any particular characteristic.

The corresponding values of r and y are then determined from equations (4.3) and (4.4).

Next, a set of continuous values of u, p, q is prescribed along some characteristic; let this set be $\{u^*, p^*, q^*\}$. In addition, a continuous value of s is prescribed and the corresponding values of r, y determined. All higher order derivatives are then determined from the differential equation and are used in Taylor's expansion to give the solution at points to, say, the left of the characteristic. Using the same set $\{u^*, p^*, q^*\}$, but assigning a different value to s and calculating the corresponding values r and y, Taylor's expansion can be used again to find the solution to the right of the characteristic.

The two solutions on the two sides of the characteristic have different forms, because of the different assigned values of s; the two sides have the same continuous values of u, p, q along the characteristic but the second order derivatives are discontinuous across the characteristic.

The indeterminancy of s arising from (4.22) also explains why the solution of (4.1) is not unique when the initial curve L in Fig. 4.1 is a characteristic.

4.4 FINITE ELEMENT SOLUTIONS

4.4.1 Introduction

The finite element method was originally developed for, and has enjoyed its greatest success with, the numerical solution of boundary value problems.

It is a natural step to consider finite element methods for time dependent problems. Many of the finite element methods for initial value problems which have appeared in the literature, have been inspired by variational principles, but most are not strictly variational. Some use variational methods in the space variables only and integrate respect to time using another approach such as finite difference methods; others use a variational principle in space and time to derive a functional that is not, in fact, stationary in the sense of Chapter 3. Often, a functional $I(U, \bar{U})$, which has an Euler-Lagrange equation of the the form $Au = f$, is developed where U is varied and $\bar{U}$ is kept fixed, then U is set equal to $\bar{U}$ at the end of the calculation. This is an example of *quasi-variational principles*.

4.4.2 Hamilton's Principle

Consider the wave equation in the form

$$\frac{\partial^2 u}{\partial x^2} = \frac{1}{C^2} \frac{\partial^2 u}{\partial t^2} \; ; \; 0 < x < 1, \; t > 0, \tag{4.23}$$

where C is the wave speed, assumed constant. Let the boundary conditions be

$$u(0, t) = u(1, t) = 0, \; t > 0 \tag{4.24}$$

and let the initial conditions be

$$u(x, 0) = f(x), \quad 0 \leqslant x \leqslant 1, \tag{4.25}$$

$$\frac{\partial u}{\partial t}(x, 0) = g(x), \quad 0 \leqslant x \leqslant 1. \tag{4.26}$$

This is the general form of the wave equation when the wave speed is constant; one of its applications is the vibration of a string fixed at both ends.

Variational principles, known more precisely as Hamiltonian variational principles, exist for many wave-equation problems. These require the value of the dependent variable at the end of the time interval over which integration is to take place and, as such, do not apply to the initial value problem given by (4.23), (4.24), (4.25), (4.26).

To use Hamiltonian variational principles, the initial condition (4.26) must be replaced by

$$u(x, T) = H(x), \quad 0 \leqslant x \leqslant 1 \tag{4.27}$$

at time $t = T$, where T is some arbitrary, but fixed, value (a 'time boundary'). Hamilton's principal states that the functional

$$I(U) = \int_0^T \int_0^1 \left\{ c^2 \left(\frac{\partial U}{\partial x} \right)^2 - \left(\frac{\partial U}{\partial t} \right)^2 \right\} dx \, dt \tag{4.28}$$

is stationary when U varies around u, the exact solution of (4.23), (4.24), (4.25), (4.27), and satisfies the essential boundary conditions

$$\begin{aligned} U(0, t) &= U(1, t) = 0, \quad 0 < t < T \\ U(x, 0) &= f(x), \quad 0 \leqslant x \leqslant 1 \\ U(x, T) &= H(x), \quad 0 \leqslant x \leqslant 1 \end{aligned} \tag{4.29}$$

It must be realized that $H(x)$ is not known, so that basis functions satisfying the third equation of (4.29) cannot be chosen. This difficulty is circumvented by carrying on as though $H(x)$ were known and, at the end of the procedured, using (4.26) to determine the parameters and constants which arise in the computation. This is well illustrated by Noble (1973) who proceeds as follows:

Suppose that the interval $0 \leqslant x \leqslant 1$ is divided into $N+1$ intervals each of width h, so that $(N+1)h = 1$, and that $\theta_i(x)$ is a function such that

$$\theta_i(x_j) = \begin{cases} 0, & i \neq j \\ 1, & i = j \end{cases}$$

where $x_j = jh$ ($j = 0, 1, \ldots, N+1$). in his example, Noble uses *hat functions*

$\theta_i(x)$ which are zero for $0 \leq x \leq x_{i-1}$ and for $x_{i+1} \leq x \leq 1$, with $\theta_i(x_i) = 1$; the function is linear in $x_{i-1} \leq x \leq x_i$ and $x_i \leq x \leq x_{i+1}$.

Now write

$$U(x, t) = \sum_{i=1}^{N} \theta_i(x)\phi_i(t) \tag{4.30}$$

so that, if hat functions are used, (4.30) is piecewise linear in x.

Assume next that the essential boundary conditions which must be satisfied by (4.50) are

$$\phi(0) = \sum_{j=1}^{N} \theta_j(x) f(x_j), \; \phi(T) = \sum_{j=1}^{N} \theta_j(x) H(x_j); \tag{4.31}$$

these are interpolated forms of (4.25) and (4.27). The variational formulation then requires the user to find the stationary values of

$$I(U) = \sum_{i=1}^{N} \sum_{j=1}^{N} \left\{ C^2 a_{ij} \int_0^T \phi_i \phi_j \, dt - b_{ij} \int_0^T \frac{d\phi_i}{dt} \frac{d\phi_j}{dt} \, dt \right\}, \tag{4.32}$$

where the $\phi_i(t)$ satisfy the boundary conditions

$$\phi_j(0) = f(x_j), \; \phi_j(T) = H(x_j). \tag{4.33}$$

The terms a_{ij}, b_{ij} are given by Noble (1973) as

$$a_{ij} = \int_0^1 \frac{d\theta_i}{dx} \frac{d\theta_j}{dx} \, dx, \; b_{ij} = \int_0^1 \theta_i \theta_j \, dx. \tag{4.34}$$

Noble also shows that for the hat functions θ_i, the a_{ij} and b_{ij} have the values

$$a_{i,i-1} = -1/h, \; a_{i,i} = 2/h, \; a_{i,i+1} = -1/h$$
$$b_{i,i-1} = h/6, \; b_{i,i} = 2h/3, \; b_{i,i+1} = h/6$$

when the x_i are evenly spaced.

The Euler-Lagrange equations for the variational formulation (4.32) are shown by Noble (1973) to be

$$\sum_{j=1}^{N} \left\{ b_{ij} \frac{d^2\phi_j}{dt^2} + C^2 a_{ij} \phi_j \right\} = 0, \; i = 1, 2, \ldots, N \tag{4.35}$$

where the $\phi_j(t)$ satisfy (4.33). As stated earlier, $H(x)$ is not known but $\partial u/\partial t = g(x)$ at time $t = 0$ and so (4.35) is solved with

$$\phi_j(0) = f(x_j), \; \phi_j'(0) = g(x_j), j = 1, \ldots, N. \tag{4.36}$$

Sec. 4.4] Finite Element Solutions 131

The numerical solution of the second order system (4.35) may be determined using linear multistep methods (Lambert, 1973) or multiderivative methods (Twizell and Khaliq, 1984). The numerical solution of the second order initial value problem will be discussed later in the chapter when finite difference methods are used to solve the simple wave equation.

An alternative approach to finding the stationary values of (4.32) is to derive a system of linear difference equations by substituting an approximation for $\phi(t)$. Noble (1973) uses the quadratic

$$\phi_j(t) = [(T^2 - t^2)f(x_j) + t^2 H(x_j) + b_j T(T-t)t]/T^2, \quad (4.37)$$

where $b_j = \phi'_j(0)$. Equation (4.37), which satisfies (4.33), is substituted into (4.32) which is then differentiated partially with respect to each $b_j (j = 1, \ldots, N)$. The necessary conditions for $I(U)$ to have a stationary value are $\partial I/\partial b_j = 0 (j = 1, \ldots, N)$. These conditions, together with the relations $b_j = \phi'_j(0)$, give a relation between the formulas of (4.33) and $\phi'_j(0)$ for $j = i-1, i, i+1$, assuming that $\phi_j(T)$ is known and $\phi'_j(0)$ is unknown ($j=i-1, i, i+1$).

Having obtained $\phi'_j(0)$, Noble (1973) shows that for the hat functions discussed above, the finite element formulation leads to the difference scheme

$$\frac{1}{4}\left(1 - \frac{9r}{20}\right)\phi_{i-1}(T) + \left(1 + \frac{9r}{20}\right)\phi_i(T) + \frac{1}{4}\left(1 - \frac{9r}{20}\right)\phi_{i+1}(T)$$

$$= \frac{1}{4}\left(1 + \frac{21r}{10}\right)\phi_{i-1}(0) + \left(1 - \frac{21r}{20}\right)\phi_i(0) + \frac{1}{4}\left(1 + \frac{21r}{20}\right)\phi_{i+1}(0)$$

$$+ T\left[\frac{1}{4}\left(1 + \frac{3r}{5}\right)\phi'_{i-1}(0) + \left(1 - \frac{3r}{10}\right)\phi'_i(0) + \frac{1}{4}\left(1 + \frac{3r}{5}\right)\phi'_{i+1}(0)\right],$$

(4.38)

where $r = C^2 T^2/h^2$ and $i = 1, 2, \ldots, N$.

The solution is obtained using the decomposition algorithm for a tridiagonal system outlined in subsection 1.2.3 of Chapter 1.

Before proceeding to the next time-step ($t = 2T$) the derivatives at time $t = T$ are required. Differentiating (4.37) gives

$$\phi'_i(T) = 2[\phi_i(T) - \phi_i(0) - T\phi'_i(0)]/T. \quad (4.39)$$

Substituting for $\phi_i(T)$ from (4.39) in (4.38) leads to a well-conditioned system of equations for $\phi'_i(T)$; (4.39) itself is unsuitable, always giving a value for $\phi'_i(T)$ which is close to zero.

4.4.3 Petrov–Galerkin methods
In the finite element solution of hyperbolic equations, Galerkin methods suffer from accuracy and stability deficiencies. In general, such methods do not achieve

optimal asymptotic error estimates for hyperbolic problems and are known to produce highly oscillatory error components. In effect, Galerkin methods give good results for elliptic problems but not for hyperbolic problems.

To remedy this situation, *Petrov-Galerkin* methods have been developed. These are distinguished from Galerkin methods by their use of weighting functions and trial solutions from different classes of functions, thus enabling optimal accuracy (according to some measure) to be attained.

The obvious potential of Petrov–Galerkin methods has led to the growth of a large literature for both second order and first order hyperbolic equations, and the interested reader is referred to the papers by Hughes *et al.* and Osher, which appear in Morton (1982), and to the numerous references cited therein.

4.5 LOW ORDER FINITE DIFFERENCE METHODS

4.5.1 A recurrence relation

It was seen in section 4.1 that finite difference methods have an important role to play in the numerical solution of second order hyperbolic partial differential equations, in that methods for equations in one space variable often generalize to equations with more than one space variable where characteristic methods are much more difficult to use. In particular, computer codes for characteristic methods are much more difficult to write, even for one-space dimension problems, than for finite difference methods.

Consider *the initial-boundary value problem* consisting of the wave equation

$$\frac{\partial^2 u}{\partial t^2} = c^2 \frac{\partial^2 u}{\partial x^2}, \tag{4.40}$$

which is defined in a region $R = [0 < x < 1] \times [t > 0]$ in the (x, t) plane with boundary ∂R, together with the boundary conditions

$$u(0, t) = u(1, t) = 0, \ t > 0 \tag{4.41}$$

and the initial conditions

$$u(x, 0) = f(x), \ 0 \leqslant x \leqslant 1 \tag{4.42}$$

$$\frac{\partial u}{\partial t}(x, 0) = g(x), 0 \leqslant x \leqslant 1. \tag{4.43}$$

In (4.40), c is the wave speed which will be assumed to be constant. At the origin, the boundary conditions $u = 0$ holds for t arbitrarily close to $t = 0$. The initial condition (4.42) with $x = 0$ holds at the origin, so that, if $f(0) \neq 0$, there is a discontinuity between the initial condition (4.42) and the boundary condition at the origin which will be propagated in the solution as t increases. Similarly, if $f(1) \neq 0$, there is a discontinuity between initial condition and boundary condition at the point $(1, 0)$ in the (x, t) plane.

The interval $0 \leq x \leq 1$ will be divided into $N+1$ subintervals each of width h, so that $(N+1)h = 1$. Superimposing a uniform grid of width h on the space variable allows the space derivative in (4.40) to be approximated by the finite difference replacement

$$\frac{\partial^2 u}{\partial x^2} = \{u(x-h, t) - 2u(x, t) + u(x+h, t)\}/h^2 + O(h^2). \tag{4.44}$$

The independent variable t will be discretized in steps of length l so that t may be written in the form $t = nl$ where n is an integer.

The region R and its boundary ∂R have thus been discretized at the points (mh, nl) where $m = 0, 1, \ldots, N+1$ and $n = 0, 1, 2, \ldots$. The value of the dependent variable at the grid point (mh, nl) is $u(mh, nl)$; this will be denoted by u_m^n or by $u_m(t)$, where $t = nl$, as convenient.

Consider now the general time level $t = nl$ at which there are N interior mesh points. At every one of these points, equation (4.60) is now applied with the space derivative replaced by (4.44). At point m on this level ($m = 1, 2, \ldots, N$), equation (4.40) becomes

$$\frac{d^2 U}{dt^2} = c^2 \{U_{m-1}(t) - 2U_m(t) + U_{m+1}(t)\}/h^2, \tag{4.45}$$

where upper case U is introduced to denote the theoretical solution of an approximating difference scheme. Due to such factors as round-off and perturbations in initial data, U itself will incur errors as time increases. The effects of these errors will be discussed further in section 4.6.

Recalling that $U_0(t) = U_{N+1}(t) = 0$ from (4.41), the N equations (4.45) may be written in the form of a system of second order *ordinary* differential equations. In matrix form this system is

$$\frac{d^2 \mathbf{U}(t)}{dt^2} = c^2 A \mathbf{U}(t), \tag{4.45}$$

where $\mathbf{U}(t) = [U_1(t), U_2(t), \ldots, U_N(t)]^T$ is the vector of approximations to the solution vector $\mathbf{u}(t)$ and A is the square matrix of order N given by

$$A = h^{-2} \begin{bmatrix} -2 & 1 & & & \\ 1 & -2 & 1 & & 0 \\ & \ddots & \ddots & \ddots & \\ & 0 & & & 1 \\ & & & 1 & -2 \end{bmatrix} \tag{4.47}$$

which has eigenvalues $\lambda_s = -4 \sin^2[s\pi/\{2(N+1)\}]/h^2$ for $s = 1, 2, \ldots, N$.

Solving (4.46) subject to the boundary conditions (4.41) and the initial conditions (4.42), (4.43) gives the analytical solution

$$\mathbf{U}(t) = \frac{1}{2} \exp(ctB)(\mathbf{f} + c^{-1}B^{-1}\mathbf{g}) + \frac{1}{2} \exp(-ctB)(\mathbf{f} - c^{-1}B^{-1}\mathbf{g}) \quad (4.48)$$

in which B is a matrix such that $B^2 = A$ and $\mathbf{f}, \mathbf{g}$ are the vectors of initial conditions. It will be seen that the elements of B need not be determined explicitly.

It is easy to show that for a time step l, equation (4.48) satisfies the recurrence relation

$$\mathbf{U}(t+l) - \{\exp(lcB) + \exp(-lcB)\}\mathbf{U}(t) + \mathbf{U}(t-l) = \mathbf{0}, \quad (4.49)$$

with $t = l, 2l, \ldots$, in which $\mathbf{0}$ is the zero vector of order N. It is clear that using (4.49) with $t = l$ requires knowledge of $\mathbf{U}(l)$ which is not contained explicitly in the initial conditions. The value of the dependent variable must therefore be estimated using (4.42), (4.43) for use with (4.49) when $t = l$ or $t = 2l$. The accuracy of this approximation to $\mathbf{U}(l)$ must not be less than that of (4.49) with the exponential functions replaced by valid approximations.

Table 4.3

(m, k) Padé	Approximant	Order of error
(0, 1)	$1 + \theta$	θ^2
(1, 1)	$(1 + \tfrac{1}{2}\theta)/(1 - \tfrac{1}{2}\theta)$	θ^3
(1, 0)	$1/(1 - \theta)$	θ^2
(0, 2)	$1 + \theta + \tfrac{1}{2}\theta^2$	θ^3
(1, 2)	$(1 + \tfrac{2}{3}\theta + \tfrac{1}{6}\theta^2)/(1 - \tfrac{1}{3}\theta)$	θ^4
(2, 2)	$(1 + \tfrac{1}{2}\theta + \tfrac{1}{12}\theta^2)/(1 - \tfrac{1}{2}\theta + \tfrac{1}{12}\theta^2)$	θ^5
(2, 1)	$(1 + \tfrac{1}{3}\theta)/(1 - \tfrac{2}{3}\theta + \tfrac{1}{6}\theta^2)$	θ^4
(2, 0)	$1/(1 - \theta + \tfrac{1}{2}\theta^2)$	θ^3

4.5.2 Rational replacements of $\exp(\pm lB)$

Consider the (m, k) Padé approximant to the exponential function e^θ, where θ is a real scalar, and m, k are non-negative integers, given by

$$e^\theta \cong R_{m,k}(\theta) = P_k(\theta)/Q_m(\theta). \quad (4.50)$$

In (4.50) P_k, Q_m are polynomials of degrees k, m, respectively, defined by

$$P_k(\theta) = p_0 + p_1\theta + p_2\theta^2 + \ldots + p_k\theta^k; \quad P_0(\theta) \equiv 1 \quad (4.51)$$

and

$$Q_m(\theta) = q_0 - q_1\theta + q_2\theta^2 - \ldots + (-1)^m q_m\theta^m; \quad Q_m(\theta) \equiv 1 \quad (4.52)$$

with $p_0 \geq p_1 > p_2 > \ldots > p_k > 0$ and $q_0 \geq q_1 > q_2 > \ldots > q_m > 0$ depending on the chosen Padé approximant. Padé (1892) published values of the terms p_i, q_j ($i = 1, \ldots, k$; $j = 1, \ldots, m$) in his *Padé table*; the first eight entries of the Padé table for e^θ are reproduced in Table 4.3. It is easy to see that $p_0 = q_0 = 1$ and that $R_{m,k}(\theta) = 1/R_{m,k}(-\theta)$ for all m, k.

On dividing the polynomial $P_k(\theta)$ by the polynomial $Q_m(\theta)$ it is found that the first $m + k + 1$ terms of the resulting power series agree with the Maclaurin expansion

$$e^\theta = 1 + \theta + \frac{\theta^2}{2!} + \frac{\theta^3}{3!} + \ldots + \frac{\theta^{m+k}}{(m+k)!} + \ldots ; \tag{4.53}$$

the term in θ^{m+k+1} is found to be in error. Hence the error in the (m, k) Padé approximant to the exponential function e^θ is written $0(\theta^{m+k+1})$.

4.5.3 A well known explicit scheme

Consider again the recurrence relation (4.49) and replace the exponential functions $\exp(lcB)$ and $\exp(-lcB)$ by their (0, 2) Padé approximants. It is easy to see that the terms in lB cancel each other and, recalling that $B^2 = A$, that (4.49) becomes

$$\mathbf{U}(t + l) - (2I + l^2 c^2 A)\mathbf{U}(t) + \mathbf{U}(t - l) = \mathbf{0}, \tag{4.54}$$

where I is the identity matrix of order N.

Equation (4.54) expresses the independent variable $\mathbf{U}$ at the advanced time $t + l$ in terms of the independent variable at the base time t and the retarded time $t - l$. The vector $\mathbf{U}(t + l)$ is not pre-multiplied by any matrix (other than the identity matrix) and the relation (4.54) is said to be *explicit*.

This explicit formula may be applied to each element of the vector $\mathbf{U}(t) \equiv \mathbf{U}(nl)$. Applied to the element $U_m(nl)$ of $\mathbf{U}(t)$, (4.54) gives for each $m = 1, 2, \ldots, N$

$$U_m^{n+1} - \{2U_m^n + l^2 c^2 (U_{m-1}^n - 2U_m^n + U_{m+1}^n)/h^2\} + U_m^{n-1} = 0$$

which, on gathering terms, becomes

$$U_m^{n+1} = 2(1 - c^2 r^2) U_m^n + c^2 r^2 (U_{m-1}^n + U_{m+1}^n) - U_m^{n-1} \tag{4.55}$$

where $r = l/h$.

The finite difference scheme (4.55) is an explicit three-level five-point difference scheme for which it is an easy matter to write computer code. It is probably the most widely used explicit scheme for solving the wave equation, but, as will be seen in section 4.6.2, the ratio r must be controlled if a meaningful numerical solution to (4.40) is to be obtained. Note that (4.55), being explicit, has only one point at the advanced time.

Equation (4.55) may also be obtained by approximating the space derivative in (4.40) by its finite difference replacement (4.44) and the time derivative by its equivalent replacement. This dual replacement may well be quicker than deriving (4.49) and then making Padé approximants. However, (4.49), which was developed by the author (Twizell, 1979), does allow a *family* of finite difference methods to be determined simply by substituting various Padé approximants; this family happens to contain the most widely used explicit scheme for solving the wave equation.

It is easy to verify from (4.53) that the Maclaurin expansion of $\{\exp(lcB) + \exp(-lcB)\}$ is

$$2I + l^2 c^2 A + \frac{1}{12} l^4 c^4 A^2 + \frac{1}{360} l^6 c^6 A^3 + \ldots, \qquad (4.56)$$

so that the error in time incurred by using the (0, 2) Padé approximant in (4.49) is $O(l^4)$ even though the (0, 2) Padé approximant itself has error $O(l^3)$. This error in time contributes to one component of the local truncation error of the method (see section 4.6.1). The local truncation error also contains terms which are due to the space discretization.

4.5.4 A well known implicit scheme

The family of difference schemes evolving from (4.49) also contains the most widely used implicit method for solving the constant coefficient wave equation (4.40). This method is obtained by approximating the exponential matrix functions in (4.49) by the (1, 1) Padé approximants giving

$$\mathbf{U}(t+l) - \{(I - \tfrac{1}{2} lcB)^{-1} (I + \tfrac{1}{2} lcB) + (I + \tfrac{1}{2} lcB)^{-1} (I - \tfrac{1}{2} lcB)\} \mathbf{U}(t) + \mathbf{U}(t-l) = \mathbf{0}. \qquad (4.57)$$

Expanding the matrix inverses and gathering together like powers of lcB in (4.57) it is easy to verify that the (1, 1) Padé approximant which, from Table 4.3, itself has error $O(l^3)$, leads to a finite difference method which is second order accurate in time.

Premultiplying the terms in $\mathbf{U}(t+l)$, $\mathbf{U}(t)$, $\mathbf{U}(t-l)$ in (4.57) by $(I - \tfrac{1}{2} lcB)$ $(I + \tfrac{1}{2} lcB)$ leads to

$$(I - \tfrac{1}{4} l^2 c^2 A) \mathbf{U}(t+l) - (2I + \tfrac{1}{2} l^2 c^2 A) \mathbf{U}(t) + (I - \tfrac{1}{4} l^2 c^2 A) \mathbf{U}(t-l) = \mathbf{0}. \qquad (4.58)$$

In (4.58) the function at the advanced time $t + l$ is premultiplied by a matrix which is not the identity matrix; the relation (4.58) is thus said to be *implicit*.

Applying (4.58) to the element $U_m(nl)$ of $\mathbf{U}(t)$, $m = 1, 2, \ldots, N$, gives

$$U_m^{n+1} - \frac{1}{4}l^2 c^2 (U_{m-1}^{n+1} - 2U_m^{n+1} + U_{m+1}^{n+1})/h^2 - \{2U_m^n + \frac{1}{2}l^2 c^2(U_{m-1}^n - 2U_m^n + U_{m+1}^n)/h^2\}$$

$$+ U_m^{n-1} - \frac{1}{4}l^2 c^2 (U_{m-1}^{n-1} - 2U_m^{n-1} + U_{m+1}^{n-1})/h^2 = 0$$

which becomes

$$-\frac{1}{4}c^2 r^2 U_{m-1}^{n+1} + (1 + \frac{1}{2}c^2 r^2)U_m^{n+1} - \frac{1}{4}c^2 r^2 U_{m+1}^{n+1}$$

$$= \frac{1}{2}c^2 r^2 U_{m-1}^n + (2 - c^2 r^2)U_m^n + \frac{1}{2}c^2 r^2 U_{m+1}^n$$

$$+ \frac{1}{4}c^2 r^2 U_{m-1}^{n-1} - (1 + \frac{1}{2}c^2 r^2)U_m^{n-1} + \frac{1}{4}c^2 r^2 U_{m+1}^{n-1}. \quad (4.59)$$

The finite difference scheme (4.59) is an implicit, three-level, ninepoint difference scheme. It is probably the most widely used implicit scheme for solving the wave equation and, as will be seen in section 4.6.2 there is no restriction on the size of r which may be used in obtaining a numerical solution. Note that, in keeping with all implicit schemes, more than one point is involved at the advanced time level.

Equation (4.59) is also a member of the family of methods obtained by rewriting (4.40) as

$$\left(\frac{\partial^2 u}{\partial t^2}\right)_m^n = ac^2 \left(\frac{\partial^2 u}{\partial x^2}\right)_m^{n+1} + (1-2a)c^2 \left(\frac{\partial^2 u}{\partial x^2}\right)_m^n + ac^2 \left(\frac{\partial^2 u}{\partial x^2}\right)_m^{n-1} \quad (4.60)$$

where $a > 0$ is a parameter to be determined (see Mitchell and Griffiths, 1980). Note that for $a = 0$, (4.60) becomes (4.40) applied at the point (mh, nl) in the (x, t) plane. Introducing the standard second order finite difference replacements, (4.60) becomes

$$-c^2 r^2 a U_{m-1}^{n+1} + (1+2c^2 r^2 a)U_m^{n+1} - c^2 r^2 a U_{m+1}^{n+1}$$

$$= c^2 r^2 (1-2a)U_{m-1}^n + 2(1+2c^2 r^2 a - c^2 r^2)U_m^n + c^2 r^2 (1-2a)U_{m+1}^n$$

$$+ c^2 r^2 a U_{m-1}^{n-1} - (1+2c^2 r^2 a)U_m^{n-1} + c^2 r^2 a U_{m+1}^{n-1}. \quad (4.61)$$

Substituting $a = 0$ in (4.61) gives (4.55) and substituting $a = \frac{1}{4}$ gives (4.59).

The computation of the numerical solution of an implicit method such as (4.58) involves the solution of a linear algebraic system whose unknowns are the elements of the vector $\mathbf{U}(t + l)$. Writing $S = I - \frac{1}{4}l^2 c^2 A$ and $D = 2I + \frac{1}{2}l^2 c^2 A$, (4.58) may be written in the form

$$S\mathbf{U}(t + l) = D\mathbf{U}(t) - S\mathbf{U}(t - l). \qquad (4.62)$$

In (4.62), the matrices D, S and the vectors $\mathbf{U}(t)$, $\mathbf{U}(t - l)$ are known so that the right-hand side is easily determined. Writing $\mathbf{U}^*(t) = D\mathbf{U}(t) - S\mathbf{U}(t - l)$, (4.82) takes the form

$$S\mathbf{U}(t + l) = \mathbf{U}^*(t) \qquad (4.63)$$

which is easily solved implicitly for $\mathbf{U}(t + l)$. The matrix A is tridiagonal, see (4.47), and so, therefore, is the matrix S. The solution of (4.63) may thus be determined economically as described in section 1.2.3 of Chapter 1.

4.5.5 Solution at the first time-step

It has been noted that the recurrence relation (4.49) is applied with $t = l, 2l, \ldots$ and that the accuracy of $\mathbf{U}$ depends on the chosen approximation to the exponential matrix functions $\exp(\pm lcB)$. The remaining difficulty then is to find a sufficiently accurate approximation to $\mathbf{U}(l)$. This is not contained explicitly in the initial conditions and it is the purpose of this subsection to derive an approximation to the dependent variable at the first time step which is sufficiently accurate to be used with a scheme such as (4.54) or (4.58).

Considering the theoretical solution given by (4.48), replacing the exponential matrix functions with their (0, 2) Padé approximants, and writing $t = l$, gives

$$\mathbf{U}(l) = \frac{1}{2}[\exp(clB) + \exp(-clB)]\mathbf{f} + \frac{1}{2}[\exp(clB) - \exp(-clB)]B^{-1}\mathbf{g}$$

$$= \frac{1}{2}[I + clB + \frac{1}{2}c^2l^2B^2 + I - clB + \frac{1}{2}c^2l^2B^2]\mathbf{f}$$

$$+ \frac{1}{2}[I + clB + \frac{1}{2}c^2l^2B^2 - I + clB - \frac{1}{2}c^2l^2B^2]B^{-1}\mathbf{g},$$

that is

$$\mathbf{U}(l) = (I + \frac{1}{2}c^2l^2A)\mathbf{f} + l\mathbf{g} + 0(l^3). \qquad (4.64)$$

Equation (4.64) is a second order accurate approximation (in time – recall that the space discretization has an error proportional to h^2). It is equation

(4.64) which is usually used with (4.54) and (4.58), both of which are third order accurate in time. Indeed, if many time-steps are to be taken with (4.54) or (4.58) then (4.64) may be used with confidence. However, an approximation to U(*l*) which is at least third order accurate in time should be used if only a few time steps are to be taken with the general scheme. In addition, (4.64) becomes unreliable for large values of *l* (see Problem 7 at the end of this chapter).

Replacing the exponential matrix functions in (4.48) with their (0, 4) Padé approximants leads to

$$U(l) = (I + \frac{1}{2}l^2 A + \frac{1}{24}l^4 A^2)\mathbf{f} + l(I + \frac{1}{6}l^2 A)\mathbf{g} + 0(l^5), \quad (4.65)$$

which may be solved explicitly to give fourth order accuracy in time. This formula also becomes unreliable for large values of *l* and the reader is referred again to Problem 7 at the end of the chapter.

4.6 ANALYSES OF FINITE DIFFERENCE METHODS

So far, two finite difference schemes have been derived for solving the constant coefficient wave equation given by (4.40). In this section of the chapter, the tests which these and other schemes must pass before they are eligible for use, will be discussed.

4.6.1 Errors and consistency

Suppose that the differential equation is written in the form $G(u) = 0$. Each of the difference schemes (4.55), (4.59) may be written in the form $F_{m,n}(U) = 0$. If U is replaced by u at the mesh points appearing in the finite difference formula, then the value of $l^{-2}F_{m,n}(u) - G(u_m^n)$ is the *local truncation error* at the mesh point (*mh*, *nl*). The value of $F_{m,n}(u)$ is found by expanding about $u(mh, nl)$ the value of u at other points appearing in the finite difference scheme. The local truncation error, which is the difference between the finite difference scheme and the differential equation it is replacing, is not to be confused with the *local discretization error* of the finite difference scheme which is the difference between the theoretical solutions of the differential and difference equation and which is given by

$$z_m^n = u_m^n - U_m^n. \quad (4.66)$$

If $l^{-2}F_{m,n}(u) - G(u_m^n)$ tends to zero as the mesh lengths h, l tend to zero the difference equation is said to be *consistent* with the differential equation.

Consider for example the five-point explicit scheme given by (4.55), then

$$F_{m,n}(u) = u_m^{n+1} - 2(1 - c^2 r^2)u_m^n - c^2 r^2(u_{m-1}^n + u_{m+1}^n) + u_m^{n-1}. \quad (4.67)$$

Expanding u_m^{n+1}, u_{m-1}^n, u_{m+1}^n, u_m^{n-1} about u_m^n and its partial derivatives (assuming u is sufficiently often differentiable with respect to both x and t) gives

$$l^{-2}F_{m,n}(u) = \left(\frac{\partial^2 u}{\partial t^2} - c^2 \frac{\partial^2 u}{\partial x^2}\right)_m^n + \frac{1}{12} h^2 \left(r^2 \frac{\partial^4 u}{\partial t^4} - c^2 \frac{\partial^4 u}{\partial x^4}\right)_m^n$$
$$+ \frac{1}{360} h^4 \left(r^4 \frac{\partial^6 u}{\partial t^6} - c^2 \frac{\partial^6 u}{\partial x^6}\right)_m^n + \ldots \qquad (4.68)$$

The first term on the right-hand side of (4.68) is the differential equation evaluated at (mh, nl). If this term is taken to the left-hand side of (4.68), the left-hand side then gives the difference at (mh, nl) between the difference equation and the differential equation which is the definition of local truncation error given by some authors, see, for instance, Smith (1978).

It is easy to show, for the constant coefficient wave equation $\partial^2 u/\partial t^2 = c^2 \partial^2 u/\partial x^2$, that $\partial^4 u/\partial t^4 = c^4 \partial^4 u/\partial x^4$, $\partial^6 u/\partial t^6 = c^6 \partial^6 u/\partial x^6$, etc. The local truncation error (4.68) thus becomes

$$h^2 c^2 \left\{\frac{1}{12}(c^2 r^2 - 1)\frac{\partial^4 u}{\partial x^4} + \frac{1}{360} h^2 (c^4 r^4 - 1) \frac{\partial^6 u}{\partial x^6} + \ldots\right\}_m^n. \qquad (4.69)$$

The term $\frac{1}{12} h^2 c^2 (c^2 r^2 - 1)\partial^4 u/\partial x^4$ is called the *principal part* of the local truncation error. It is easy to see that the local truncation error vanishes when $c^2 r^2 = 1$ and the explicit five point scheme (4.55) takes the simple form

$$U_m^{n+1} = U_{m-1}^n + U_{m+1}^n - U_m^{n-1}, \qquad (4.70)$$

which is therefore an *exact* representation of the wave equation (4.40). For the simple wave equation ($c^2 = 1$) the condition for the local truncation error to vanish implies that the finite difference mesh is square ($r=l/h=1$). The principal part of the local truncation error (4.69) is seen to be $O(h^2 + l^2)$; the first part $O(h^2)$ is related to the space discretization, and the second part $O(l^2)$ is related to use use of the (0, 2) Padé approximant in (4.49) which gives rise to (4.54).

The local truncation error (4.69) is now denoted by $T_{m,n}(u)$. It is seen that $T_{m,n} \to 0$ as $h, l \to 0$ showing that (4.55) is a consistent finite difference scheme for solving the wave equation (4.40). The implicit difference scheme (4.59) is also consistent with the wave equation. In fact, all except the (0, 1) and (1, 0) Padé approximants yield consistent finite difference schemes from (4.49).

The local *order of accuracy* of (4.55) is $O(l^2 + h^2)$, that is, it is second order accurate with respect to both time and space. This must not be confused with the global accuracy of the method which gives an indication of the accuracy of the method over the whole of the region R.

4.6.2 Stability

The error, $u_m^n - U_m^n$, at the mesh point (mh, nl) in using a finite difference scheme to solve the differential equation must be bounded as $n \to \infty$ for fixed h, l and as h, $l \to 0$ for a fixed value of nl. In both cases the number of applications of the difference scheme becomes infinite in the limit and there is a possibility of unbounded amplification of errors. The concept of stability is concerned with the boundedness of the solution of the finite-difference equations and this is examined by finding conditions under which

$$Z_m^n = U_m^n - \tilde{U}_m^n \tag{4.71}$$

remains bounded as n increases for fixed h, l. In (4.71) U_m^n is, as before, the theoretical solution of the finite difference scheme and $\tilde{U}_m^n$ is the solution of the scheme which is actually obtained, so that $\tilde{U}_m^n$ contains rounding errors. The analysis considers the growth of perturbations in initial data or the growth of errors introduced at mesh points at a given time level.

There are three common methods of investigating stability: the *energy method*, the *von Neumann* or *Fourier method*, and the *matrix method*. The energy method is a powerful tool in dealing with particular equations or particular classes of equations. Its application can, unfortunately, become rather messy and each problem to which it is applied requires a different treatment. The successful application of the method will be due in no small part to the ingenuity of the user. The strength of the method lies in its ability to deal effectively with boundary conditions, variable coefficients and non-linear problems. Besides proving the stability of a finite difference scheme, the energy method can indicate the correct choice of method. However, the method provides only sufficient conditions for stability which may be far removed from what is necessary in certain initial-boundary value problems.

The method calculates the sum of the squares of the errors Z_m^n ($m=1,\ldots,M$; $n=1, 2, \ldots$) at time level nl. This sum of squared errors is called the *energy* from which the method gets its name. It must be noted, however, that the conserved quantity (the energy) is not the physical energy of the system, a point which can cause confusion.

The explicit finite difference scheme (4.55) and the implicit scheme (4.59) are both three(time)-level schemes. The energy method is very difficult to use with three-level schemes and, as the problem is properly posed with constant coefficients, the wave equation (4.40) can be written as a system of two *first* order hyperbolic equations (as in section 4.9) with the numerical solution obtained using a two-level difference scheme. Stability can then be considered more easily using the energy method. A stability analysis for an explicit difference scheme used to solve the one-dimensional heat flow problem, using the energy method, is given in Chapter 5. For a theoretical discussion of the application of the energy method to three-level difference schemes, the reader is referred to Richtmeyer and Morton (1967, Chapter 7).

The Fourier method or von Neumann method, developed by J. von Neumann in the early 1940s, expresses the initial line of errors Z_m^0 in terms of a finite Fourier series and determines the criterion governing the growth of a function which reduces to this Fourier series for $t = 0$.

The Fourier series can be written in terms of sines and cosines in the usual way but it is soon seen that the algebraic manipulation involved is reduced if the exponential form is used. The errors at the $N + 2$ mesh points along the initial line $t = 0$ are written in terms of the error function

$$Z_m^0 = \sum_{r=0}^{N+1} A_r \exp(i\beta_r mh), \qquad (4.72)$$

where $m = 0, 1, \ldots, N+1$, $i = +\sqrt{-1}$, and the $A_r (r = 0, 1, \ldots, N+1)$ are the Fourier coefficients. The $N + 2$ equations in (4.72) are sufficient to determine the $N + 2$ unknowns $A_r (r = 0, 1, \ldots, N+1)$ uniquely.

Assuming that the finite difference method to be analyzed as linear, so that separate solutions will be additive, it is only necessary to consider the propagation of the error due to a single, typical term. Consider the term associated with the typical frequency $|\beta_s| = |\beta|$. The coefficient A_s is constant and can be neglected.

To examine the propagation of this single, typical error as $t \to \infty$, the solution of the finite difference scheme must be found which reduces to $\exp(i\beta mh)$ when $t = 0$. Let such a solution be

$$e^{\alpha t} e^{i\beta x} = e^{\alpha nl} e^{i\beta mh}$$

where $\alpha = \alpha(\beta)$ is complex. The original error component $e^{i\beta mh}$ will not grow with time (as n increases, that is) if

$$|e^{\alpha l}| \leq 1$$

for all α. This is *von Neumann's criterion for stability*. Often the notation $\xi = e^{\alpha l}$ is introduced, where ξ is known as the *amplification factor*.

In deriving this condition, a number of assumptions were made and so a number of important points must be emphasized before the condition is used. The first of these points is that the Fourier method applies only if the coefficients of the linear difference scheme are constant. If the difference scheme has variable coefficients, the method can still be applied at every point, individually, of the mesh (this is referred to as examining the local stability of the variable coefficient difference scheme) and it is reasonable to assume that the difference scheme is stable globally if it is stable at *every* mesh point. The von Neumann condition is necessary only for three level difference schemes such as (4.55) and (4.59); for two level difference schemes such as those to be met in sections

4.11, 4.12 and in Chapter 5, the condition is sufficient as well as necessary. This also applies to problems with more than one space variable. Strictly speaking, the Fourier or von Neumann method applies only to pure initial value problems with periodic initial data. In practice, however, it is used to analyze finite difference schemes applied to initial-boundary value problems also.

As an example, consider the stability of the three-level, explicit finite difference scheme (4.55) when applied to the wave equation (4.40) in the region $R = [0 < x < 1] \times [t > 0]$. A harmonic decomposition is made of the error Z at mesh points at a given time level, as in the above description of the Fourier method. It is easy to show that the error Z_m^n at the mesh point (mh, nl) defined by (4.71) satisfies the difference scheme (4.55). Thus

$$Z_m^{n+1} = 2(1 - c^2 r^2) Z_m^n + c^2 r^2 (Z_{m-1}^n + Z_{m+1}^n) - Z_m^{n-1} \tag{4.73}$$

and substituting $Z_m^n = e^{\alpha n l} e^{i\beta m h} \neq 0$ gives

$$e^{\alpha(n+1)l} e^{i\beta mh} = 2(1 - c^2 r^2) e^{\alpha n l} e^{i\beta mh}$$
$$+ c^2 r^2 e^{\alpha n l}(e^{i\beta(m-1)h} + e^{i\beta(m+1)h}) - e^{\alpha(n-1)l} e^{i\beta mh}$$

Cancelling by $e^{\alpha n l} e^{i\beta mh}$ and writing $\xi = e^{\alpha l}$ leads to

$$\xi^2 - \{2(1 - c^2 r^2) + c^2 r^2 (e^{-i\beta h} + e^{i\beta h})\}\xi + 1 = 0.$$

Using Euler's relation $e^{i\theta} = \cos\theta + i \sin\theta$ and the half-angle formula $\cos\theta = 1 - 2\sin^2 \tfrac{1}{2}\theta$ for any real θ, gives

$$\xi^2 - 2(1 - 2c^2 r^2 \sin^2 \tfrac{1}{2} \beta h)\xi + 1 = 0. \tag{4.74}$$

This is the stability equation of the explicit difference scheme (4.55) and von Neumann's necessary condition for stability requires both values of ξ to satisfy $|\xi| \leq 1$; the condition on the mesh ratio $r = l/h$ to ensure stability is now determined.

In general, three-level difference schemes, explicit and implicit, are of the form

$$a\xi^2 - 2b\xi + a = 0, \tag{4.75}$$

where a, b are functions of r, h. The zeros of (4.75) are easily seen to be $\xi = [b \pm (b^2 - a^2)^{1/2}]/a$, so that $|\xi| \leq 1$ when $|b| \leq |a|$. In the case of (4.74) this gives the three-part inequality

$$-1 \leq 1 - 2c^2 r^2 \sin^2 \tfrac{1}{2} \beta h \leq 1. \tag{4.76}$$

The right-hand side of (4.76) is satisfied for all values of r (recall that c, β, h are all real); the left-hand side is satisfied only for $c^2 r^2 \leq 1$ since $0 \leq \sin^2 \frac{1}{2}\beta h \leq 1$. The right-hand side of (4.76) suggests that the finite difference scheme is *unconditionally stable* ($c^2 r^2 \geq 0$) but the left-hand side imposes the condition $c^2 r^2 \leq 1$. The *stability criterion* for the method is the intersection of the two ranges of values namely $0 < c^2 r^2 \leq 1$ (obviously, the possibility $c^2 r^2 = 0$ may be ignored).

The matrix method of analysis is applicable to initial boundary value problems, but it can only be used for hyperbolic equations when the difference scheme has been expressed in two-time level form. Suppose that the difference scheme is in the general form

$$D\mathbf{U}^{n+1} = B\mathbf{U}^n + C\mathbf{U}^{n-1} + \mathbf{b}^n \tag{4.77}$$

where $\mathbf{b}^n$ is a vector which depends on the boundary conditions and D, B, C are square matrices of order N (recall, N is the number of mesh points at each time level). In the case of a differential equation with constant coefficients, the matrices D, B, C are constant; in the case of a variable coefficient problem, the matrices D, B, C are evaluated at times $(n+1)l$, nl, $(n-1)l$ respectively. In constant coefficient problems the matrices D and C are usually equal and non-singular.

Writing (4.77) in the form

$$\mathbf{U}^{n+1} = D^{-1}B\mathbf{U}^n + D^{-1}C\mathbf{U}^{-1} + D^{-1}\mathbf{b}^n,$$

it follows that a pertubation $\mathbf{Z}^0$ of the initial conditions will satisfy

$$\mathbf{Z}^{n+1} = D^{-1}B\mathbf{Z}^n + D^{-1}C\mathbf{Z}^{n-1}, \tag{4.78}$$

This may be written

$$\begin{bmatrix} \mathbf{Z}^{n+1} \\ \mathbf{Z}^n \end{bmatrix} = \begin{bmatrix} D^{-1}B & D^{-1}C \\ I & 0 \end{bmatrix} \begin{bmatrix} \mathbf{Z}^n \\ \mathbf{Z}^{n-1} \end{bmatrix} \tag{4.79}$$

which is of the form

$$\mathbf{E}^{n+1} = W\mathbf{E}^n \tag{4.80}$$

where $\mathbf{E}^{n+1} = [\mathbf{Z}^{n+1}, \mathbf{Z}^n]^T$; W is called the *amplification matrix*. It follows that

$$\|\mathbf{E}^{n+1}\| \leq \|W\| \cdot \|\mathbf{E}^n\|. \tag{4.81}$$

where $\|\cdot\|$ denotes a suitable norm. Equation (4.80) is a two-time level equation and the von Nuemann necessary condition for the stability of a difference scheme such as (4.77), using a constant time step and letting $n \to \infty$, is

$$\|W\| \leq 1. \tag{4.82}$$

Sec. 4.6]   Analyses of Finite Difference Methods

The success of the matrix method for second order hyperbolic equations thus depends on being able to obtain a suitable estimate for $\|W\|$. The spectral norm, which is equal to the spectral radius when W is a symmetric matrix, is the usual choice of norm. The spectral radius is usually used to approximate the spectral norm when W is non-symmetric, leading occasionally to an erroneous stability interval (see subsection 4.12.1). Often, the matrices $B, C, D, D^{-1}B, D^{-1}C$ have the same system of linearly independent eigenvectors and the eigenvalues of W are given by

$$\det \begin{bmatrix} \alpha_k^{-1}\beta_k - \mu & \alpha_k^{-1}\gamma_k \\ 1 & -\mu \end{bmatrix} = 0$$

where $\alpha_k, \beta_k, \gamma_k$ ($k = 1, \ldots, N$) are the eigenvalues of D, B, C.

It is an easy task to verify that the matrix method gives the same stability criterion for the explicit scheme (4.55) as that determined by the Fourier method. Taking the scheme in its matrix form (4.54), it may be written in two-time level form as

$$\mathbf{U}^{n+1} = (2I + l^2 c^2 A)\mathbf{U}^n - \mathbf{V}^n, \tag{4.83}$$
$$\mathbf{V}^{n+1} = \mathbf{U}^n$$

so that the error vector $\mathbf{E}$ satisfies

$$\mathbf{E}^{n+1} = \begin{bmatrix} 2I + l^2 c^2 A & -I \\ I & 0 \end{bmatrix} \mathbf{E}^n. \tag{4.84}$$

The $2N$ eigenvalues of the amplification matrix are determined from the N quadratic equations

$$\det \begin{bmatrix} 2 - 4c^2 r^2 \sin^2 [s\pi/\{2(N+1)\}] - \mu & -1 \\ 1 & -\mu \end{bmatrix} = 0$$

for $s = 1, 2, \ldots, N$ (the eigenvalues of the matrix A were given with equation (4.47)). The eigenvalues of the amplification matrix are thus given by

$$\mu_s^2 - 2(1 - 2c^2 r^2 \sin^2 \theta_s)\mu_s + 1 = 0 \tag{4.85}$$

for $s = 1, 2, \ldots, N$, where $\theta_s = s\pi/\{2(N+1)\}$. Clearly, equation (4.85) is analogous to (4.74), and it follows from (4.82) that the stability criterion is $0 < c^2 r^2 \leq 1$, as obtained by the Fourier method.

Turning now to the implicit method (4.59), which is based on the (1, 1) Padé approximant, the matrix method for stability proceeds from (4.58) and leads to the amplification matrix

$$W = \begin{bmatrix} (I - \tfrac{1}{4}l^2 c^2 A)^{-1}(2I + \tfrac{1}{2}l^2 c^2 A) & -I \\ I & 0 \end{bmatrix} \tag{4.86}$$

which has $2N$ eigenvalues found from the N quadratic equations

$$\det \begin{bmatrix} (1 - \tfrac{1}{4}c^2 r^2 \lambda_s)^{-1}(2 + \tfrac{1}{2}l^2 c^2 \lambda_s) - \mu_s & -1 \\ 1 & -\mu_s \end{bmatrix} = 0.$$

These give the stability equations

$$(1 - \tfrac{1}{4}c^2 r^2 \lambda_s)\mu_s^2 - 2(1 + \tfrac{1}{4}c^2 r^2 \lambda_s)\mu_s + (1 - \tfrac{1}{4}c^2 r^2 \lambda_s) = 0, \qquad (4.87)$$

where λ_s ($s = 1, 2, \ldots, N$) are the eigenvalues of the matrix A, given after (4.47). Equation (4.87) is of the form (4.75) for each $s = 1, 2, \ldots, N$, and it follows, for stability, that

$$-1 + \tfrac{1}{4}c^2 r^2 \lambda_s \leqslant 1 + \tfrac{1}{4}c^2 r^2 \lambda_s \leqslant 1 - \tfrac{1}{4}c^2 r^2 \lambda_s \qquad (4.88)$$

must be satisfied. The left-hand side of (4.88) is trivially satisfied for all values of $c^2 r^2$; the right-hand side gives $\tfrac{1}{2}c^2 r^2 \lambda_s \leqslant 0$ for each $s = 1, 2, \ldots, N$ which is always satisfied since the eigenvalues $\lambda_s (s = 1, 2, \ldots, N)$ of A are all negative. The implicit method (4.59) is therefore unconditionally stable.

The unconditional stability of (4.59) can also be verified by the Fourier method; this is left to the reader as an excercise.

The treatment of stability considered so far, namely that a method is stable if perturbations in the initial conditions are not magnified, is the more conventional one. Other definitions of stability, based on how a finite difference method treats unimportant components of a solution, have been developed for two time level difference schemes. Three-level schemes such as (4.55) and (4.59) could threfore be treated in this way, but further discussion of this approach to stability will be left until Chapter 5.

4.6.3 Convergence

The problem of convergence of a finite difference scheme for solving a partial differential equation (elliptic, hyperbolic or parabolic) consists of finding the criteria under which the local discretization error (4.66) at *a fixed mesh point*, tends to zero uniformly as the mesh is refined. For a second order hyperbolic equation such as (4.40), this refinement means that $h, l \to 0$ and $m, n \to \infty$. In carrying out the convergence analysis, it may be convenient to assume that h, l do not tend to zero independently but according to a relationship of the form $l = rh^\alpha$, where r is a constant and $\alpha \geqslant 1$ is some parameter.

As an example of a convergence analysis for a finite difference scheme, consider the five-point explicit scheme (4.55). It is easy to see from (4.67), (4.68), (4.40) and (4.66) that the local discretization error z_m^n at the fixed mesh point (mh, nl) satisfies

$$z_m^{n+1} = 2(1 - c^2 r^2)z_m^n + c^2 r^2 (z_{m-1}^n + z_{m+1}^n) - z_m^{n-1} + 0(l^2 h^2 + l^4).$$

Sec. 4.6]   **Analyses of Finite Difference Methods**   147

It was seen in subsection 4.6.2 that the stability criterion for (4.55) is $0 < c^2 r^2 \leq 1$. Assuming, therefore, that the stability criterion is satisfied, it follows that

$$|z_m^{n+1}| \leq 2(1 - c^2 r^2)|z_m^n| + c^2 r^2 |z_{m-1}^n| + c^2 r^2 |z_{m+1}^n| + |z_m^{n-1}| + k^{(n)}(l^2 h^2 + l^4)$$

$$\leq 2\hat{z}^{(n)} + \hat{z}^{(n-1)} + k^{(n)}(l^2 h^2 + l^4),$$

where $\hat{z}^{(n)} = \max |z_m^n|$ and $k^{(n)} = \max |\partial^4 u / \partial x^4|_m^n$ for $m = 1, 2, \ldots, N$. Thus

$$\hat{z}^{(n+1)} \leq 2\hat{z}^{(n)} + \hat{z}^{(n-1)} + k^{(n)}(l^2 h^2 + l^4)$$

and, assuming that the same initial conditions are used with both differential equation and difference equation ($\hat{z}^{(0)} = 0$), and further that $\hat{z}^{(1)} = 0$, it follows that

$$\hat{z}^{(n+1)} \leq (2\bar{z}^{(n)} + \bar{z}^{(n-1)} + 1)k^{(0)}(l^2 h^2 + l^4), \tag{4.89}$$

where $\bar{z}^{(n)}$ is the coefficient of $k^{(n)}(l^2 h^2 + l^4)$ in $\hat{z}^{(n)} = \bar{z}^{(n)} k^{(n)} (l^2 h^2 + l^4)$. Inequality (4.89) leads to

$$\hat{z}^{(n+1)} \leq l^2 (h^2 + l^2) k^{(0)} (2\bar{z}^{(n)} + \bar{z}^{(n-1)} + 1)$$

$$\to 0, \text{ as } h, l \to 0.$$

The reader will have noted that the expression $2\bar{z}^{(n)} + \bar{z}^{(n-1)} + 1$ grows as n increases. The local discretization error does, however, tend to zero as the mesh is refined and convergence is thus assured.

An alternative approach for establishing convergence is to assume that the theoretical solution of the difference scheme at the fixed mesh point (mh, nl) has the form

$$U_m^n = e^{im\theta} e^{in\psi} \neq 0, \tag{4.90}$$

where θ is real and ψ is complex, and to derive the condition, or conditions, under which such a solution remains bounded. The arguments used are very similar to those used in the Fourier method for examining the stability of a linear finite difference scheme with constant coefficients.

Substituting (4.90) into the explicit scheme (4.55), for instance, and dividing by $e^{im\theta} e^{in\psi}$, gives

$$e^{i\psi} = 2(1 - c^2 r^2) + c^2 r^2 (e^{-i\theta} + e^{-i\theta}) - e^{-i\psi}. \tag{4.91}$$

Using the Euler formulas $e^{\pm i\theta} = \cos\theta \pm i\sin\theta$ and the half-angle formula $\cos\theta = 1 - 2\sin^2\tfrac{1}{2}\theta$, equation (4.91) becomes

$$\sin^2 \tfrac{1}{2}\psi = c^2 r^2 \sin^2 \tfrac{1}{2}\theta. \tag{4.92}$$

Now c, r, θ are all real and so $0 \leq \sin^2 \tfrac{1}{2}\psi \leq 1$ whenever $0 \leq c^2 r^2 \leq 1$ since $0 \leq \sin^2 \tfrac{1}{2}\theta \leq 1$. This implies that ψ is real whenever $c^2 r^2 \leq 1$ which, in turn, implies that U_m^n is bounded whenever $c^2 r^2 \leq 1$. It is assumed that u_m^n is bounded everywhere in the region R and it follows that the local discretization error is bounded at the fixed grid point (mh, nl) provided $c^2 r^2 \leq 1$. This condition for convergence coincides with the von Neumann necessary condition for the stability of (4.55).

It is easy to verify that the solution of (4.55) does not remain bounded at the mesh point (mh, nl) if $c^2 r^2 > 1$, for then $\sin^2 \tfrac{1}{2}\psi > 1$ for some value or values of θ. Hence ψ occurs in complex conjugate pairs $\psi_1 = \mu + i\nu$ and $\psi_2 = \mu - i\nu$, say, where μ and ν ($\nu \neq 0$) are real, and:

for ψ_1, $U_m^n = e^{im\theta} e^{in(\mu + i\nu)} = e^{im\theta} e^{in\mu} e^{-n\nu}$;

for ψ_2, $U_m^n = e^{im\theta} e^{in(\mu - i\nu)} = e^{im\theta} e^{in\mu} e^{n\nu}$.

This means that for one of ψ_1 or ψ_2, U_m^n grows exponentially as $n \to \infty$ since either $e^{-n\nu}$ or $e^{n\nu}$ grows exponentially as $n \to \infty$ for real values of ν ($\nu \neq 0$, otherwise ψ is real). This exponential growth of U_m^n occurs whenever $c^2 r^2 > 1$, irrespective of the initial conditions $U_m^0 = e^{im\theta}$ which may be arbitrarily small.

It must be concluded therefore that U_m^n is bounded only for $0 \leq c^2 r^2 \leq 1$ which is the criterion for convergence (the possibility $c^2 r^2 = 0$ can be discounted as in subsection 4.6.2). Since u_m^n is bounded, the local discretization error $z_m^n = u_m^n - U_m^n$ is not bounded if $c^2 r^2 > 1$; z_m^n does not then tend to zero as the mesh is refined and convergence does not take place.

It is important at this point to consider the *Courant-Friedrichs-Lewy (CFL) condition* for second order hyperbolic equations. This condition states that the *domain of dependence of an explicit difference scheme must include the domain of dependence of the differential equation*. This means that the characteristics through the advanced point, when $t = (n+1)l$, must intersect the line through the points used by the scheme at the base time level ($t = nl$) within the range of those points, otherwise the difference scheme is not convergent.

The slope of the characteristics of (4.40) through $(mh, (n+1)l)$ are $+1/c$ and $-1/c$ and it is thus easy to verify geometrically that the CFL condition for (4.55), for instance, is $0 < cr \leq 1$.

In this section of the chapter, the concepts of consistency, stability and convergence of a finite difference scheme have been discussed, and it has been seen that for the consistent explicit scheme (4.55) at least, the criteria for stability and convergence coincide. For pure initial value problems the three concepts are connected more formally by the *Lax Equivalence Theorem* which states that 'for a properly posed linear initial value problem and a linear finite difference approximation to it that satisfies the consistency condition, stability is the necessary and sufficient condition for convergence'. In this context, a problem is properly posed if

(i) the solution $u(x, t)$ is unique whenever it exists;
(ii) the solution depends continuously on the initial conditions;
(iii) a solution always exists for initial conditions which are arbitrarily close to initial conditions for which no solution exists.

4.7 HIGHER ORDER METHODS

In this section, higher order approximations to the matrix exponential function will be used in the recurrence relation (4.49). This will have the effect of raising the order of that part of the local truncation error which is a power of the time increment l only. That component of the error which relates to the space discretization will remain unchanged unless a different approximation to the space derivative in (4.40) is used.

In addition to giving a more favourable truncation error, the use of higher order approximants to the matrix exponential functions in (4.49) improves the stability properties of explicit finite difference schemes yielded by (4.49).

If, for instance, the (0, 4) Padé approximant is used instead of the (0, 2) approximant, $\exp(lcB) + \exp(-lcB) = 2I + l^2c^2B^2 + \frac{1}{12}l^4c^4B^4 + O(l^6) = 2I + l^2c^2A + \frac{1}{12}l^4c^4A^2 + O(l^6)$ and (4.49) becomes

$$\mathbf{U}(t + l) - (2I + l^2c^2A + \frac{1}{12}l^4c^4A^2)\mathbf{U}(t) + \mathbf{U}(t - l) + O(l^6) = \mathbf{0}. \quad (4.93)$$

Squaring the matrix A, given by (4.47), gives

$$A^2 = h^{-4} \begin{bmatrix} 5 & -4 & 1 & & & & \\ -4 & 6 & -4 & 1 & & & 0 \\ 1 & -4 & 6 & -4 & 1 & & \\ & & & & & & \\ 0 & & & & 1 & -4 & 5 \end{bmatrix}$$

and applying (4.9.3) to the mesh point (mh, nl) gives the explicit, seven-point, three-level scheme

$$U_m^{n+1} = 2(1 - \frac{1}{2}c^2r^2)^2 U_m^n + c^2r^2(1 - \frac{1}{3}c^2r^2)(U_{m-1}^n + U_{m+1}^n)$$

$$+ \frac{1}{12}c^4r^4(U_{m-2}^n + U_{m+2}^n) - U_m^{n-1} \quad (4.94)$$

for $m \neq 1, N$. This finite difference scheme was introduced in Twizell (1979).

Examining the form of the matrix A^2 it is clear that, for $m = 1$, (4.94) must be modified to

$$U_1^{n+1} = 2(1 - c^2r^2 + \frac{5}{24}c^4r^4)U_1^n + c^2r^2(1 - \frac{1}{3}c^2r^2)U^n$$

$$+ \frac{1}{12}c^4r^4 U_3^n - U_1^{n-1}, \qquad (4.95)$$

and for $m = N$ to

$$U_N^{n+1} = \frac{1}{12}c^4r^4 U_{N-2}^n + c^2r^2(1 - \frac{1}{3}c^2r^2)U_{N-1}^n$$

$$+ 2(1 - c^2r^2 + \frac{5}{24}c^4r^4)U_N^n - U_N^{n-1}. \qquad (4.96)$$

The reader should check, by first calculating $F_{m,n}(U)$, that the general form (4.94) of the seven-point scheme is consistent with the differential equation (4.40) and that the local truncation error is

$$-\frac{1}{12}c^2h^2\left(\frac{\partial^4 u}{\partial x^4}\right)_m^n + \frac{1}{360}\{l^4 - c^2h^4 - 5c^4l^2h^2\}\left(\frac{\partial^6 u}{\partial x^6}\right)_m^n \ldots, \qquad (4.97)$$

which represents an improvement when compared with (4.69) in that the term $\frac{1}{12}c^4l^2\partial^4u/\partial x^4$ has vanished; the term $T = \frac{1}{360}l^4$ relates to the use of the (0, 4) Padé approximant.

Perhaps the most significant benefit in using the seven-point explicit scheme is in its extended stability range. It follows from section 4.6 that a perturbation $\mathbf{Z}^0$ of a vector of initial conditions is propagated in such a way that

$$\mathbf{Z}^{n+1} = (2I + l^2c^2A + \frac{1}{12}l^4c^4A^2)\mathbf{Z}^n - \mathbf{Z}^{n-1} \qquad (4.98)$$

for $n = 0, 1, \ldots$. The amplification matrix W of (4.80) is thus

$$W = \begin{bmatrix} 2I + l^2c^2A + \dfrac{1}{12}l^4c^4A^2 & -I \\ I & 0 \end{bmatrix}$$

which has $2N$ eigenvalues found by solving the N quadratic equations

$$\det\begin{bmatrix} 2 + l^2c^2\lambda_s + \dfrac{1}{12}l^4c^4\lambda_s^2 - \mu_s & -1 \\ 1 & -\mu_s \end{bmatrix} = 0 \qquad (4.99)$$

Sec. 4.7] Higher Order Methods 151

for $s = 1, 2, \ldots, N$; $\lambda_s = -4 \sin^2 [s\pi/\{2(N+1)\}]/h^2$, with $s = 1, 2, \ldots, N$, are the eigenvalues of the matrix A. For each $s = 1, 2, \ldots, N$, (4.99) becomes

$$\mu_s^2 - 2(1 - 2c^2r^2\sigma_s^2 + \frac{2}{3}c^4r^4\sigma_s^4)\mu_s + 1 = 0 \qquad (4.100)$$

where $\sigma_s^2 = \sin^2[s\pi/\{2(N+1)\}]$. Equation (4.100) is the stability equation for the method and the necessary condition for stability is $|\mu_s| \leq 1$ for $s = 1, 2, \ldots, 2N$. Following the discussion of equation (4.75), the inequalities

$$-1 \leq 1 - 2c^2r^2\sigma_s^2 + \frac{2}{3}c^4r^4\sigma_s^4 \leq 1 \qquad (4.101)$$

must be satisfied for each $s = 1, \ldots, N$ and it is from (4.101) that the restriction on cr will be found. It is convenient to write $q_s = c^2r^2\sigma_s^2$ (each q_s is therefore non-negative) and to consider (4.101) in the form $-1 \leq 1 - 2q_s + \frac{2}{3}q_s^2 \leq 1$. The left-hand side leads to the requirement $-1 + q_s - \frac{1}{3}q_s^2 \leq 0$ for each $s = 1, \ldots, N$ and it is very easy to show that the quadratic function $-1 + q_s - \frac{1}{3}q_s$ has a maximum value of -0.25 where $q_s = 1.5$. This suggests that the seven-point scheme may be unconditionally stable but the right-hand side of (4.101) quickly dispels this possibility. The right-hand side gives $-q_s + \frac{1}{3}q_s^2 \leq 0$; this leads to $0 \leq q_s \leq 3$, that is $0 \leq c^2r^2\sigma_s^2 \leq 3$ and, since $0 \leq \sigma_s^2 \leq 1$, it follows that $0 \leq c^2r^2 \leq 3$. The possibility $c^2r^2 = 0$ can be ignored, as before, and the stability criterion for the seven-point explicit scheme (4.94) is $0 < cr \leq \sqrt{3}$.

Convergence can be verified as in subsection 4.6.2 by assuming a solution of (4.94), at the fixed mesh point (mh, nl), to be of the form $U_m^n = e^{in\psi}e^{im\theta}$, where θ is real and ψ is complex. It is left to the reader to show that ψ, θ, cr are connected by the equation

$$\sin^2\frac{1}{2}\psi = \frac{1}{3}r^2(3 - r^2)\sin^2\frac{1}{2}\theta$$

and to deduce from this that the seven-point formula converges for $0 < cr \leq \sqrt{3}$.

The advantages in choosing a higher order Padé approximant to the matrix exponential functions in (4.49) should now be apparent. To this end, Twizell and Tirmizi (1984) have examined the numerical solution of (4.40) using (4.49) with the (1, 2), (2, 2), (2, 1) and (2, 0) Padé approximants to the matrix exponential functions. For these four approximants, the findings of Twizell and Tirmizi may be summarized as follows.

4.7.1 The (1, 2) Padé approximant
Here, the recurrence relation (4.49) becomes

$$(I - \frac{1}{9}r^2A)\mathbf{U}(t + l) - (2I + \frac{7}{9}r^2A)\mathbf{U}(t) + (I - \frac{1}{9}r^2A)\mathbf{U}(t - l) = 0 \qquad (4.102)$$

which, when applied to the mesh point (mh, nl), gives a consistent nine-point, three-level, implicit finite difference scheme with truncation error $0(h^2 + l^2)$ which has $T = -\frac{1}{36} l^2$. The stability restriction on r is given by $0 < cr \leqslant 3\sqrt{5}/5 \cong 1.34$. In view of the value of T_4 and the stability restriction on r, there appears to be little to be gained in using the finite difference scheme derived from (4.102) in preference to any of the schemes based on the (0, 2), (1, 1) or (0, 4) Padé approximants.

4.7.2 The (2, 2) Padé approximant
Using the (2, 2) Padé approximant, (4.49) becomes

$$(I - \frac{1}{12} r^2 A + \frac{1}{144} r^4 A^2)\mathbf{U}(t + l) - (2I + \frac{5}{6} r^2 A + \frac{1}{72} r^4 A^2)\mathbf{U}(t)$$

$$+ (I - \frac{1}{12} r^2 A + \frac{1}{144} r^4 A^2)\mathbf{U}(t - l) = \mathbf{0} \quad (4.103)$$

which, when applied to the mesh point (mh, nl), gives a consistent fifteen-point, three-level, implicit finite difference scheme which is unconditionally stable. Its truncation error is $0(h^2 + l^4)$ with $T = \frac{1}{360} l^4$.

4.7.3 The (2, 1) Padé approximant
Here, the recurrence relation (4.49) becomes

$$(I - \frac{1}{9} r^2 A + \frac{1}{36} r^4 A^2)\mathbf{U}(t + l) - (2I + \frac{7}{9} r^2 A)\mathbf{U}(t)$$

$$+ (I - \frac{1}{9} r^2 A + \frac{1}{36} r^4 A^2)\mathbf{U}(t - l) = \mathbf{0}, \quad (4.104)$$

and applying (4.104) to the mesh point (mh, nl) gives a consistent thirteen-point, three-level scheme which is implicit and unconditionally stable. Its local truncation error is $0(h^2 + l^2)$ with $T = -\frac{1}{36} l^2$.

4.7.4 The (2, 0) Padé approximant
Using this approximant, (4.49) becomes

$$(I + \frac{1}{4} r^4 A^2)\mathbf{U}(t + l) - (2I + r^2 A)\mathbf{U}(t) + (I + \frac{1}{4} r^4 A^2)\mathbf{U}(t - l) = \mathbf{0}, \quad (4.105)$$

and applying (4.105) to the mesh point (mh, nl) gives a consistent thirteen-point, three-level implicit scheme which is unconditionally stable. The local truncation error of the method is $0(h^2 + l^2)$ with $T_4 = \frac{7}{12} l^2$. This high value of T_4 indicates the numerical results obtained using this method will be less accurate than the other methods in this section.

Sec. 4.7] Higher Order Methods

A more accurate estimate of the solution at the first time step than that afforded by (4.64) or (4.65), may be obtained for use with (4.93), (4.102), (4.103), (4.104) by replacing the exponential matrix functions in (4.48) with their (0, 6) Padé approximants, giving

$$U(l) = (I + \frac{1}{2} l^2 A + \frac{1}{24} l^4 A^2 + \frac{1}{720} l^6 A^3)\mathbf{f} + l(I + \frac{1}{6} l^2 A + \frac{1}{120} l^4 A^2)\mathbf{g} + O(l^7).$$

This formula may be used explicitly to give $U(l)$ to sixth-order accuracy in time. Like (4.64) and (4.65), however, it, too, becomes unreliable for large values of l.

To illustrate the benefits of using higher order Padé approximations in (4.49), the resulting finite difference methods were tested in Twizell and Tirmizi (1984) on the simple model problem

$$\frac{\partial^2 u}{\partial t^2} = \frac{\partial^2 u}{\partial x^2}; \quad 0 < x < 1, t > 0$$

$$u(0, t) = u(1, t) = 0; \quad t > 0$$

$$u(x, 0) = \sin \pi x, \quad \partial u(x, 0)/\partial t = 0; \quad 0 \leqslant x \leqslant 1.$$

The theoretical solution of this problem is $u(x, t) = \sin \pi x \cos \pi t$.

The errors $u - \tilde{U}$ in the computed solution of this problem for $x = 0$ (the mid-point of the range $0 \leqslant x \leqslant 1$ where the errors are greatest) at time $t = 1.0$, are given in Table 4.4; the notation $-0.49(-4)$, for instance, means -0.49×10^{-4}. The theoretical solution at this mesh point is $u(0.5, 1.0) = -1$. A space increment $h = 0.1$ is used and r is given the values 0.5, 1.0, 2.0 corresponding to time steps $l = 0.05, 0.1, 0.2$, respectively. It is not immediately obvious from Table 4.4 where instability occurs as the methods which are unstable for $r = 2$ continue to give reasonable results for this value of r. The errors are largely in keeping with the theory; the reader will recall that the method based on (0, 2) Padé approximant is exact for $r = 1$.

Table 4.4

Method (Padé)	$r = 0.5$	Errors $r = 1.0$	$r = 2.0$
(0, 2)	−0.49(−4)	−0.22(−7)	−0.90(−3)
(1, 1)	−0.18(−3)	−0.63(−3)	−0.38(−2)
(0, 4)	−0.83(−4)	−0.84(−4)	−0.97(−4)
(1, 2)	−0.97(−4)	−0.14(−3)	−0.32(−3)
(2, 2)	−0.83(−4)	−0.84(−4)	−0.89(−4)
(2, 1)	−0.71(−4)	−0.44(−4)	−0.15(−5)
(2, 0)	−0.32(−4)	−0.20(−2)	−0.24(−1)

4.8 TWO SPACE VARIABLES

4.8.1 Introduction

Any homogeneous linear second order hyperbolic partial differential equation in two space variables x, y can be transformed to the form

$$\frac{1}{c^2}\frac{\partial^2 u}{\partial t^2} - \left(A\frac{\partial^2 u}{\partial x^2} + B\frac{\partial^2 u}{\partial x \partial y} + C\frac{\partial^2 u}{\partial y^2}\right) + D + Eu = 0, \quad (4.106)$$

where $c > 0$ is the constant wave speed, A, B, C, D are functions of x, y, t, $\partial u/\partial x$, $\partial u/\partial y$, $\partial u/\partial t$, but not of the function $u(x, y, t)$ nor of its second derivatives, and E is a function of x, y, t; in addition $A > 0$, $C > 0$ and $4AC - B^2 > 0$.

The simple second order hyperbolic equation

$$\frac{\partial^2 u}{\partial t^2} - c^2 \left(\frac{\partial^2 u}{\partial x^2} + \frac{\partial^2 u}{\partial y^2}\right) = 0, \quad (4.107)$$

together with the initial conditions $u(x, y, 0) = f(x, y)$ and $\partial u(x, y, 0)/\partial t = g(x, y)$ for $-\infty < x, y < \infty$, form the Cauchy initial value problem in two space variables. The domain of dependence of the Cauchy problem in the (x, y) plane for a point P whose x, y, t, coordinates are X, Y, T is the circle

$$(x - X)^2 + (y - Y)^2 \leq c^2 T^2.$$

This circle is cut from the (x, y) plane by the circular cone with apex angle $\tan^{-1}(1/c)$, vertex at P, and axis parallel to the x-axis. This cone is known as the *characteristic cone* for the two-dimensional wave.

It was noted at the beginning of section 4.5 that the method of characterstics does not generalize easily to two-dimensional problems. For this reason only finite difference methods will be considered in this section.

Consider the two-dimensional initial-boundary value problem consisting of the wave equation (4.107), defined in a region $[0 < x, y < 1] \times [t > 0]$ in (x, y, t) space, together with the boundary conditions

$$u(x, y, t) = 0; \quad \forall (x, y) \in \partial R, \; t > 0, \quad (4.108)$$

where ∂R is the boundary of the square region $R = \{(x, y): 0 < x, y < 1\}$, and the initial conditions

$$u(x, y, 0) = f(x, y), \; \partial u(x, y, 0)/\partial t = g(x, y); \; 0 \leq x, y \leq 1. \quad (4.109)$$

This is the problem of a vibrating square membrane fixed round its edges.

As in section 4.5, there will be a discontinuity between initial condition and boundary conditions if $f(x, y) \neq 0$ for any point $(x, y) \in \partial R$. It will be assumed that $u(x, y, t)$ is bounded everywhere in $R \times [t \geq 0]$.

Two Space Variables

The intervals $0 \leqslant x \leqslant 1$ and $0 \leqslant y \leqslant 1$ will each be divided into $N+1$ subintervals each of width h, so that $(N+1)h = 1$. Geometrically, this means that a square mesh of width h has been superimposed on the unit square R bounded by the lines $x=0, y=0, x=1, y=1$, by drawing lines parallel to the x, y axes; this discretization gives N^2 mesh points within the square and $N+2$ equally spaced points along each of its sides.

The independent variable t will be discretized in steps of length l, as before, so that $t = nl$ with $n = 0, 1, \ldots$.

The value of the dependent variable u at the mesh point (kh, mh, nl) in $R \times [t \geqslant 0]$ is $u(kh, mh, nl)$ where $m = 0, 1, \ldots, N+1$ and $n = 0, 1, \ldots$; this will be denoted by $u_{k,m}^n$ which is a simple extension of the notation used in section 4.5. The theoretical solution of a finite difference scheme approximating a two-dimensional hyperbolic equation will thus be denoted by $U_{k,m}^n$; the vector of such values at the interior mesh points of R at time $t = nl$ ($n = 0, 1, \ldots$) will be written in the form

$$\mathbf{U}^n = (U_{1,1}^n, U_{2,1}^n, \ldots, U_{N,1}^n; U_{1,2}^n, U_{2,2}^n, \ldots, U_{N,2}^n; \ldots;$$
$$U_{1,N}^n, U_{2,N}^n, \ldots, U_{N,N}^n)^T. \quad (4.10)$$

Superimposing the uniform mesh of width h on the space variable allows the space derivatives to be approximated by the finite difference replacements

$$\frac{\partial^2 u}{\partial x^2} = \{u(x-h, y, t) - 2u(x, y, t) + u(x, y+h, t)\}/h^2 + O(h^2). \quad (4.111)$$

and

$$\frac{\partial^2 u}{\partial y^2} = \{u(x, y-h, t) - 2u(x, y, t) + u(x, y+h, t)\}/h^2 + O(h^2). \quad (4.112)$$

Consider now the time level $t = nl$ at which there are N^2 interior mesh points. At every one of these points, the differential equation is applied with the space derivatives approximated by (4.111) and (4.112). As in section 4.5 these applications result in a system of second order ordinary differential equations, now of order N^2, of the form (4.46), that is $d^2\mathbf{U}(t)/dt^2 = c^2 A \mathbf{U}(t)$.

Recalling that $U = 0$ everywhere on the boundary of R, the matrix A is now block tridiagonal of order N^2 and is given by

$$A = h^{-2} \begin{bmatrix} C & I & & & \\ I & C & I & & 0 \\ & & \ddots & & \\ & 0 & & I & C & I \\ & & & & I & C \end{bmatrix} \quad (4.113)$$

In (4.113) I is the itentity matrix of order N and C is the tridiagonal matrix

$$C = \begin{bmatrix} -4 & 1 & & & \\ 1 & -4 & 1 & & 0 \\ & \ddots & \ddots & \ddots & \\ & 0 & 1 & -4 & 1 \\ & & & 1 & -4 \end{bmatrix} \qquad (4.114)$$

of order N. The eigenvalues of A are given by

$$\lambda_{i,j} = -4h^{-2}\left(\sin^2 \frac{i\pi}{2(N+1)} + \sin^2 \frac{j\pi}{2(N+1)}\right) \qquad (4.115)$$

for $i, j = 1, \ldots, N$.

Solving (4.107) subject to the boundary conditions (4.108) and initial conditions (4.109) the analytical solution is again found to be of the form (4.48), which satisfies the recurrence relation (4.49) for $t = l, 2l, \ldots$. As in section 4.5, Padé approximants to the exponential matrix function can be made in (4.49) and estimates for $\mathbf{U}(l)$ may be obtained using equations (4.64), (4.65) or (4.105) as appropriate.

4.8.2 An explicit method

Using the (0, 2) Padé approximant in (4.49) again leads to (4.59) where, now, the matrix A is given by (4.113). The finite difference scheme which arises from the application of (4.54) to the mesh point (kh, mh, nl), where $k, m = 1, \ldots, N$ and $n = 1, 2, \ldots$, is given by

$$U_{k,m}^{n+1} - 2U_{k,m}^n - c^2l^2\left(U_{k,m-1}^n + U_{k-1,m}^n - 4U_{k,m}^n + U_{k+1,m}^n + U_{k,m+1}^n\right)/h^2$$

$$+ U_{k,m}^{n-1} = 0.$$

Gathering together terms and writing $r = l/h$ gives the seven-point three-level, explicit scheme

$$U_{k,m}^{n+1} = 2(1 - 2c^2r^2)U_{k,m}^n$$

$$- c^2r^2\left(U_{k,m-1}^n + U_{k+1,m}^n + U_{k,m+1}^n + U_{k-1,m}^n\right) - U_{k,m}^{n-1}. \qquad (4.116)$$

Equation (4.116) is probably the most widely used explicit scheme for solving

(4.107) and, like its one-dimensional analog, can also be derived by substituting (4.111), (4.112) and the approximation

$$\partial^2 u/\partial t^2 = \{u(x, y, t-l) - 2u(x, y, t) + u(x, y, t+l)\}/l^2 + 0(l^2)$$

directly into (4.107).

Expanding (4.116) about the mesh point (kh, mh, nl) by Taylor's series, it may be shown that the principal part of the local truncation error is given by

$$\frac{1}{12} h^2 \left\{ (c^2 r^2 - 1) \left(\frac{\partial^4 u}{\partial x^4} + \frac{\partial^4 u}{\partial y^4} \right) + 2c^2 r^2 \frac{\partial^4 u}{\partial x^2 \partial y^2} \right\}_m^n \quad (4.117)$$

so that the scheme is consistent with the differential equation. Unlike the one-dimensional case, the local truncation error does not vanish for any value of cr, so using the $(0, 2)$ Padé approximation in (4.49) does not yield a finite difference scheme which is an exact representation of the two-dimensional wave equation.

Investigating stability using, say, the matrix method, the amplification matrix of order $2N^2$ is of the form

$$W = \begin{bmatrix} 2I + c^2 l^2 A & -I \\ I & 0 \end{bmatrix}. \quad (4.118)$$

The $2N^2$ eigenvalues of W in (4.118) are determined by solving the N^2 quadratic equations

$$\det \begin{bmatrix} 2 + l^2 c^2 \lambda_{i,j} - \mu_{i,j} & -1 \\ 1 & -\mu_{i,j} \end{bmatrix} = 0 \quad (4.119)$$

for $i, j = 1, \ldots, N$. This leads to

$$\mu_{i,j}^2 - 2(1 + \frac{1}{2} c^2 l^2 \lambda_{i,j}) \mu_{i,j} + 1 = 0. \quad (4.120)$$

Comparing (4.120) with (4.75), it follows that, for stability, the inequalities

$$-1 \leq 1 + \frac{1}{2} c^2 l^2 \lambda_{i,j} \leq 1 \quad (4.121)$$

must be satisfied. It is clear from (4.115) that each λ_{ij} is negative $-8h^{-2} < \lambda_{i,j} < 0$ so that the right-hand side of (4.121) is trivially satisfied. The left-hand side, however, gives $-\frac{1}{2} c^2 l^2 \lambda_{i,j} \leq 2$ and thus $\lambda_{i,j} \leq 4/c^2 l^2$ from which the reader can easily deduce that $0 \leq c^2 r^2 \leq \frac{1}{2}$. Discounting the possibility $cr = 0$, it follows that the necessary condition for the stability of (4.116) is $0 < cr \leq \frac{1}{2} \sqrt{2} \approx 0.71$.

This restriction is more severe than the analogous restriction for the one-dimensional wave equation. Both, in fact, are examples of the criterion $0 < cr\sqrt{s}/s$

for the stability of the appropriate finite difference scheme derived from (5.45) for the wave equation with s space dimensions.

The convergence of (4.116) is established by following either of the procedures outlined in subsection 4.6.3. Suppose that the theoretical solution of (4.136) has the form $U_{k,m}^n = e^{ik\phi} e^{im\theta} e^{in\psi}$ where θ, ϕ are real and ψ is complex. Substituting into (4.116) leads to

$$\sin^2 \frac{1}{2}\psi = c^2 r^2 (\sin^2 \frac{1}{2}\phi + \sin^2 \frac{1}{2}\theta),$$

from which it may be deduced tht (4.116) converges to the solution of the differential equation at the fixed mesh point (kh, mh, nl) provided $0 < cr \leq \frac{1}{2}\sqrt{2}$.

A higher order explicit finite difference scheme may be obtained by using the $(0, 4)$ Padé approximant in (4.49). This scheme requires $0 < cr \leq \frac{1}{2}\sqrt{6}$ for stability. Its implementation is no more difficult than the method based on the $(0, 2)$ method, even though it does require the square of the matrix A given by (4.113), (4.114), but it will not be discussed further.

4.8.3 An implicit method

Probably the most widely used implicit method for solving the two-dimensional wave equation (4.107) is the two-dimensional analog of (4.59) which is obtained by replacing the matrix exponential functions in (4.49) by their $(1, 1)$ Padé approximants. This leads to (4.58) which, when applied to the mesh point (kh, mh, nl), where $k, m = 1, \ldots, N$ and $n = 1, 2, \ldots$, gives the fifteen-point, three-level, implicit scheme

$$(1 + c^2 r^2) U_{k,m}^{n+1} - \frac{1}{4} c^2 r^2 (U_{k,m-1}^{n+1} + U_{k-1,m}^{n+1} + U_{k+1,m}^{n+1} + U_{k,m+1}^{n+1})$$

$$= 2(1 - c^2 r^2) U_{k,m}^n + \frac{1}{2} c^2 r^2 (U_{k,m-1}^n + U_{k-1,m}^n + U_{k+1,m}^n + U_{k,m+1}^n)$$

$$- (1 + c^2 r^2) U_{k,m}^{n-1} + \frac{1}{4} c^2 r^2 (U_{k,m-1}^{n-1} + U_{k-1,m}^{n-1} + U_{k+1,m}^{n-1} + U_{k,m+1}^{n-1}).$$
(4.122)

This scheme has local truncation error $O(h^2 + l^2)$ as does (4.116); in addition it is unconditionally stable. However, it does have the disadvantage of requiring the solution of a linear system at each time step and, if a fine space discretization is used, solving the system at each one of a large number of time steps can become expensive.

The user must then strongly consider using an explicit method such as (4.116), or that method based on the $(0, 4)$ Padé approximant. The conditional stability criteria of such methods are not too restrictive and so allow the dependent variable to be calculated at reasonably high values of t without using a prohibitively large number of time steps.

4.9 FIRST ORDER EQUATIONS

In analysing the stability of a three-time level difference replacement of a constant coefficient second order hyperbolic equation, it was noted in subsection 4.6.2 that the scheme could be written as a two-time level scheme. In an analogous manner, the second order hyperbolic equation itself can be written as a system of two *first* order hyperbolic equations and it is, therefore, appropriate to develop and analyze numerical methods for the solution of first order hyperbolic equations. This will largely be done for a single equation but the development and analysis carries over to a system.

Given an $M \times M$ real matrix A with real eigenvalues and M linearly independent eigenvectors, and a vector $\mathbf{u}(x, t)$ with M components, the general first order hyperbolic system is given by

$$\frac{\partial \mathbf{u}}{\partial t} + A\mathbf{u} = \mathbf{0}. \tag{4.123}$$

The matrix A is not necessarily symmetric. Given, additionally, the initial conditions

$$\mathbf{u}(x, 0) = \mathbf{g}(x); \quad -\infty < x < \infty, \tag{4.124}$$

(4.123) together with (4.124) form the *Cauchy initial value problem* for a first order hyperbolic system.

Explicit finite difference methods of solution are used to solve initial value problems of the type (4.123) and (4.124). The solution is computed in a triangular domain and may be obtained when A has positive, negative, or a mixture of positive and negative eigenvalues.

If, on the other hand, the problem is an initial-boundary value problem, a solution can only be obtained when the problem is well-posed. There are three types of initial-boundary value problem; they are related to the properties of the eigenvalues of A as follows:

Type 1 A positive definite (the outflow problem).
Here the problem is given in the form

$$\frac{\partial \mathbf{u}}{\partial t} + A\frac{\partial \mathbf{u}}{\partial t} = \mathbf{0}; \quad x > 0, t > 0$$

$$\mathbf{u}(0, t) = \mathbf{v}(t); \quad t > 0 \tag{4.125}$$

$$\mathbf{u}(x, 0) = \mathbf{g}(x); \quad x > 0$$

and the solution is determined in the first quarter plane. Explicit or implicit methods may be used to obtain the solution.

Type 2 A negative definite (the inflow problem).

Type 2 A negative definite (the inflow problem).

Here the problem is given in the form

$$\frac{\partial \mathbf{u}}{\partial t} + A \frac{\partial \mathbf{u}}{\partial x} = \mathbf{0}; \ x < 0, t > 0$$

$$\mathbf{u}(0, t) = \mathbf{v}(t); \ t > 0 \tag{4.126}$$

$$\mathbf{u}(x, 0) = \mathbf{g}(x); \ x < 0$$

and the solution is computed in the second quarter plane. This problem may also be solved numerically using explicit or implicit finite difference methods.

Type 3 A neither positive nor negative definite.

Here the initial-boundary value problem is much more difficult to solve. The solution is obtained in an open rectangle and implicit methods are used in the computation. Suppose that k of the M eigenvalues of A are positive and $M-k$ are negative. The problem must be given in the form

$$\frac{\partial \mathbf{u}}{\partial t} + A \frac{\partial \mathbf{u}}{\partial x} = \mathbf{0}; \ L_0 < x < L_1, t > 0$$

$$\mathbf{u}(x, 0) = \mathbf{g}(x); \ L_0 \leqslant x \leqslant L_1 \tag{4.127}$$

any k components of $\mathbf{u}$ given on the line $x = L_0, t > 0$; the other $M-k$ components of $\mathbf{u}$ given on the line $x = L_1, t > 0$.

The numerical solution is obtained in the open rectangle $L_0 < x < L_1, t > 0$.

In each of the these three types of problem a total of $2M$ boundary and/or initial conditions are specified, thus matching the total number of derivatives (M with respect to each of x, t) in the hyperbolic system (4.123). It will be found in the following pages that certain higher order implicit methods require all components of the vector $\mathbf{u}(x, t)$ to be given on the boundaries $x = L_0$, $x = L_1$ with $t > 0$. This raises the number of specified conditions to $3M$ and, as the total number of derivatives in (4.123) is still only $2M$, the problem becomes over-posed. If all $3M$ conditions are not available the initial-boundary value problem can usually be solved using a lower order method.

It was noted that hyperbolic systems may be solved using two-step finite difference methods; that is, the solution $\mathbf{u}$ is obtained at time $t = (n+1)l$ in terms of the solution at time $t = nl$, where l is a convenient time step and $n = 0, 1, 2, \ldots$. In cases where the matrix A in (4.123) is constant, the finite difference methods to be developed in the following subsections of this chapter, may be applied to a system of the form (4.123) with only minor alterations.

To a certain extent this is also true when the matrix A is a function of x or a function of x, t. In these cases the matrix A is evaluated at the mesh point to which the finite difference is being applied before proceeding with the computation. Analyses in problems when A is variable are carried out locally; the concept of a local stability criterion, for instance, was introduced in subsection 4.6.2.

Systems in which the matrix A is constant, or a function of x, or a function of x and t, are examples of *linear* first order hyperbolic systems. *Non-linear* hyperbolic systems arise in conservation problems where the system is in the *conservation form*

$$\frac{\partial \mathbf{u}}{\partial t} + \frac{\partial \mathbf{f}(\mathbf{u})}{\partial x} = \mathbf{0} \tag{4.128}$$

where $\mathbf{f}$ has M components.

The non-linear system (4.128) can be written in a form analogous to (4.123) by writing

$$A(\mathbf{u}) \equiv \frac{\partial \mathbf{f}}{\partial \mathbf{u}} = \frac{\partial (f_1, f_2, \ldots, f_M)}{\partial (u_1, u_2, \ldots, u_M)}. \tag{4.129}$$

The matrix $A(\mathbf{u})$ in (4.129) is the Jacobian of $\mathbf{f}$ with respect to $\mathbf{u}$. System (4.128) can now be written in the form

$$\frac{\partial \mathbf{u}}{\partial t} + A(\mathbf{u})\frac{\partial \mathbf{u}}{\partial x} = \mathbf{0} \tag{4.130}$$

and finite difference methods applied and analyzed as for variable coefficient problems in which $A = A(x)$ or $A = A(x, t)$.

It is not intended to consider finite difference methods for variable coefficient or non-linear problems in this text; for details of such problems the reader is referred to the texts of Mitchell and Griffiths (1980) or Richtmyer and Morton (1967). Finite difference methods for a single equation (that is, $M=1$) will be developed and analyzed; such methods are easily adapted to problems in which $M > 1$. Furthermore, only single equation problems derived from (4.123) with A positive definite will be considered; generalization to other problems is then immediate.

4.10 THE ROLE OF CHARACTERISTICS

As in the case of second order hyperbolic equations, a study of finite difference methods for solving first order hyperbolic equations should not be undertaken before the role of characteristics is undertsood.

In this section, the general first order hyperbolic equation

$$a \frac{\partial u}{\partial x} + b \frac{\partial u}{\partial t} = c \tag{4.131}$$

will be considered where a, b, c are functions of x, t, u but not of $\partial u/\partial x$ or $\partial u/\partial t$. The abbreviations p, q for $\partial u/\partial x$, $\partial u/\partial t$ will be retained so that (4.131) may be written in the form

$$ap + bq = c. \tag{4.132}$$

The total differential formula is

$$du = \frac{\partial u}{\partial x} dx + \frac{\partial u}{\partial t} dt = p\, dx + q\, dt \tag{4.133}$$

and eliminating p from (4.132), (4.133) gives

$$du = \frac{(c - bq)}{a} dx + q\, dt$$

which can be written in the form

$$q(a\, dt - b\, dx) + (c\, dx - a\, du) = 0. \tag{4.134}$$

It is possible to derive an equation independent of p because a, b, c are independent of p; this equation, (4.134), can further be made independent of q by stipulating that it hold only along some curve C in the x, t plane which satisfies the ordinary differential equation

$$a\, dt - b\, dx = 0. \tag{4.135}$$

It follows from (4.134), (4.135) that, along C, the relation

$$c\, dx - a\, du = 0 \tag{4.136}$$

holds also, and, since a, b are uniquely determined at any point (x, t), dt/dx has only one value at (x, t). This means that the curve C is the only curve which passes through (x, t); this curve is called the *characteristic curve* or *characteristic* through (x, t). From (4.135), (4.136), it follows that

$$\frac{dx}{a} = \frac{dt}{b} = \frac{du}{c} \tag{4.137}$$

along C. The reader will recall from section 4.1 that, for a second order equation, there are two characteristic curves through every point (x, t).

The Role of Characteristics

Suppose now that the value of u is known at N points $T_m^{(1)}(m=1,\ldots,N)$ on some initial curve L in the (x, t) plane; the curve L must not be a characteristic curve. Then through the point $T_m^{(1)}$ the characteristic C_m may be drawn. The *Method of Characteristics* for first order hyperbolic equations, which provides a procedure for integrating along the characteristics, approximates C_m by the straight lines $T_m^{(k)}T_m^{(k+1)}(k=1, 2, \ldots)$ where the points $T_m^{(k)}(k=1, 2, \ldots)$ lie on C_m; these points need not be equally spaced along C_m. The notation $x_m^{(k)}$ will be used to denote the value of x at the point $T_m^{(k)}(k=1, 2, \ldots)$ on C_m ($m=1, 2, \ldots, N$), with similar notations for t, u, a, b, c and, when required, for $p, q, f \equiv dt/dx = b/a$ at that point.

The algorithm advances the integration from $T_m^{(k)}$ to $T_m^{(k+1)}(k=1, 2, \ldots)$ on C_m ($m=1, \ldots, N$) as follows:

(1) a, b, c at $T_m^{(k)}$ are evaluated using the forms in which they appear in the differential equation (4.131);
(2) assume that $x_m^{(k+1)}$ is known, then a predicted estimate of $t_m^{(k+1)}$ may be obtained by adapting (4.135) to give

$$a_m^{(k)}(t_m^{(k+1)} - t_m^{(k)}) - b_m^{(k)}(x_m^{(k+1)} - x_m^{(k)}) = 0, \quad (4.138)$$

in which $t_m^{(k+1)}$ is the only unknown;
(3) a predicted value of $u_m^{(k+1)}$ is obtained by adapting (4.136) to give

$$c_m^{(k)}(x_m^{(k+1)} - x_m^{(k)}) - a_m^{(k)}(u_m^{(k+1)} - u_m^{(k)}) = 0, \quad (4.139)$$

in which $u_m^{(k+1)}$ is the only unknown;
(4) a, b, c at $T_m^{(k+1)}$ are evaluated using $x_m^{(k+1)}, t_m^{(k+1)}, u_m^{(k+1)}$ in the forms in which a, b, c appear in the differential equation (4.131);
(5) a corrected value of $t_m^{(k+1)}$ is obtained by writing (4.138) in the form

$$\frac{1}{2}(a_m^{(k+1)} + a_m^{(k)})(t_m^{(k+1)} - t_m^{(k)}) - \frac{1}{2}(b_m^{(k+1)} + b_m^{(k)})(x_m^{(k+1)} - x_m^{(k)}) = 0 \quad (4.140)$$

in which $t_m^{(k+1)}$ is the only unknown;
(6) a corrected value of $u_m^{(k+1)}$ is obtained by writing (4.139) in the form

$$\frac{1}{2}(c_m^{(k+1)} + c_m^{(k)})(x_m^{(k+1)} - x_m^{(k)}) - \frac{1}{2}(a_m^{(k+1)} + a_m^{(k)})(u_m^{(k+1)} - u_m^{(k)}) = 0 \quad (4.141)$$

in which $u_m^{(k+1)}$ is the only unknown.

An iterative procedure is set up using stages (4), (5), (6) of the algorithm and is carried out until each of $t_m^{(k+1)}, u_m^{(k+1)}$ converge to a predetermined accuracy.

In the event that q, p, f are required at $T_m^{(k+1)}$, they may now be calculated from (4.134), (4.132) and the equation $f = b/a$ respectively.

It was noted in section 4.2 that certain discontinuities can arise across the characteristics of second order hyperbolic equations. Similar problems can arise with first order hyperbolic equations, as will now be seen.

The first such difficulty arises with discontinuities in initial data and is easily demonstrated by reference to the simple equation for which $a = b = c \neq 0$ in (4.131). The characteristics for this problem are the straight lines $t - x =$ constant (see Fig. 4.6).

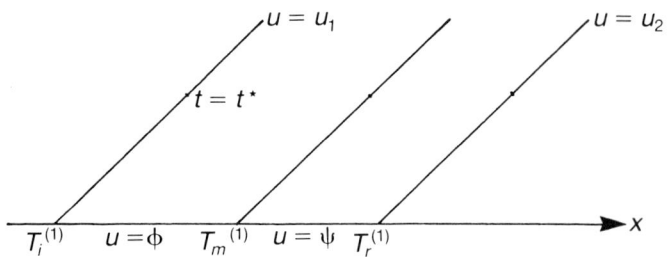

Fig. 4.6 – Discontinuities in the solution.

Suppose that the initial curve L is the x-axis and that there is a discontinuity in the function u at the initial point $T_m^{(1)}$ on L in such a way that

$$u(x, 0) = \phi(x); \quad -\infty < x < x_m^{(1)} \qquad (4.142)$$

and

$$u(x, 0) = \psi(x); \quad x_m^{(1)} < x < \infty, \qquad (4.143)$$

with $\phi(x_m^{(1)}) \neq \psi(x_m^{(1)})$. The characteristic through $T_m^{(1)}$ is the line $t - x = x_m^{(1)}$.

Consider now the initial point $T_i^{(1)}$ at which $x_i^{(1)} < x_m^{(1)}$ and the initial point $T_r^{(1)}$ at which $x_m^{(1)} < x_r^{(1)}$; the characteristics through these initial points are the lines $t - x = x_i^{(1)}$ and $t - x = x_r^{(1)}$, respectively. From (4.137) it follows that the solution along the characteristic $t - x = x_i^{(1)}$ is $u(x, t) = \phi(x_i^{(1)}) + t$, and that the solution along $t - x = x_r^{(1)}$ is $u(x, t) = \psi(x_r^{(1)}) + t$; let these solutions be u_1 and u_2, respectively.

Hence for the two points on these two characteristics which have the same t coordinate $t = t^*$,

$$u_1 - u_2 = \phi(x_i^{(1)}) - \psi(x_r^{(1)}). \qquad (4.144)$$

Suppose now, that $x_i^{(1)} \to x_m^{(1)}$ and $x_r^{(1)} \to x_m^{(1)}$. Then for the value t^*, $u_1 \neq u_2$ since $\phi(x_m^{(1)}) \neq \psi(x_m^{(1)})$. There is thus a discontinuity in the function u along

Sec. 4.10]   **The Role of Characteristics**   165

the characteristic through the initial point $T_m^{(1)}$, at which there is a discontinuity in initial conditions. The initial discontinuity has been propagated along the characteristic and since the right-hand side of (4.144) is constant for any t^*, the discontinuity is not damped as t increases.

The second difficulty arises with discontinuities in one, or both, of the first derivatives at an initial point on L. Suppose, as before, that the hyperbolic equation reduces to $\partial u/\partial x + \partial u/\partial t = 1$, and that the curve L is the x-axis. Let the initial conditions be

$$u(x, 0) = 0; \quad -\infty < x \leq 0, \tag{4.145}$$

$$u(x, 0) = x; \quad 0 < x < \infty, \tag{4.146}$$

then, recalling that the differential equation is $p + q = 1$, it follows that $p = 0, q = 1$ for $-\infty < x \leq 0, t = 0$ and $p = 1, q = 0$ for $0 < x < \infty, t = 0$. There is thus a discontinuity in both first derivatives at the origin even though the initial distributions of the function itself make u continuous at the origin.

The characteristic through any initial point $T_m^{(1)}$ is $t - x = x_m^{(1)}$ and from (4.137) it is seen that $u - u_m^{(1)} = t = x - x_m^{(1)}$ along this characteristic. Consider, therefore, the initial point $T_i^{(1)}$ for which $x_i^{(1)} \leq 0$ and the initial point $T_r^{(1)}$ for which $0 < x_r^{(1)}$, and consider the two points on these characteristics which have the same time value t (see Fig. 4.7).

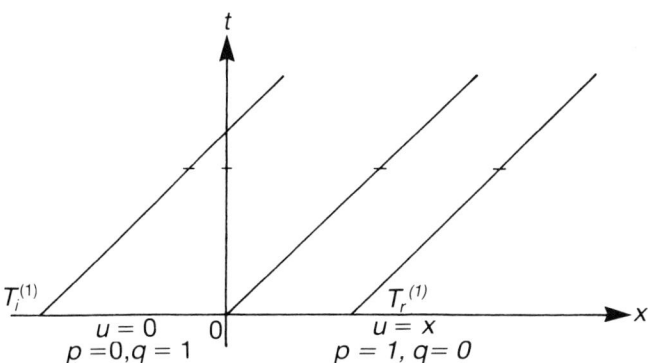

Fig. 4.7 – Discontinuities in the first derivatives.

The solution at this point on the characteristic through $T_i^{(1)}$ is from (4.137), (4.145), $u - u_i^{(1)} = t - t_i^{(1)}$, that is

$$u = t; \quad -\infty < x \leq 0, \; t \geq 0 \tag{4.147}$$

and the solution at the analogous point on the characteristic through $T_r^{(1)}$ is, from (4.137), (4.146), $u - u_r^{(1)} = x - x_r^{(1)}$, that is

$$u = x; \; 0 < x < \infty, \; t \geq 0 \tag{4.148}$$

It is clear that, as $x_i^{(1)} \to 0^-$, $u = 0$, $p = 0$, $q = 1$, and that, as $x_r^{(1)} \to 0^+$, $u \to 0$, $p = 1$, $q = 0$ for any points on the characteristic through $T_i^{(1)}$, $T_r^{(1)}$ which have the same value of t. It may be concluded from this that the discontinuities in the first derivatives at the origin are propagated along the characteristic $t = x$ and that, since p and q have constant values to the left and (different) constant values to the right of this characteristic, these discontinuities are not damped as t increases.

It has been seen that any discontinuity in the initial conditions of a first order hyperbolic equation is propagated along the characteristic through the point of discontinuity. It is reasonable to conclude therefore that, for such problems, the method of characteristics should be used to integrate along the characteristics. For systems of hyperbolic equations, or for a single equation with more than one space variable, however, it is much more difficult to write computer codes and the user may well be advised to use finite difference methods to obtain a numerical solution.

4.11 CENTRAL DIFFERENCE METHODS FOR FIRST ORDER EQUATIONS

4.11.1 Development of the methods

Returning now to the single equation form of (4.127), namely

$$\frac{\partial u}{\partial t} + a \frac{\partial u}{\partial x} = 0; \; a > 0 \tag{4.149}$$

defined in a region $R = [0 < x < 1] \times [t > 0]$, with initial and boundary conditions given by

$$u(x, 0) = g(x); \; 0 \leq x \leq 1, \tag{4.150}$$

$$u(0, t) = v(t); \; t > 0, \tag{4.151}$$

$$u(1, t) = w(t); \; t > 0. \tag{4.152}$$

With boundary conditions given on both $x = 0$ and $x = 1$, the problem is overposed; but, as noted in section 4.9, some commonly used finite difference methods require both these sets of boundary conditions, and so they must be specified. If $g(0) \neq v(0)$ and/or $g(1) \neq w(0)$ there are discontinuities between initial and boundary conditions.

Suppose that the interval $0 \leq x \leq 1$ is divided into $N+1$ equal parts each of width h, then $(N+1)h = 1$ and the x-coordinates of the N points of the

Sec. 4.11] Central Difference Methods for First Order Equations

discretization are $x_m = mh(m = 1, \ldots, N)$. Let $\mathbf{u}(t) = [u_1(t), u_2(t), \ldots, u_N(t)]^T$, T denoting transpose, be the solution vector at the N points of the discretization at time t.

Replacing the space derivative in (4.149) with the central difference formula

$$\frac{\partial u}{\partial x} = \{u(x+h, t) - u(x-h, t)\}/2h + O(h^2)$$

leads to the system of first order ordinary differential equations

$$\frac{d\mathbf{U}(t)}{dt} = -\frac{1}{2}aB\mathbf{U}(t) + \frac{1}{2}a\mathbf{b}_t. \qquad (4.153)$$

In (4.153), upper case U has been introduced to denote the theoretical solution of an approximating finite difference scheme. The matrix B is of order N and is given by

$$hB = \begin{bmatrix} 0 & 1 & & & & \\ -1 & 0 & 1 & & & \\ 0 & -1 & 0 & 1 & & \\ & & & \ddots & & \\ 0 & & & -1 & 0 & 1 \\ & & & & -1 & 0 \end{bmatrix}, \qquad (4.154)$$

and $\mathbf{b}_t = [v_t, 0, \ldots, 0, -w_t]^T/h$ is an N-component vector with first element the numerical (frozen) value of the boundary condition at $x = 0$ and time $t = nl(n = 0, 1, 2, \ldots$ and l is a constant time step); and last element minus the numerical value of the boundary condition at $x = 1$ and time $t = nl$. The region R has clearly been covered by a rectangular mesh.

The solution of (4.153) may be shown to be

$$\mathbf{U}(t) = B^{-1}\mathbf{b}_t + \exp(-\frac{1}{2}atB) \{\mathbf{g} - B^{-1}\mathbf{b}_t\}, \qquad (4.155)$$

where $\mathbf{g}$ is the vector of initial values. It is easy to show that (4.155) satisfies the recurrence relation

$$\mathbf{U}(t+l) = B^{-1}\mathbf{b}_t + \exp(-\frac{1}{2}alB) \{\mathbf{U}(t) - B^{-1}\mathbf{b}_t\} \qquad (4.156)$$

and it is this formula which may be used to derive a family of finite difference methods for the numerical solution of (4.149) subject to (4.150), (4.151), (4.152).

Replacing the exponential matrix function in (4.156) with its (0, 1) Padé approximant gives

$$U(t + l) = B^{-1} b_t + (I - \frac{1}{2} alB) \{U(t) - B^{-1} b_t\} + 0(l^2)$$

which, when applied to the point (mh, nl), gives the scheme

$$U_m^{n+1} = U_m^n - \frac{1}{2} ar (U_{m+1}^n - U_{m-1}^n).$$

Unfortunately, this explicit scheme is unconditionally unstable and should never be used.

Using the (1, 0) Padé approximant to the exponential matrix function in (4.156) gives

$$U(t + l) = B^{-1} b_t + (I + \frac{1}{2} alB)^{-1} \{U(t) - B^{-1} b_t\} + 0(l^2). \qquad (4.157)$$

Writing (4.157) in implicit form gives

$$(I + \frac{1}{2} alB)U(t + l) - \frac{1}{2} alB_{t+l} = U(t) \qquad (4.158)$$

where $b_{t+l} = [v_{t+l}, 0, \ldots, 0, -w_{t+l}]^T/h$. The reader will note the switch of the vector of boundary conditions from the base time level $t = nl$ to the advanced time level $t = (n+1)l$; this is quite legitimate and is, indeed, essential, as may be seen by applying (4.158) to any point (mh, nl) at time level t. Writing $r = l/h$, the resulting finite difference scheme (Twizell, 1980) is

$$U_m^{n+1} + \frac{1}{2} ar(U_{m-1}^{n+1} - U_{m-1}^{n+1}) = U_m^n. \qquad (4.159)$$

The scheme is reliable over the whole range $0 \leq x \leq 1$ even where there are discontinuities between initial conditions and boundary conditions.

Replacing the exponential matrix function in (4.156) with its (1, 1) Padé approximant gives

$$U(t + l) = B^{-1} b_t + (I + \frac{1}{4} alB)^{-1} (I - \frac{1}{4} alb) \{U(t) - B^{-1} b_t\} + 0(l^3). \qquad (4.160)$$

Writing (4.160) implicitly gives

$$(I + \frac{1}{4} alB)U(t + l) - \frac{1}{4} alb_{t+l} = (I - \frac{1}{4} alB)U(t) + \frac{1}{4} al b_t, \qquad (4.161)$$

Sec. 4.11]   **Central Difference Methods for First Order Equations**   169

which, when applied to the mesh point (mh, nl), yields the well known Crank-Nicolson type implicit scheme

$$-\frac{1}{4}arU_{m-1}^{n+1} + U_m^{n+1} + \frac{1}{4}arU_{m+1}^{n+1} = \frac{1}{4}arU_{m-1}^n + U_m^n - \frac{1}{4}arU_{m+1}^n. \qquad (4.162)$$

This method is unreliable near the boundaries when there are discontinuities between initial and boundary conditions.

4.11.2 Analyses of the central difference methods

As in subsection 4.6.1, the notation $G(u) = 0$ will be used for the differential equation (4.149). The finite difference schemes (4.159), (4.162) may both be written in the form $F_{m,n}(U) = 0$. If U is replaced by u at the mesh points appearing in the finite difference formula, then the local truncation error at the mesh point (mh, nl) is given by $l^{-1}F_{m,n}(U) - G(u_m^n)$; the Taylor expansion is used to obtain expressions for u at mesh points other than (mh, nl) in terms of $u(mh, nl)$ and its derivatives. If $l^{-1}F_{m,n}(u) - G(u_m^n)$ tends to zero as the mesh lengths tend to zero, the difference equation is consistent with the differential equation.

Consider first of all the finite difference scheme (4.159). Clearly

$$F_{m,n}(u) = u_m^{n+1} - \frac{1}{2}ar(u_{m+1}^{n+1} - u_{m-1}^{n+1}) - u_m^n$$

and the reader can easily verify that the local truncation error is

$$\left(-\frac{1}{2}l\frac{\partial^2 u}{\partial t^2} + \frac{1}{6}ah^2\frac{\partial^3 u}{\partial x^3} + \text{higher order terms}\right)_m^n \qquad (4.163)$$

and thus conclude that (4.159) is consistent with (4.149).

Following the study of stability in subsection 4.6.2, a perturbation $\mathbf{Z}^0$ of the initial conditions will satisfy

$$\mathbf{Z}^{n+1} = (I + \frac{1}{2}alB)^{-1}\mathbf{Z}^{(n)}. \qquad (4.164)$$

The eigenvalues of the matrix B are given by $\lambda_s = 2i \cos\{s\pi/(N+1)\}/h$ where $i = \sqrt{-1}$ and $s = 1, 2, \ldots, N$; the N eigenvalues of the amplification matrix $(I + \frac{1}{2}alB)^{-1}$ are therefore $\mu_s = 1/(1+iar\cos\theta_s)$ where $r = l/h$ and $\theta_s = s\pi/(N+1)$ for $s = 1, 2, \ldots, N$. Thus,

$$\mu_s = (1 - iar\cos\theta_s)/(1 + a^2r^2\cos^2\theta_s)$$

from which it follows that $|\mu_s| = 1$ for each $s = 1, 2, \ldots, N$. The finite difference scheme (4.159) is therefore unconditionally stable.

In examining the convergence of (4.159), the local discretization error $z_m^n = u_m^n - U_m^n$ is considered at a fixed mesh point (mh, nl). As in subsection 4.6.3, z_m^n satisfies

$$z_m^{n+1} + \frac{1}{2} ar(z_{m-1}^{n+1} - z_{m-1}^{n+1}) = z_m^n + k(lh^2 + l^2) \qquad (4.165)$$

where k depends on the upper bounds of $|\partial^2 u/\partial t^2|$ and $|\partial^3 u/\partial x^3|$. Denoting by $\hat{z}^{(n)}$ the maximum value of $|z_m^n|$, $m = 1, \ldots, N$, (4.165) leads to

$$(1 + ar)\hat{z}^{(n+1)} \leq \hat{z}^{(n)} + kl(h^2 + l), \qquad (4.166)$$

and so, if the same initial conditions are used for differential equation and difference equation, $\hat{z}^{(0)} = 0$ and (4.166) gives

$$z^{(n+1)} \leq kl(h^2+l)\left[\frac{1}{1+ar} + \frac{1}{(1+ar)^2} + \ldots + \frac{1}{(1+ar)^n}\right]$$

$$= nl \cdot \frac{k(h^2+l)}{(1+ar)^n}\left[1 + \frac{(n-1)ar}{2} + \ldots + \frac{(ar)^{n-1}}{n}\right] \qquad (4.167)$$

(using the formula $S_v = A(1-R^v)/(1-R)$ for the sum of geometric series $A + AR + \ldots + AR^v$). Clearly $\hat{z}^{(n+1)} \to 0$ as $h, l \to 0$ at the fixed mesh point (mh, nl) and the method is convergent.

Turning now to the finite difference scheme (4.162),

$$F_{m,n}(U) = \frac{1}{4} ar(U_{m+1}^{n+1} - U_{m-1}^{n+1}) + U_m^{n+1} + \frac{1}{4} ar(U_{m+1}^n - U_{m-1}^n) - U_m^n$$

from which the local truncation error

$$\left(\frac{1}{6} a h^2 \frac{\partial^3 u}{\partial x^3} - \frac{1}{12} l^2 \frac{\partial^3 u}{\partial t^3} + \text{higher order terms}\right)_m^n \qquad (4.168)$$

may be deduced.

A perturbation $\mathbf{Z}^0$ of the initial conditions will satisfy

$$\mathbf{Z}^{n+1} = (I + \frac{1}{4} alB)^{-1}(I - \frac{1}{4} alB)\mathbf{Z}^n \qquad (4.169)$$

and this perturbation will not be magnified if no eigenvalue of the amplification

Sec. 4.11]   **Central Difference Methods for First Order Equations**   171

matrix $W = (I + \frac{1}{4} alB)^{-1} (I - \frac{1}{4} alB)$ exceeds unity in modulus. The eigenvalues μ_s of W may easily be shown to be

$$\mu_s = (1 - \frac{1}{4} a^2 r^2 \cos^2 \theta_s - iar \cos \theta_s)/(1 + \frac{1}{4} a^2 r^2 \cos^2 \theta_s)^2 \quad (4.170)$$

where, as before, $\theta_s = s\pi/(N+1)$ for $s = 1, 2, \ldots, N$. Hence $|\mu_s| = 1$ and the finite difference method (4.162) is unconditionally stable.

To investigate convergence this time, it will be assumed that the theoretical solution of the difference scheme (4.162) at the fixed mesh point (mh, nl) is of the form (4.90), namely $U_m^n = e^{im\theta} e^{in\psi}$, where θ is real and ψ is complex, and to determine the restriction on r (if any) under which such a solution is bounded and, consequently, under which the local discretization error is bounded, as the mesh is refined.

Substituting (4.90) into (4.162) and dividing by $e^{im\theta} e^{in\psi}$ gives

$$\frac{1}{4} ar(e^{i\theta} - e^{-i\theta})e^{i\psi} + e^{i\psi} = \frac{1}{4} ar(e^{-i\theta} - e^{i\theta}) + 1$$

from which it follows, using Euler's relations $e^{\pm i\theta} = \cos \theta \pm i \sin \theta$, that

$$\cos \psi + i \sin \psi = \frac{1 - \frac{1}{4} a^2 r^2 \sin^2 \theta}{1 + \frac{1}{4} a^2 r^2 \sin^2 \theta} - i \frac{ar \sin \theta}{1 + \frac{1}{4} a^2 r^2 \sin^2 \theta}. \quad (4.171)$$

Since a, r, θ are real, it follows from (4.171) that $-1 \leq \cos \psi, \sin \psi \leq 1$ for all non-negative values of r. The finite difference scheme (4.162) is therefore convergent for all $r > 0$ (the possibility $r = 0$ may be discounted), agreeing with the finding that the scheme is unconditionally stable.

4.11.3 A note on local truncation errors

The principal parts of the local truncation errors of the finite difference schemes (4.159), (4.162) are each in two parts. In (4.163), for instance, which relates to (4.159), the term $\frac{1}{6} a h^2 \partial^3 u/\partial x^3$ arises from the use of the central difference replacement of $\partial u/\partial x$ in the differential equation, while the term $-\frac{1}{2} l\partial^2 u/\partial t^2$ arises from the use of the (1, 0) Padé approximant to the exponential matrix function in (4.156). In (4.168), the term $\frac{1}{6} a h^2 \partial^3 u/\partial x^3$ is again present, while the term $-\frac{1}{12} l^2 \partial^3 u/\partial t^3$ arises from the use of the (1, 1) Padé approximant to the exponential matrix function in (4.156).

The term $\frac{1}{6} a h^2 \partial^3 u/\partial x^3$ will always be present in the local truncation error of a finite difference scheme yielded by (4.156). In addition there will be a term of the form $C_q l^{q-1} \partial^q u/\partial t^q$, where $q = m+k+1$, which depends on the Padé approximant used in (4.156); this observation is made in Khaliq (1983). The constants C_q are contained in Twizell and Khaliq (1981) and are reproduced in

Table 4.5 for the finite difference schemes yielded by the first eight entries of the Padé table; it is seen that, for m odd, $C_q < 0$ and for m zero or even, $C_q > 0$.

Table 4.5

Method (Padé)	C_q	Extrapolation $\mathbf{U}^{(E)}$	E_q
(0, 1)	$C_2 = \dfrac{1}{2}$	$2\mathbf{U}^{(1)} - \mathbf{U}^{(2)}$	$E_3 = \dfrac{4}{3}$
(1, 1)	$C_3 = -\dfrac{1}{12}$	$(4\mathbf{U}^{(1)} - \mathbf{U}^{(2)})/3$	$E_5 = \dfrac{1}{10}$
(1, 0)	$C_2 = -\dfrac{1}{2}$	$2\mathbf{U}^{(1)} - \mathbf{U}^{(2)}$	$E_3 = \dfrac{4}{3}$
(0, 2)	$C_3 = \dfrac{1}{6}$	$(4\mathbf{U}^{(1)} - \mathbf{U}^{(2)})/3$	$E_4 = \dfrac{1}{3}$
(1, 2)	$C_4 = -\dfrac{1}{72}$	$(8\mathbf{U}^{(1)} - \mathbf{U}^{(2)})/7$	$E_5 = -\dfrac{8}{945}$
(2, 2)	$C_5 = \dfrac{1}{720}$	$(16\mathbf{U}^{(1)} - \mathbf{U}^{(2)})/15$	$E_7 = -\dfrac{1}{1890}$
(2, 1)	$C_4 = \dfrac{1}{72}$	$(8\mathbf{U}^{(1)} - \mathbf{U}^{(2)})/7$	$E_5 = -\dfrac{8}{945}$
(2, 0)	$C_3 = \dfrac{1}{6}$	$(4\mathbf{U}^{(1)} - \mathbf{U}^{(2)})/3$	$E_4 = -\dfrac{1}{3}$

4.11.4 Extrapolation of the schemes

The component of the local truncation error due to the approximation of the derivative by its central difference replacement will not change; however, the component due to the chosen Padé approximant can be improved by at least one power of l by the process of *extrapolation*. This process is illustrated now for the finite difference schemes based on the $(1,0)$ and $(1,1)$ Padé approximants.

Consider the method based on the $(1, 0)$ Padé approximant in the form (4.157). Written over two single time steps this gives

$$\mathbf{U}^{(1)}(t+2l) = B^{-1}\mathbf{b}_t + (I + \tfrac{1}{2} alB)^{-1}(I + \tfrac{1}{2} alB)^{-1} \{\mathbf{U}(t) - B^{-1}\mathbf{b}_t\} + 0(l^2),$$

$$= B^{-1}\mathbf{b}_t + \{I - alB + \tfrac{3}{4} a^2 l^2 B^2 - \tfrac{1}{2} a^3 l^3 B^3 + \tfrac{5}{16} a^4 l^4 B^4 + 0(l^5)\} \cdot$$

$$\{\mathbf{U}(t) - B^{-1}\mathbf{b}_t\}, \qquad (4.172)$$

Sec. 4.11]   **Central Difference Methods for First Order Equations**   173

and written over a double time step $2l$ it gives

$$\mathbf{U}^{(2)}(t+2l) = B^{-1}\mathbf{b}_t + (I + alB)^{-1}\{\mathbf{U}(t) - B^{-1}\mathbf{b}_t\} + 0(l^2),$$

$$= B^{-1}\mathbf{b}_t + \{I - alB + a^2l^2B^2 - a^3l^3B^3 + a^4l^4B^4 + 0(l^5)\}\,.$$

$$\{\mathbf{U}(t) - B^{-1}\mathbf{b}_t\}. \qquad (4.173)$$

Comparing (4.172) and (4.173) with the Maclaurin expansion

$$\mathbf{U}^{(M)}(t+2l) = B^{-1}\mathbf{b}_t + \{I - alB + \frac{1}{2}a^2l^2B^2 - \frac{1}{6}a^3l^3B^3 + \frac{1}{24}a^4l^4B^4$$

$$- \frac{1}{120}a^5l^5B^5 + \ldots\}\{\mathbf{U}(t) - B^{-1}\mathbf{b}_t\}, \qquad (4.174)$$

it is verified that both (4.172) and (4.173) each have error $0(l^2)$, and each is therefore only first order accurate, but that $\mathbf{U}^{(E)} = \mathbf{U}^{(E)}(t+2l)$ defined by

$$\mathbf{U}^{(E)} = 2\mathbf{U}^{(1)} - \mathbf{U}^{(2)} = \mathbf{U}^{(M)} + 0(l^3) \qquad (4.175)$$

is second order accurate. The first order (in time) formula (4.157) has been extrapolated to give second order accuracy.

To extrapolate the method based on the (1, 1) Padé approximant, it is considered in the form (4.140). Written over two single time steps it becomes

$$\mathbf{U}^{(1)}(t+2l) = B^{-1}\mathbf{b}_t + (I + \frac{1}{4}alB)^{-1}(I - \frac{1}{4}alB)(I + \frac{1}{4}alB)^{-1}(I - \frac{1}{4}alB).$$

$$\{\mathbf{U}(t) - B^{-1}\mathbf{b}_t\} + 0(l^3),$$

$$= B^{-1}\mathbf{b}_t + \{I - alB + \frac{1}{2}a^2l^2B^2 - \frac{3}{16}a^3l^3B^3 + \frac{1}{16}a^4l^4B^4$$

$$- \frac{5}{256}a^5l^5B^5 + 0(l^6)\}\{\mathbf{U}(t) - B^{-1}\mathbf{b}_t\}, \qquad (4.176)$$

and over a double time step $2l$ it gives

$$\mathbf{U}^{(2)}(t+2l) = B^{-1}\mathbf{b}_t + (I + \frac{1}{2}alB)^{-1}(I - \frac{1}{2}alB)\{\mathbf{U}(t) - B^{-1}\mathbf{b}_t\} + 0(l^3),$$

$$= B^{-1}\mathbf{b}_t + \{I - alB + \frac{1}{2}a^2l^2B^2 - \frac{1}{4}a^3l^3B^3 + \frac{1}{8}a^4l^4B^4$$

$$- \frac{1}{16}a^5l^5B^5 + 0(l^5)\}\{\mathbf{U}(t) - B^{-1}\mathbf{b}_t\}. \qquad (4.177)$$

Comparing (4.176) and (4.177) with the Maclaurin expansion (4.174), it is easily seen that each is second order accurate in time (error in time = $0(l^3)$) but that $\mathbf{U}^{(E)} = \mathbf{U}^{(E)}(t+2l)$ defined by

$$\mathbf{U}^{(E)} = \frac{4}{3}\mathbf{U}^{(1)} - \frac{1}{3}\mathbf{U}^{(2)} = \mathbf{U}^{(M)} + 0(l^5) \tag{4.178}$$

is fourth order accurate in time.

In extrapolating the scheme based on the (1, 0) Padé approximant, one extra order of accuracy in time was obtained, and in extrapolating the scheme based on the (1, 1) Padé approximant, two extra orders were obtained. This second, bonus, order of accuracy is always obtained when a method based on a Padé approximant which has numerator and denominator of the same degree, is extrapolated.

The coefficients of $\mathbf{U}^{(1)}$ and $\mathbf{U}^{(2)}$ in the definition of $\mathbf{U}^{(E)}$ can easily be calculated from the following formulas (Twizell and Khaliq, 1981) which also give the order of the error in the $\mathbf{U}^{(E)}$:

$$\mathbf{U}^{(E)} = (2^{m+k}\mathbf{U}^{(1)} - \mathbf{U}^{(2)})/(2^{m+k} - 1) + 0(l^{m+k+2}) \tag{4.179}$$

for $m \neq k$, and

$$\mathbf{U}^{(E)} = (2^{2m}\mathbf{U}^{(1)} - \mathbf{U}^{(2)})/(2^{2m} - 1) + 0(l^{2m+3}) \tag{4.180}$$

for $m = k$.

The extrapolation formulas (4.175), (4.178) are contained in Table 4.5. The term $\frac{1}{6} a h^2 \partial^3 u/\partial x^3$ will still be present in the local truncation error of the extrapolated form of each finite difference method. In addition there will now be a term of the form $E_q l^{q-1} \partial^q u/\partial t^q$ ($q=m+k+2$ for $m \neq k$, $q=2m+3$ for $m=k$). The constants E_q (Twizell and Khaliq, 1981) are also contained in Table 4.5.

4.11.5 Higher order methods

As an alternative to extrapolating in time, higher order Padé approximants may be used in the recurrence relation (4.156) to improve accuracy. The higher order (in time) finite difference methods so obtained may themselves be extrapolated to give even higher accuracy in time. If a large value of h is used in the space discretization, the anticipated improvement in accuracy of high order methods in time may be dulled, as the component of the principal part of the local truncation error relating to the space discretization is unaffected by the choice of Padé approximant to be used in (4.156). The user must carefully weigh the benefits of improved accuracy against the costs of extrapolating or using a method which offers high order in time.

Replacing the matrix exponential function in (4.156) by the (0, 2) and (1, 2) Padé approximants leads to two finite difference methods which are unconditionally unstable. Khaliq and Twizell (1982) developed, analyzed, extrapolated and tested finite difference schemes based on the (2, 0), (2, 1), (2, 2) Padé approximants.

4.11.6 The Lax-Wendroff method
This method cannot be derived from (4.156) but may be used explicitly to solve (4.149) with (4.150), (4.151), (4.152) over the open rectangular region R. The method is given by

$$U_m^{n+1} = \frac{1}{2} ar(1+ar)U_{m-1}^n + (1-a^2 r^2)U_m^n - \frac{1}{2} ar(1-ar)U_{m+1}^n$$

and has the same local truncation error as the implicit scheme based on the (2, 0) Padé approximant, namely

$$\left(\frac{1}{6} a h^2 \frac{\partial^3 u}{\partial x^3} + \frac{1}{6} l^2 \frac{\partial^3 u}{\partial t^3} + \ldots\right)_m^n. \tag{4.181}$$

The method is stable and convergent for $0 < ar \leq 1$ and, because it is an explicit method, is an attractive alternative to the implicit methods derived from (4.156) provided the ratio r can be controlled.

To examine the behaviour of the schemes developed and extrapolated in this section, each is now applied to (4.149) with $a = 1$, that is $\partial u/\partial t + \partial u/\partial x = 0$, in the region $R = [0 < x < 1] \times [t > 0]$ with two different sets of initial conditions and boundary conditions as follows:

(i) $u(x, 0) = 1 + x;\ 0 \leq x \leq 1$
$u(0, t) = t,\ u(1, t) = 2 - t;\ t > 0.$

The theoretical solution of this problem in the region R is given by

$u(x, t) = 1 + x - t,\ x \geq t$
$u(x, t) = t - x,\ x < t$

so that there exists a discontinuity in the solution across the line $t = x$ in the (x, t) plane

(ii) $u(x, 0) = e^x;\ 0 \leq x \leq 1$
$u(0, t) = e^t,\ u(1, t) = e^{1-t};\ t > 0.$

The theoretical solution of this problem in the region R is given by

$u(x, t) = e^{x-t},\ x \geq t$
$u(x, t) = e^{t-x},\ x < t$

so that there exist discontinuities in the first derivatives across the line $t = x$.

The errors in the computed solution for the first problem are given in Table 4.6 and for the second problem in Table 4.7; for each problem $h = 0.1$ and $l = 0.2$ so that $r = 2$. The errors are given for $t = 0.4(0.4)2.0$ at the mesh points where $x = 0.5$.

Examining Tables 4.6 and 4.7 it is seen that all the methods incur their worst errors at points near the lines of discontinuities. The methods, in their raw and extrapolated forms, give better results when $t > x$.

The method based on the (1, 0) Padé approximant, which does not use the initial discontinuity point at the origin, portrays a smooth behaviour throughout the region.

Table 4.6

Method	Time				
	0.4	0.8	1.2	1.6	2.0
(1, 0) raw	1.4(−1)	3.4(−1)	4.6(−2)	2.1(−1)	3.4(−1)
extrapolated	2.3(−2)	1.3(−1)	2.2(−2)	1.4(−2)	1.2(−2)
(1, 1) raw	5.4(−2)	1.1(−1)	1.4(−1)	2.6(−2)	3.8(−1)
extrapolated	7.2(−3)	1.3(−2)	6.4(−2)	6.3(−3)	3.0(−2)

Table 4.7

Method	Time				
	0.4	0.8	1.2	1.6	2.0
(1, 0) raw	1.8(−1)	6.5(−2)	1.0(−1)	2.0(−1)	2.1(−1)
extrapolated	1.0(−1)	3.7(−2)	6.4(−2)	1.5(−2)	3.7(−2)
(1, 1) raw	4.2(−2)	9.1(−2)	4.1(−2)	3.5(−2)	4.5(−2)
extrapolated	2.4(−2)	3.7(−2)	5.9(−3)	7.5(−3)	1.2(−3)

4.12 BACKWARD DIFFERENCE METHODS FOR FIRST ORDER EQUATIONS

4.12.1 Development, analysis and extrapolation of low order methods

Backward difference methods may be used to compute numerical solutions of the properly posed initial-boundary value problem

$$\frac{\partial u}{\partial t} + a \frac{\partial u}{\partial x} = 0; \quad a > 0, x > 0, t > 0 \tag{4.182}$$

Sec. 4.12] Backward Difference Methods for First Order Equations

(the *outflow* problem) with initial conditions given by

$$u(x, 0) = g(x); \quad x \geq 0 \tag{4.183}$$

and boundary conditions by

$$u(0, t) = v(t); \quad t > 0. \tag{4.184}$$

Equation (4.182) is the single equation form of (4.125) and the solution is obtained in the first quarter-plane $x > 0$, $t > 0$. If $a < 0$, $x < 0$ in (4.182) the initial-boundary value problem is the single equation form of (4.126) and the solution is obtained in the second quarter plane $x < 0$, $t > 0$, the initial conditions (4.183) being given for $x \leq 0$. If $g(0) \neq v(0)$ there is a discontinuity between initial and boundary conditions at the origin.

Suppose that the solution of (4.182) is sought in some arbitrary region $R = [0 < x \leq X, t > 0]$ of the first quarter plane. The interval $0 \leq x \leq X$ is divided into N equal parts each of width h so that $Nh = X$. The coordinates of the N points of this discretization are $x_m = mh$ ($m = 1, 2, \ldots, N$); clearly $x_0 = 0$ is the x-coordinate of every point on the t-axis.

As in section 4.11 the vector $\mathbf{u}(t) = [u_1(t), \ldots, u_N(t)]^T$, T denoting transpose, will represent the solution vector at the N points of the discretization at time $t > 0$, and $\mathbf{U}(t)$ will denote the theoretical solution of a consistent, approximating finite difference scheme. The time variable t will again be discretized by a constant time step l such that $t = nl (n = 0, 1, \ldots)$.

Replacing the space derivative in (4.182) with the low order backward difference formula

$$\frac{\partial u}{\partial x} = \{u(x, t) - u(x - h, t)\}/h + O(h) \tag{4.185}$$

and applying (4.182) with (4.185) to every one of the N discrete points at time level t, leads to the system of first order ordinary differential equations

$$\frac{d\mathbf{U}(t)}{dt} = -a C \mathbf{U}(t) + a \mathbf{c}_t. \tag{4.186}$$

In (4.236), C is a square matrix of order N given by

$$hC = \begin{bmatrix} 1 & 0 & & & \\ -1 & 1 & 0 & & 0 \\ 0 & -1 & 1 & & \\ & & & \ddots & \\ 0 & & & -1 & 1 \end{bmatrix} \tag{4.187}$$

and $\mathbf{c}_t$ is a vector with N components given by

$$h\mathbf{c}_t = [v_t, 0, \ldots, 0]^T. \tag{4.188}$$

In (4.188), v_t is the numerical (frozen) value of the boundary condition at time $t = nl$ ($n = 1, 2, \ldots$); for $n = 0$, let $v_t = g(0)$. The N eigenvalues of the matrix C each have the value $1/h$.

The solution of (4.186) with (4.183) is

$$\mathbf{U}(t) = C^{-1}\mathbf{c}_t + \exp(-atC)\{\mathbf{g} - C^{-1}\mathbf{c}_t\}, \tag{4.189}$$

where $\mathbf{g} = [g(h), g(2h), \ldots, g(Nh)]^T$ is the vector of initial conditions. It is easy to show that this vector $\mathbf{U}(t)$ satisfies

$$\mathbf{U}(t+l) = C^{-1}\mathbf{c}_t + \exp(-alC)\{\mathbf{U}(t) - C^{-1}\mathbf{c}_t\}. \tag{4.190}$$

Using the (0, 1) Padé approximant to the exponential matrix function in (4.190) leads to

$$\mathbf{U}(t+l) = C^{-1}\mathbf{c}_t + (I - alC)\{\mathbf{U}(t) - C^{-1}\mathbf{c}_t\} + O(l^2), \tag{4.191}$$

where I is the identity matrix of order N. Applying (4.191) to the mesh point (mh, nl) in the (x, t) plane ($m = 1, 2, \ldots, N; n = 0, 1, \ldots$), and using U_m^n to denote the theoretical solution of (4.191) at this mesh point, gives the three-point explicit scheme

$$U_m^{n+1} = (1 - ar)U_m^n + arU_{m-1}^n \tag{4.192}$$

where $r = l/h$, as before. For $m = 1, n = 0$, equation (4.192) uses the origin as one of its mesh points so that a discontinuity between initial and boundary conditions will be propagated in the numerical solution yielded by (4.192).

It is instructive at this point to consider the *Courant-Friedrichs-Lewy (CFL) condition* for first order hyperbolic equations. The condition states that *the characteristic through the advanced point of an explicit difference scheme must intersect the line through the points used by the scheme at the base time level within the range of those points.*

From section 4.10, the slope of the characteristic through $(mh, (n+1)l)$ is $1/a$ and it is thus easy to verify geometrically that the CFL condition for (4.192) is $0 \leq ar \leq 1$.

Turning now to the local truncation error and taking

$$F_{m,n}(u) = u_m^{n+1} - (1-ar)u_m^n - aru_{m-1}^n,$$

it is easy to show that the local truncation error of (4.192) is

$$\left(-\frac{1}{2}ah\frac{\partial^2 u}{\partial x^2} + \frac{1}{2}l\frac{\partial^2 u}{\partial t^2} + \ldots\right)_m^n \tag{4.193}$$

from which it is seen that (4.192) is consistent with (4.182). In (4.193), the component $-\frac{1}{2}ah\partial^2 u/\partial x^2$ is due to the finite difference replacement (4.185) and will be included in the local truncation error of each finite difference

Sec. 4.12] Backward Difference Methods for First Order Equations

method derived from (4.190). The component $\frac{1}{2} l \, \partial^2 u/\partial t^2$ is due to the use of the (0, 1) Padé approximant in (4.190); the constant $C_2 = \frac{1}{2}$ in this component is given in Table 4.5.

It is clear that a pertubation $\mathbf{Z}^0$ of the initial conditions satisfies

$$\mathbf{Z}^{n+1} = (I - alC)\mathbf{Z}^n \tag{4.194}$$

from (4.191) and since every eigenvalue of C has the value $1/h$, each eigenvalue of the amplification matrix $I - alC$ has the value $1 - ar$. The stability criterion of (4.192) thus satisfies

$$-1 \leqslant 1 - ar \leqslant 1$$

from which it follows that $0 \leqslant ar \leqslant \frac{1}{2}$. This restriction on ar is, however, too severe; it certainly provides a sufficient condition for stability, but the condition is by no means necessary (see Morton, 1980).

Turning, therefore, to the Fourier method, it will be assumed that $Z_m^n = e^{\alpha nl} e^{i\beta mh}$ (α complex, β real), which represents the difference between the theoretical and actual computed solution of the difference scheme (4.192) satisfies

$$Z_m^{n+1} = (1-ar)Z_m^n + ar \, Z_{m-1}^n \tag{4.195}$$

Substituting for Z_m^{n+1}, Z_m^n, Z_{m-1}^n and dividing by $e^{\alpha nl} e^{i\beta mh}$ gives

$$\xi = 1 - ar + are^{-i\beta h} \tag{4.196}$$

where $\xi = e^{\alpha l}$ is the amplification factor. This leads to

$$\xi = 1 - 2ar \sin^2 \tfrac{1}{2} \beta h + iar \sin \beta h$$

so that $|\xi|^2 = 1 - 4ar(1-ar)\sin^2 \tfrac{1}{2} \beta h$. The reader can then verify that the von Neumann necessary condition for stability, $|\xi| \leqslant 1$, is satisfied for $0 < ar \leqslant 1$ (the possibility $ar = 0$ may be ignored).

This restriction on ar is in agreement with the CFL condition for the method; it may also be obtained by carrying out a convergence analysis of the finite difference scheme along the lines outlined in subsection 4.6.3. The reader has seen the most commonly used illustration of the failure of the matrix method of analyzing stability.

The explicit scheme yielded by the (0, 1) Padé approximant may be extrapolated, as in section 4.11, to improve the component of the local truncation error due to the Padé approximant. The extrapolated form of (4.191) is obtained from (4.179) and is given in Table 4.5. The local truncation error (4.193) is improved to

$$\left(-\frac{1}{2} a h \frac{\partial^2 u}{\partial x^2} + \frac{4}{3} l^2 \frac{\partial^3 u}{\partial t^3} + \ldots \right)_m^n . \tag{4.197}$$

Using the (1, 0) Padé approximant to the exponential matrix function in (4.190) gives

$$\mathbf{U}(t+l) = C^{-1}\mathbf{c}_t + (I+alc)^{-1}\{\mathbf{U}(t) - C^{-1}\mathbf{c}_t\} + 0(l^2). \quad (4.198)$$

Premultiplying all terms in (4.198) by $(I+alC)$ and writing the term alc_t on the right-hand side as alc_{t+l} on the left-hand side, gives the implicit scheme

$$(I+alC)\mathbf{U}(t+l) - alc_{t+l} = \mathbf{U}(t). \quad (4.199)$$

Applying (4.199) to the mesh point (mh, nl) gives the three-point formula

$$(1+ar)U_m^{n+1} - arU_{m-1}^{n+1} = U_m^n. \quad (4.200)$$

Equation (4.200) uses two points at the advanced time level $t+l$ and would, therefore, be normally classed as implicit. The reader will remember, however, that the boundary condition associated with the differential equation being solved is given along the axis $t = 0$, so that, for any n, applying (4.200) to the mesh point (h, nl) means that the mesh points (h, nl), $(0, (n+1)l)$, $(h,(n+1)l)$ are used. The only unknown in the application of (4.200) to (h, nl) is, therefore, U_1^{n+1}; this means that (4.200) is bing used explicitly after all. Obviously, for all other $m = 2, 3, \ldots, N$ the scheme is also used explicitly. The CFL condition may not, however, be used in analyzing the scheme. An attractive feature of the method is that any discontinuities between initial and boundary conditions at the origin are ignored by the method.

The local truncation error of the scheme may be shown to be

$$\left(-\frac{1}{2}ah\frac{\partial^2 u}{\partial x^2} - \frac{1}{2}l\frac{\partial^2 u}{\partial t^2} + \ldots\right)_m^n \quad (4.201)$$

from which the finite difference method is seen to be consistent.

The matrix method may be used with confidence, this time, in examining stability. A perturbation $\mathbf{Z}^0$ of the vector of initial conditions satisfies

$$\mathbf{Z}^{n+1} = (I+alC)^{-1}\mathbf{Z}^n \quad (4.202)$$

and, since the eigenvalues of C are all $1/h$, the eigenvalues of the amplification matrix $(I+alC)^{-1}$ are all $1/(1+ar)$, and thus never exceed unity for non-negative values of ar. The initial perturbation is therefore damped as t increases, from which it is concluded that the scheme is unconditionally stable. The reader can quickly verify that the scheme is also convergent for any value of $ar > 0$.

The finite difference scheme (4.200) has an obvious advantage over (4.192) in that it is unconditionally stable, although both have similar local truncation errors. Moreover (4.200) allows large time steps to be used for a given value of h, for then the amplification factor $1/(1+ar)$ decreases and any perturbation of the initial conditions is quickly damped.

Sec. 4.12] Backward Difference Methods for First Order Equations

The scheme may be extrapolated in the now familiar way, the extrapolation formula being derived from (4.179) and appearing in Table 4.5. The local truncation error of the extrapolated form is

$$\left(-\frac{1}{2}ah\frac{\partial^2 u}{\partial x^2} + \frac{4}{3}l^2\frac{\partial^3 u}{\partial t^3} + \ldots\right)_m^n. \tag{4.203}$$

Using the (1, 1) Padé approximant to $\exp(-alC)$ in (4.190) gives

$$\mathbf{U}(t+l) = C^{-1}\mathbf{c}_t + (I + \frac{1}{2}alC)^{-1}(I - \frac{1}{2}alC)\{\mathbf{U}(t) - C^{-1}\mathbf{c}_t\} + 0(l^3).$$

Premultiplying all terms by the matrix $(I+\frac{1}{2}alC)$ and splitting the term $al\mathbf{c}_t$ into $\frac{1}{2}al\mathbf{c}_{t+l} + \frac{1}{2}al\mathbf{c}_t$, leads to

$$(I + \frac{1}{2}alC)\mathbf{U}(t+l) - \frac{1}{2}al\mathbf{c}_{t+l} = (I - \frac{1}{2}alC)\mathbf{U}(t) + \frac{1}{2}al\mathbf{c}_t \tag{4.204}$$

which, when applied to the mesh point (mh, nl) gives the four point scheme

$$(1 + \frac{1}{2}ar)U_m^{n+1} - \frac{1}{2}arU_{m-1}^{n+1} = (1 - \frac{1}{2}ar)U_m^n + \frac{1}{2}arU_{m-1}^n \tag{4.205}$$

(Twizell, 1980). Like (4.200), this scheme may also be applied explicitly in the first quadrant of the (x, t) plane.

The local truncation error of the scheme may be written down with the aid of Table 4.5; it is

$$\left(-\frac{1}{2}ah\frac{\partial^2 u}{\partial x^2} - \frac{1}{12}l^2\frac{\partial^3 u}{\partial t^3} + \ldots\right)_m^n \tag{4.206}$$

from which it is seen that the scheme is consistent and, of course, more accurate than (4.200).

The numerical stability of the scheme is established by the matrix method which shows that a perturbation $\mathbf{Z}^0$ of the vector of initial conditions satisfying

$$\mathbf{Z}^{(n+1)} = (I + \frac{1}{2}alC)^{-1}(I - \frac{1}{2}alC)\mathbf{Z}^n \tag{4.207}$$

does not increase in magnitude as time increases. This is verified by noting that the eigenvalues $(1-\frac{1}{2}ar)/(1+\frac{1}{2}ar)$ of the amplification matrix $(I+\frac{1}{2}alC)^{-1}(I-\frac{1}{2}alC)$ are less than $+1$ for small positive values of ar and greater than -1 for large positive values of ar. An initial perturbation $\mathbf{Z}^0$ will not grow as time increases, but for $ar > 2$ such a perturbation will oscillate about $\mathbf{0}$. It is easy to verify that this method, too, is convergent for all $ar > 0$ (discounting the possibility $ar = 0$).

Extrapolation of the scheme gives two more powers of accuracy in time. The extrapolating algorithm is obtained from (4.180) and is given in Table 4.5. The local truncation error of the extrapolated form is

$$\left(-\frac{1}{2}ah\frac{\partial^2 u}{\partial x^2} + \frac{1}{10}l^4\frac{\partial^5 u}{\partial t^5} + \ldots\right)^n_m. \tag{4.208}$$

4.12.2 High order backward difference methods

It was seen in subsection 4.12.1, that extrapolation in time does, indeed, bring about an improvement in the principal parts of the local truncation errors of the finite difference schemes, which are based on the $(0,1), (1,0), (1,1)$ Padé approximants to the exponential matrix function in the recurrence relation (4.190).

Further improvement in the component of the local truncation error due to the chosen Padé approximant can be obtained by replacing $\exp(-alC)$ in (4.190) with its $(0,2), (1,2), (2,2), (2,1)$ or $(2,0)$ Padé approximant. The components, the extrapolated forms of the five algorithms, and the improved components of the local truncation errors are contained in Table 4.5. Stability analyses show, however, that the method based on the $(0, 2)$ Padé approximant is stable only for $0 < ar \leq 2$ and that the method based on the $(1, 2)$ Padé approximant is stable only for $0 < ar \leq \sqrt{6}$; the three methods based on the $(2,0) (2,1), (2,2)$ approximants to $\exp(-alC)$ are all unconditionally stable.

The improvements in accuracy brought about by these methods may not, unfortunately, be sufficient to justify their use for larger values of h. This is because the component of the local truncation error due to approximating $\partial u/\partial x$ in (4.182), with the first order replacement (4.185), is still present, and tends to overshadow any improvement brought about by extrapolating in time.

To a certain extent, this difficulty can be overridden by introducing a second order backward difference approximant to $\partial u/\partial x$ at the mesh points (mh, nl) for $m = 2, 3, \ldots, N$ and $n = 0, 1, \ldots$, whilst retaining the first order approximant (4.185) at the points (h, nl) adjacent to the boundary. This mixture of first and second order approximants to $\partial u/\partial x$ was used successfully by Oliger (1974) and justified in the convergence theorems of Gustafsson (1975). The resulting difference schemes can all be used explicitly, though some enjoy the stability properties of implicit schemes, as in subsection 4.12.1. The only irritation in their implementation is the use of a different formula for $m = 1$, as is now seen.

Consider the second order replacement

$$\frac{\partial u}{\partial x} = \{u(x-2h, t) - 4u(x-h, t) + 3u(x, t)\}/2h + O(h^2) \tag{4.209}$$

to the space derivative in the differential equation (4.182). This replacement uses three mesh points at any time $t = nl$, so that any attempt to use it at the mesh points where $x = h$, requires the value of u to be given at the points

Sec. 4.12] Backward Difference Methods for First Order Equations

$(-h, nl)$ in the second quadrant of the (x, t) plane. Of course, such values of u are not available and, in effect, (4.209) can only be used at the mesh points (mh, nl) where $m = 2, \ldots, N$ for any $n = 0, 1, \ldots$. At the mesh points (h, nl), equation (4.185), written for convenience as

$$\frac{\partial u}{\partial x} = \{2u(x, t) - 2u(x-h, t)\}/2h + O(h), \tag{4.210}$$

is retained.

Applying (4.182) with (4.210) or (4.209), as appropriate, to the N mesh points at time level $t = nl$, leads to the system of first order ordinary differential equations

$$\frac{d\mathbf{U}(t)}{dt} = -\frac{1}{2}aD\mathbf{U}(t) + \frac{1}{2}a\mathbf{d}_t. \tag{4.211}$$

In (4.211) D is the matrix given by

$$hD = \begin{bmatrix} 2 & & & & \\ -4 & 3 & & & 0 \\ 1 & -4 & 3 & & \\ & \ddots & \ddots & \ddots & \\ 0 & & 1 & -4 & 3 \end{bmatrix} \tag{4.212}$$

and $\mathbf{d}_t$ is the N-component vector given by

$$h\mathbf{d}_t = [2v_t, -v_t, 0, \ldots, 0]^T, \tag{4.213}$$

where v_t is, as previously, the numerical (frozen) value of the boundary condition at time t, and T denotes transpose. One eigenvalue of the matrix D has the value $2/h$ and the other $N-1$ eigenvalues have the value $3/h$.

The solution of (4.211) with (4.183) is

$$\mathbf{U}(t) = D^{-1}\mathbf{d}_t + \exp(-\frac{1}{2}atD)\{\mathbf{g} - D^{-1}\mathbf{d}_t\}, \tag{4.214}$$

where $\mathbf{g}$ is the vector of initial conditions. It may be shown that (4.214) satisfies the recurrence relation

$$\mathbf{U}(t+l) = D^{-1}\mathbf{d}_t + \exp(-\frac{1}{2}alD)\{\mathbf{U}(t) - D^{-1}\mathbf{d}_t\} \tag{4.215}$$

and replacing the exponential matrix function with its (0, 1) Padé approximant leads to

$$\mathbf{U}(t+l) = (I - \frac{1}{2}alD)\mathbf{U}(t) + \frac{1}{2}al\mathbf{d}_t + O(l^2). \tag{4.216}$$

Applying (4.216) to the mesh point (mh, nl) gives

$$U_1^{n+1} = (1-ar)U_1^n + arv_t, \qquad (4.217)$$

$$U_m^{n+1} = -\frac{1}{2}arU_{m-2}^n + 2arU_{m-1}^n + (1-\frac{3}{2}ar)U_m^n; \quad m \neq 1 \qquad (4.218)$$

recalling that, for $m = 2$, the term U_{m-2}^n is a boundary condition and has the value v_t for $n \neq 0$ or $g(0)$ for $n = 0$.

Equation (4.217) is identical with (4.192) applied to the mesh point (h, nl) so that its local truncation error is given by (4.193). The component $-\frac{1}{2} ah \partial^2 u/\partial x^2$ of (4.193) will therefore be present in the local truncation error of each point (h, nl) adjacent to the boundary $x = 0$ of any finite difference scheme yielded by (4.215).

The local truncation error of (4.218) is given by

$$\left(\frac{1}{2} l \frac{\partial^2 u}{\partial t^2} - \frac{1}{3} a h^2 \frac{\partial^3 u}{\partial x^3} + \ldots \right)_m^n ; \qquad (4.219)$$

the component $\frac{1}{2} l \partial^2 u/\partial t^2$ is due to using the $(0, 1)$ Padé approximant in (4.215), while the component $-\frac{1}{3} ah^2 \partial^3 u/\partial x^3$ is due to replacing the space derivative $\partial u/\partial x$ in the differential equation by the second order formula (4.209). The second term in (4.219) will occur in the local truncation error of any finite difference scheme, originating from (4.215), applied to the general mesh point (mh, nl) for $m = 2, \ldots, N$.

Equations (4.217), (4.218) must not be investigated individually for stability; instead the *whole* scheme (4.216) must be analyzed. This is done in the usual way be determining the criterion under which a perturbation Z^0 in the initial data does not grow as time increases. The behaviour of such a perturbation is governed as time increases by the relation

$$Z^{n+1} = (I - \frac{1}{2} alD)Z^n \qquad (4.220)$$

and a necessary condition for stability may be found by first determining the eigenvalues of the amplification matrix $I-alD$ (these are $1 - ar$ and $1 - \frac{3}{2} ar$) and, in the usual way, using that eigenvalue which gives the smallest stability interval. It is easy to show that this interval is obtained from $-1 \leqslant -\frac{3}{2} ar \leqslant 1$, leading to $0 < ar \leqslant \frac{4}{3}$ for stability. The reader can verify that convergence also occurs for this interval.

In view of the conditional stability of the method based on the $(0, 1)$ Padé approximant, and on the fact that the method propagates discontinuities between initial and boundary conditions at the origin, its usefulness is limited.

Sec. 4.12] Backward Difference Methods for First Order Equations

The same order of accuracy as the method based on the (0, 1) Padé approximant can be attained by using the (1, 0) approximant in (4.215) to give

$$\mathbf{U}(t+l) = D^{-1}\mathbf{d}_t + (I + \frac{1}{2}alD)^{-1}\{\mathbf{U}(t) - D^{-1}\mathbf{d}_t\} + 0(l^2). \quad (4.221)$$

Premultiplying (4.221) by the matrix $(I + \frac{1}{2} alD)$, and writing the term $\frac{1}{2} al\mathbf{d}_t$ on the right-hand side of the resulting equation as $-\frac{1}{2} al\mathbf{d}_{t+l}$ on the left-hand side gives

$$(I + \frac{1}{2}alD)\mathbf{U}(t+l) - \frac{1}{2}al\mathbf{d}_{t+l} = \mathbf{U}(t), \quad (4.222)$$

which is first order accurate in time.

Applying (4.222) to the mesh point (mh, nl) gives

$$(1+ar)U_1^{n+1} - arv_{t+l} = U_1^n, \quad (4.223)$$

$$(1 + \frac{3}{2}ar)U_2^{n+1} - 2ar\, U_1^{n+1} + \frac{1}{2}arv_{t+l} = U_2^n, \quad (4.224)$$

$$(1 + \frac{3}{2}ar)U_m^{n+1} - 2ar\, U_{m-1}^{n+1} + \frac{1}{2}arU_{m-2}^{n+1} = U_m^n;\ m \neq 1, 2. \quad (4.225)$$

It is thus seen that, although (4.222) is implicit in nature, it can be applied explicitly, U_1^{n+1} being the only unknown in (4.223), U_2^{n+1} therefore being the only unknown in (4.224), and U_m^{n+1} ($m = 3, 4, \ldots, N$) being the only unknown in the $N-2$ successive applications of (4.225). Note that the origin, where there may be a discontinuity between initial and boundary conditions, is not a mesh point.

The local truncation error of (4.223) is the same as that of (4.200) and is given by (4.201); the local truncation error of each of (4.254), (4.225) is easily written down by referring to Table 4.5. It is

$$\left(-\frac{1}{3} a h^2 \frac{\partial^3 u}{\partial x^3} - \frac{1}{2} l \frac{\partial^2 u}{\partial t^2} + \cdots \right)_m^n \quad (4.276)$$

and the finite difference algorithm arising from the (1, 0) Padé approximant is seen to be consistent with (4.182).

The stability of (4.222) is easily verified in the, by now, familiar way. The amplification matrix is $(I+\frac{1}{2}alD)^{-1}$, which has one eigenvalue $1/(1+ar)$ and $N-1$ eigenvalues equal to $1/(1+\frac{3}{2} ar)$. It is an easy exercise to deduce that (4.222) yields an unconditionally stable algorithm and, further, to verify convergence.

In view of its favourable stability properties, and that it is used explicitly, it is worthwhile to extrapolate (4.222). The extrapolated formula, derived from (4.179) and given in Table 4.5, has local truncation error given by (4.203) for $m = 1$, and

$$\left(-\frac{1}{3}ah^2\frac{\partial^3 u}{\partial x^3} + \frac{4}{3}l^2\frac{\partial^3 u}{\partial t^3} + \dots\right)_m^n; \quad m \neq 1. \tag{4.227}$$

Further improvement in accuracy may be achieved by using the (1, 1) Padé approximant to the matrix exponential function in (4.215) to give

$$\mathbf{U}(t+l) = D^{-1}\mathbf{d}_t + (I + \tfrac{1}{4}alD)^{-1}(I - \tfrac{1}{4}alD)\{\mathbf{U}(t) - D^{-1}\mathbf{d}_t\} + 0(l^3). \tag{4.228}$$

Premultiplying (4.228) by $(I+\tfrac{1}{4} alD)$ and taking one of the two $\tfrac{1}{4} al\mathbf{d}_t$ terms on the right-hand side of the resulting equation, to its left-hand side as $-\tfrac{1}{4} al\mathbf{d}_{t+l}$, gives

$$(I + \tfrac{1}{4}alD)\mathbf{U}(t+l) - \tfrac{1}{4}al\mathbf{d}_{t+l} = (I - \tfrac{1}{4}alD)\mathbf{U}(t) + \tfrac{1}{4}al\mathbf{d}_t \tag{4.229}$$

which is second order accurate in time.

Applying (4.229) to the mesh point (h, nl) gives

$$(1 + \tfrac{1}{2}ar)U_1^1 - \tfrac{1}{2}arv_l = (1 - \tfrac{1}{2}ar)U_1^0 + \tfrac{1}{2}arg(0); n = 0, \tag{4.230}$$

$$(1 + \tfrac{1}{2}ar)U_1^{n+1} - \tfrac{1}{2}arv_{t+l} = (1 - \tfrac{1}{2}ar)U_1^n + \tfrac{1}{2}arv_t; \quad n \neq 0. \tag{4.231}$$

Applying (4.229) to the mesh point $(2h, nl)$ gives

$$(1 + \tfrac{3}{4}ar)U_2^{n+1} - arU_1^{n+1} + \tfrac{1}{4}arv_{t+l} = (1 - \tfrac{3}{4}ar)U_2^n + arU_1^n - \tfrac{1}{4}arv_t, \tag{4.232}$$

and to the mesh points (mh, nl), $m = 3, \dots, N$ gives

$$(1 + \tfrac{3}{4}ar)U_m^{n+1} - arU_m^{n+1} + \tfrac{1}{4}arU_{m-2}^{n+1} = (1 - \tfrac{3}{4}ar)U_m^n + arU_{m-1}^n - \tfrac{1}{4}arU_{m-2}^n. \tag{4.233}$$

Equations (4.230) and (4.231) with $n = 0$ are the same equation if $g(0) = v_0 = v(0)$, that is, if there is no discontinuity between initial and boundary condition at the origin. If there is such a discontinuity, however, it is clear that it will be propagated in the numerical solution.

Sec. 4.12] Backward Difference Methods for First Order Equations

Clearly, there is only one unknown in (4.230) which may be solved to give U_1^1 explicitly; also, there is only one unknown in (4.231) which may be solved explicitly to give U_1^{n+1} ($n \neq 0$). Whatever the value of n, equation (4.232) then has only one unknown, U_2^{n+1}, which may be determined explicitly. Finally, for any n, (4.233) is explicit also; its unknown, U_m^{n+1}, being determined explicitly for any $m = 3, 4, \ldots, N$. Overall, then, the method based on the (1, 1) Padé approximant may be applied explicitly even though in its vector form (4.229) it is implicit in nature.

The local truncation errors of (4.230), (4.231) are the same as (4.205) and are given by (4.206). The local truncation errors of (4.232), (4.233) are both

$$\left(-\frac{1}{3} a h^2 \frac{\partial^3 u}{\partial x^3} - \frac{1}{12} l^2 \frac{\partial^3 u}{\partial t^3} + \ldots \right)_m^n. \qquad (4.234)$$

The stability of the method is examined by considering the eigenvalue of the amplification matrix $(I+\frac{1}{4} alD)^{-1} (I-\frac{1}{4} alD)$. This matrix has one eigenvalue equal to $(1-\frac{1}{2} ar)/(1+\frac{1}{2} ar)$ and $N - 1$ eigenvalues equal to $(1-\frac{3}{4} ar)/(1+\frac{3}{4} ar)$ neither of which exceeds unity in modulus for any non-negative value of ar. It must be noted, however, that, for $ar > 4/3$, a perturbation $\mathbf{Z}^0$ of the initial conditions will oscillate about $\mathbf{0}$. The convergence of the method can easily be verified.

Looking at the local truncation error (4.234), it may reasonably be concluded that, for problems without discontinuities between initial and boundary conditions at least, the method based on the (1, 1) Padé approximant is to be preferred to the extrapolated form of the method based on the (1, 0) approximant. Furthermore, the method given by (4.229) can itself be extrapolated to give two extra orders of accuracy in time. This is done by referring to equation (4.180) and by consulting Table 4.5; the local truncation error of the extrapolated form of the method is seen to be

$$\left(-\frac{1}{3} a h^2 \frac{\partial^3 u}{\partial x^3} + \frac{1}{10} l^4 \frac{\partial^5 u}{\partial x^5} + \ldots \right)_m^n. \qquad (4.235)$$

The advantages in choosing a higher order Padé approximant to the matrix exponential function, when a reasonably high order approximant to the space derivative in the differential equation has been used, should now be evident to the reader. It should be equally apparent that further methods derived from the recurrence relation (4.215) will be explicit in nature or may be applied explicitly even though they are implicit in nature. It is reasonable, therefore, to discard methods which are not unconditionally stable. Two such methods are those based on the (0, 2) (1, 2) Padé approximants which have finite stability intervals.

Khaliq (1983) and Twizell and Khaliq (1984) have developed and tested the algorithms based on the (2, 0), (2, 1), (2, 2) Padé approximants to exp

188   **Hyperbolic Equations**   [Ch. 4

$(-\frac{1}{2} alD)$ in (4.215). These algorithms give very accurate numerical results for large values of ar and small values of h. Using, first of all, the $(2, 0)$ Padé approximant in (4.215), see Table 4.3, gives

$$\mathbf{U}(t+l) = D^{-1}\mathbf{d}_t + (I + \frac{1}{2}alD + \frac{1}{8}a^2l^2D^2)^{-1}\{\mathbf{U}(t) - D^{-1}\mathbf{d}_t\} + 0(l^3). \quad (4.236)$$

Premultiplying (4.236) by $(I+\frac{1}{2} alD + \frac{1}{8} a^2 l^2 D^2)$, then taking the term $(\frac{1}{2} alI + \frac{1}{8} a^2 l^2 D)\mathbf{d}_t$ on the right side to the left-hand side and writing it as $-(\frac{1}{2} alI + \frac{1}{8} a^2 l^2 D)\mathbf{d}_{t+l}$, gives

$$(I + \frac{1}{2}alD + \frac{1}{8}a^2l^2D^2)\mathbf{U}(t+l) - (\frac{1}{2}alI + \frac{1}{8}a^2l^2D)\mathbf{d}_{t+l} = \mathbf{U}(t). \quad (4.237)$$

The matrix D^2 is given by

$$h^2 D^2 = \begin{bmatrix} 4 & & & & & & 0 \\ -20 & 9 & & & & & \\ 21 & -24 & 9 & & & & \\ -8 & 22 & -24 & 9 & & & \\ 1 & -8 & 22 & -24 & 9 & & \\ & \ddots & \ddots & \ddots & \ddots & \ddots & \\ 0 & 1 & -8 & 22 & -24 & 9 \end{bmatrix} \quad (4.238)$$

and has one eigenvalue equal to $4/h^2$ and $N - 1$ eigenvalues equal to $9/h^2$. Applying (4.237) to the mesh points (mh, nl) leads to a linear system which may be written in matrix form as

$$E\mathbf{U}(t+l) = \boldsymbol{\phi}^n. \quad (4.239)$$

The matrix E is of order N and is of the lower triangular form

$$E = \begin{bmatrix} e_6 & & & & & & 0 \\ e_7 & e_1 & & & & & \\ e_8 & e_2 & e_1 & & & & \\ e_4 & e_3 & e_2 & e_1 & & & \\ e_5 & e_4 & e_3 & e_2 & e_1 & & \\ 0 & \ddots & \ddots & \ddots & \ddots & \ddots & \\ & & e_5 & e_4 & e_3 & e_2 & e_1 \end{bmatrix}, \quad (4.240)$$

where

$$e_1 = 1 + \frac{3}{2}ar + \frac{9}{8}a^2r^2, \quad e_2 = -2ar - 3a^2r^2, \quad e_3 = \frac{1}{2}ar + \frac{11}{4}a^2r^2,$$

Sec. 4.12] Backward Difference Methods for First Order Equations

$$e_4 = -a^2 r^2, \quad e_5 = \frac{1}{8}a^2 r^2, \quad e_6 = 1 + ar + \frac{1}{2}a^2 r^2,$$

$$e_7 = -2ar - \frac{5}{2}a^2 r^2, \quad e_8 = \frac{1}{2}ar + \frac{21}{8}a^2 r^2, \quad (4.241)$$

and the vector $\boldsymbol{\phi}^n = (\phi_1^n, \phi_2^n, \ldots, \phi_N^n)^T$ has elements

$$\phi_1^n = U_1^n + ar(1 + \frac{1}{2}ar)v_{t+l}, \quad \phi_2^n = U_2^n - ar(\frac{1}{2} + \frac{11}{8}ar)v_{t+l},$$

$$\phi_3^n = U_3^n + \frac{3}{4}a^2 r^2 v_{t+l}, \quad \phi_4^n = U_4^n - \frac{1}{8}a^2 r^2 v_{t+l},$$

$$\phi_m^n = U_m^n \quad (m = 5, 6, \ldots, N). \quad (4.242)$$

The value of U at the origin is not used by this six-point finite difference scheme.

Clearly the vector $\mathbf{U}(t+l)$ is obtained explicitly from (4.239) by the method of forward substitution; that is, $U_1^{n+1} = \phi_1^n/e_6$ is determined from the first equation of the system, then U_2^{n+1} is obtained from the second equation of the system, and so on.

For $m = 4, \ldots, N$ the local truncation error is

$$\left(-\frac{1}{3} a h^2 \frac{\partial^3 u}{\partial x^3} + \frac{1}{6} l^2 \frac{\partial^3 u}{\partial t^3} + \ldots \right)_m^n, \quad (4.243)$$

but for $m = 1, 2, 3$ this accuracy is not achieved. The coefficient $C_3 = \frac{1}{6}$ in (4.243) is greater in modulus than the corresponding term in (4.234) relating to the (1, 1) Padé approximant. It may reasonably be concluded therefore that, for problems without a discontinuity between initial and boundary conditions at the origin, the method based on the (1, 1) Padé approximant is to be preferred to that based on the (2, 0) approximant, particularly in view of the loss of accuracy by the method based on the (1, 1) approximant at only one point adjacent to the boundary compared with three points by the method based on the (2, 0) approximant. Where discontinuities are present, however, this observation is unlikely to be true.

The stability of (4.237) is easily established by examining, in the usual way, the propagation of a perturbation $\mathbf{Z}^0$ in the vector of initial conditions. Such a perturbation satisfies

$$\mathbf{Z}^{(n+1)} = (I + \frac{1}{2} alD + \frac{1}{8} a^2 l^2 D^2)^{-1} \mathbf{Z}^{(n)} \quad (4.244)$$

and it is easy to show that one eigenvalue of the amplification matrix $(I + \frac{1}{2} alD + \frac{1}{8} a^2 l^2 D^2)^{-1}$ is $1/(1 + ar + \frac{1}{2} a^2 r^2)$ while the others are all $1/(1 + \frac{3}{2} ar + \frac{9}{8} a^2 r^2)$.

All N eigenvalues are thus between zero and unity for positive values of ar and the scheme given in vector form by (4.237) is unconditionally stable.

Assuming that the user has decided in favour of the method based on the (2, 0) Padé approximant for a problem with discontinuities between initial and boundary conditions, the method may be extrapolated, as given in Table 4.5, using formula (4.179), after which the local truncation error for $m = 4,\ldots,N$ becomes

$$\left(-\frac{1}{3}a\,h^2\,\frac{\partial^3 u}{\partial x^3} - \frac{1}{3}l^3\,\frac{\partial^4 u}{\partial t^4} + \ldots\right)^n_m. \tag{4.245}$$

Turning next to the (2, 1) Padé approximant to the matrix exponential function, given in Table 4.3, the recurrence relation (4.215) becomes

$$\mathbf{U}(t+l) = D^{-1}\mathbf{d}_t + (I + \frac{1}{3}alD + \frac{1}{24}a^2 l^2 D^2)^{-1}(I - \frac{1}{6}alD)\{\mathbf{U}(t) - D^{-1}\mathbf{d}_t\} + O(l^4).$$

After premultiplying by $(I + \frac{1}{3}alD + \frac{1}{24}a^2 l^2 D^2)$ the resulting equation may be written in the from

$$(I + \frac{1}{3}alD + \frac{1}{24}a^2 l^2 D^2)\mathbf{U}(t+l) - (\frac{1}{3}alI + \frac{1}{24}a^2 l^2 D)\mathbf{d}_{t+l}$$

$$= (I - \frac{1}{6}alD)\mathbf{U}(t) + \frac{1}{6}al\mathbf{d}_t. \tag{4.246}$$

Applying (4.246) to the mesh points (mh, nl) leads to a linear system of the form (4.239) so that the elements of $\mathbf{U}(t+l)$ may, again, be found explicitly by forward substitution. The elements of the lower triangular matrix E are now

$$e_1 = 1 + ar + \frac{3}{8}a^2 r^2,\quad e_2 = -\frac{4}{3}ar - a^2 r^2,\quad e_3 = \frac{1}{3}ar + \frac{11}{12}a^2 r^2,$$

$$e_4 = -\frac{1}{3}a^2 r^2,\quad e_5 = \frac{1}{24}a^2 r^2,\quad e_6 = 1 + \frac{2}{3}ar + \frac{1}{6}a^2 r^2,$$

$$e_7 = -\frac{4}{3}ar - \frac{5}{6}a^2 r^2,\quad e_8 = \frac{1}{3}ar + \frac{7}{8}a^2 r^2, \tag{4.247}$$

while the elements of the vector $\boldsymbol{\phi}^n$ are

$$\phi_1^n = (1 - \frac{1}{3}ar)U_1^n + ar(\frac{2}{3} + \frac{1}{6}ar)v_{t+l} + \frac{1}{3}arv_t;\quad n \neq 0,$$

$$\phi_2^n = \frac{2}{3}arU_1^n + (1 - \frac{1}{2}ar)U_2^n - ar(\frac{1}{3} - \frac{11}{24}ar)v_{t+l} - \frac{1}{6}arv_t;\quad n \neq 0$$

Sec. 4.12] Backward Difference Methods for First Order Equations

$$\phi_3^n = -\frac{1}{6}arU_1^n + \frac{2}{3}arU_2^n + (1 - \frac{1}{2}ar)U_3^n + \frac{1}{4}a^2r^2v_{t+l},$$

$$\phi_4^n = -\frac{1}{6}arU_2^n + \frac{2}{3}arU_3^n + (1 - \frac{1}{2}ar)U_4^n - \frac{1}{24}a^2r^2v_{t+l},$$

$$\phi_m^n = -\frac{1}{6}arU_{m-2}^n + \frac{2}{3}arU_{m-1}^n + (1 - \frac{1}{2}ar)U_m^n \quad (m = 5, \ldots, N). \quad (4.248)$$

For $n = 0$, changes to ϕ_1^n, ϕ_2^n in (4.248) are needed; these arise because the boundary condition (4.184) applies for $t > 0$ while the initial condition (4.183) applies at $x = 0, t = 0$. The terms become

$$\phi_1^0 = (1 - \frac{1}{3}ar)U_1^0 + ar(\frac{2}{3} + \frac{1}{6}ar)v_l + \frac{1}{3}\mathrm{arg}(0),$$

$$\phi_2^0 = \frac{2}{3}arU_1^0 + (1 - \frac{1}{2}ar)U_2^0 - ar(\frac{1}{3} + \frac{11}{24}ar)v_l - \frac{1}{6}\mathrm{arg}(0). \quad (4.249)$$

For $m = 4, \ldots, N$ the local truncation error is

$$\left(-\frac{1}{3}ah^2\frac{\partial^3 u}{\partial x^3} + \frac{1}{72}l^3\frac{\partial^4 u}{\partial t^4} + \ldots\right)_m^n, \quad (4.250)$$

but this accuracy is not achieved for $m = 1, 2, 3$. The need to introduce special formulations for ϕ_1^0, ϕ_2^0, given by (4.249), shows that the origin is used in the implementation of the eight-point scheme (4.246), so that a discontinuity between initial and boundary conditions at this point will be propagated by the scheme, and the accuracy of the scheme may not be achieved at further mesh points (mh, nl). This is an example of the difficulties encountered by increasing the order from an even order of accuracy, formula (4.237), to one higher order by increasing the number of points at the base time level $t = nl$, discussed by Abarbanel et al. (1975).

The stability of (4.246) is examined by considering the propagation of the initial error vector $\mathbf{Z}^0$, which satisfies

$$\mathbf{Z}^{n+1} = (I + \frac{1}{3}alD + \frac{1}{24}a^2l^2D^2)^{-1}(I - \frac{1}{6}alD)\mathbf{Z}^n. \quad (4.251)$$

One eigenvalue of the amplification matrix $(I + \frac{1}{3}alD + \frac{1}{24}a^2l^2D^2)^{-1}(I - \frac{1}{6}alD)$ is $(1 - \frac{1}{3}ar)/(1 + \frac{2}{3}ar + \frac{1}{6}a^2r^2)$ and the other $N - 1$ eigenvalues are all $(1 - \frac{1}{2}ar)/(1 + ar + \frac{3}{8}a^2r^2)$; all eigenvalues are thus less than unity in modulus for any positive value of ar, and the scheme is therefore unconditionally stable.

Extrapolation of (4.246) improves the local truncation error for all mesh points $(mh, nl), m = 4, 5, \ldots, 6$, to

$$\left(-\frac{1}{3}ah^2 \frac{\partial^3 u}{\partial x^2} - \frac{8}{945}l^4 \frac{\partial^5 u}{\partial t^5} + \ldots\right)_m^n. \qquad (4.252)$$

The extrapolation procedure arises from equation (4.179) and is given in Table 4.5.

The final method of the family arising from (4.215) to be considered in this subsection is that obtained by replacing the exponential matrix function with its (2, 2) Padé approximant given in Table 4.3. The recurrence relation (4.215) then becomes

$$\mathbf{U}(t+l) = D^{-1}\mathbf{d}_t + (I + \frac{1}{4}alD + \frac{1}{48}a^2 l^2 D^2)^{-1}(I - \frac{1}{4}alD + \frac{1}{48}a^2 l^2 D^2).$$

$$\{\mathbf{U}(t) - D^{-1}\mathbf{d}_t\} + 0(l^5),$$

which, after premultiplying by $(I + \frac{1}{4}alD + \frac{1}{48}a^2 l^2 D^2)$ and rearranging gives

$$(I + \frac{1}{4}alD + \frac{1}{48}a^2 l^2 D^2)\mathbf{U}(t+l) - (\frac{1}{4}alI + \frac{1}{48}a^2 l^2 D)\mathbf{d}_{t+l}$$

$$= (I - \frac{1}{4}alD + \frac{1}{48}a^2 l^2 D^2)\mathbf{U}(t) + (\frac{1}{4}alI - \frac{1}{48}a^2 l^2 D)\mathbf{d}_t \qquad (4.253)$$

which is fourth order accurate in time.

Applying (4.253) to each mesh point (mh, nl) leads to a linear system whose unknowns are the N elements of the vector $\mathbf{U}(t+l)$. This system may be written in matrix form as (4.236) where, now,

$$e_1 = 1 + \frac{3}{4}ar + \frac{3}{16}a^2 r^2, \quad e_2 = -ar - \frac{1}{2}a^2 r^2, \quad e_3 = \frac{1}{4}ar + \frac{11}{24}a^2 r^2,$$

$$e_4 = -\frac{1}{6}a^2 r^2, \quad e_5 = \frac{1}{48}a^2 r^2, \quad e_6 = 1 + \frac{1}{2}ar + \frac{1}{12}a^2 r^2,$$

$$e_7 = -ar - \frac{5}{12}a^2 r^2, \quad e_8 = \frac{1}{4}ar + \frac{7}{16}a^2 r^2 \qquad (4.254)$$

and the elements of the vector $\boldsymbol{\phi}^n$ are

$$\phi_1^n = (1 - \frac{1}{2}ar + \frac{1}{12}a^2 r^2)U_1^n + ar(\frac{1}{2} + \frac{1}{12}ar)v_{t+l} + ar(\frac{1}{2} - \frac{1}{12}ar)v_t; n \neq 0,$$

$$\phi_1^0 = (1 - \frac{1}{12}ar + \frac{1}{12}a^2 r^2)U_1^0 + ar(\frac{1}{2} + \frac{1}{12}ar)v_l + ar(\frac{1}{2} - \frac{1}{12}ar)g(0),$$

Sec. 4.12] Backward Difference Methods for First Order Equations

$$\phi_2^n = ar(1 - \frac{5}{12}ar)U_1^n + (1 - \frac{3}{4}ar + \frac{3}{16}a^2r^2)U_2^n - ar(\frac{1}{4} + \frac{11}{48}ar)v_{t+1}$$

$$- ar(\frac{1}{4} - \frac{11}{48}ar)v_t; n \neq 0,$$

$$\phi_2^0 = ar(1 - \frac{5}{12}ar)U_1^0 + (1 - \frac{3}{4}ar + \frac{3}{16}a^2r^2)U_2^0 - ar(\frac{1}{4} + \frac{11}{48}ar)v_l$$

$$- ar(\frac{1}{4} - \frac{11}{48}ar)g(0),$$

$$\phi_3^n = ar(-\frac{1}{4} + \frac{7}{16}ar)U_1^n + ar(1 - \frac{1}{2}ar)U_2^n + (1 - \frac{3}{4}ar + \frac{3}{16}a^2r^2)U_3^n$$

$$+ \frac{1}{8}a^2r^2 v_{t+1} - \frac{1}{8}a^2r^2 v_t; n \neq 0,$$

$$\phi_3^0 = ar(-\frac{1}{4} + \frac{7}{16}ar)U_1^0 + ar(1 - \frac{1}{2}ar)U_2^0 + (1 - \frac{3}{4}ar + \frac{3}{16}a^2r^2)U_3^0$$

$$+ \frac{1}{8}a^2r^2 v_l - \frac{1}{8}a^2r^2 g(0),$$

$$\phi_4^n = -\frac{1}{6}a^2r^2 U_1^n + ar(-\frac{1}{4} + \frac{11}{24}ar)U_2^n + ar(1 - \frac{1}{2}ar)U_3^n$$

$$+ (1 - \frac{3}{4}ar + \frac{3}{16}a^2r^2)U_4^n - \frac{1}{48}a^2r^2 v_{t+1} + \frac{1}{48}a^2r^2 v_t; n \neq 0,$$

$$\phi_4^0 = -\frac{1}{6}a^2r^2 U_1^0 + ar(-\frac{1}{4} + \frac{11}{24}ar)U_2^0 + ar(1 - \frac{1}{2}ar)U_3^0$$

$$+ (1 - \frac{3}{4}ar + \frac{3}{16}a^2r^2)U_4^0 - \frac{1}{48}a^2r^2 v_l + \frac{1}{48}a^2r^2 g(0),$$

$$\phi_m^n = \frac{1}{48}a^2r^2 U_{m-4}^n - \frac{1}{6}a^2r^2 U_{m-3}^n + ar(-\frac{1}{4} + \frac{11}{48}ar)U_{m-2}^n$$

$$+ ar(1 - \frac{1}{2}ar)U_{m-1}^n + (1 - \frac{3}{4}ar + \frac{3}{16}a^2r^2)U_m^n; m = 5,\ldots,N.$$

(4.255)

Like (4.237) and (4.246), the ten-point scheme yielded by equation (4.253) is implicit in nature but may be used explicitly, the vector $\mathbf{U}(t+l)$ being determined using forward substitution to solve (4.236) with (4.254), (4.255). The need to give separate expressions for ϕ_m^0 ($m = 1, 2, 3, 4$) in (4.255) arises because of the use of the origin as a mesh point in the implementation of (4.253) with $n = 0$. As with the methods based on the (1, 1), (2, 1) Padé approximants, a discontinuity between initial and boundary conditions at the origin may lead to considerable loss of accuracy in U near the boundary $t = 0$.

The local truncation error of (4.253) for $m = 4, \ldots, N$ is

$$\left(-\frac{1}{3} a h^2 \frac{\partial^3 u}{\partial x^3} + \frac{1}{720} l^4 \frac{\partial^5 u}{\partial t^5} + \ldots \right)_m^n, \quad (4.256)$$

with less accuracy being achieved for $m = 1, 2, 3$.

The stability of the method is easily established by showing that no criterion need be placed on ar. An initial error vector $\mathbf{Z}^0$ satisfies the equation

$$\mathbf{Z}^{n+1} = (I + \frac{1}{4} alD + \frac{1}{48} a^2 l^2 D^2)^{-1} (I - \frac{1}{4} alD + \frac{1}{48} a^2 l^2 D^2) \mathbf{Z}^n, \quad (4.257)$$

and it is trivial to show that one eigenvalue of the amplification matrix $(I + \frac{1}{4} alD + \frac{1}{48} a^2 l^2 D^2)^{-1} (I - \frac{1}{4} alD + \frac{1}{48} a^2 l^2 D^2)$ is $(1 - \frac{1}{2} ar + \frac{1}{12} a^2 r^2)/(1 + \frac{1}{2} ar + \frac{1}{12} a^2 r^2)$, while the other $N-1$ eigenvalues all have the value $(1 - \frac{3}{4} ar + \frac{3}{16} a^2 r^2)/(1 + \frac{3}{4} ar + \frac{3}{16} a^2 r^2)$. All eigenvalues of the amplification matrix are therefore positive for real, positive values of ar, but none exceed unity in magnitude, and (4.253) is therefore unconditionally stable.

The method may be extrapolated by putting $m = 2$ in (4.180); the extrapolation formula is given in Table 4.5. Two extra powers of accuracy in time are achieved by the extrapolation; the local truncation error is given by

$$\left(-\frac{1}{3} a h^2 \frac{\partial^3 u}{\partial x^3} - \frac{1}{1890} l^6 \frac{\partial^7 u}{\partial t^7} + \ldots \right)_m^n \quad (4.258)$$

except for the three points adjacent to the boundary $t = 0$ where this accuracy may not be achieved.

To examine the behaviour of the higher order (2, 0), (2, 1), (2, 2) schemes developed in this subsection, they are now applied to (4.149) with $a = 1$, that is $\partial u/\partial t + \partial u/\partial x = 0$ with $x > 0, t > 0$. The initial conditions are taken to be

$$g(x) = \sin 2k\pi x; \quad x \geq 0$$

and the boundary conditions to be

$$v(t) = -\sin 2k\pi t; \quad t > 0$$

where k is an integer. The theoretical solution of this problem is

$$u(x, t) = \sin 2k\pi (x - t)$$

Sec. 4.12] Backward Difference Methods for First Order Equations

and the numerical solution will be calculated for x in the interval $0 < x \leq 1$. The integer k gives the number of complete waves in the interval $0 \leq x \leq 1$.

The boundedness of the solution and the build-up of error may be examined with reference to two norms. Let $Z_m^n = u(mh, nl) - U_m^n$, as in (4.66), with $n = 0, 1, \ldots$ and $m = 0, 1, \ldots, N$, so that $\mathbf{Z}^n$ is the vector of such errors and has $N+1$ elements. Define two norms by

$$\|\mathbf{U}^n\|_2^2 = h \sum_{m=0}^{N} |U_m^n|^2 \qquad (4.259)$$

and

$$\|\mathbf{U}^n\|_\infty = \max_m |U_m^n|. \qquad (4.260)$$

It must be noted that the vector $\mathbf{U}^n$ used in (4.259) and (4.260) now has $N+1$ elements, as it includes the boundary condition U_0^n at any time $t = nl$. The quantity $\|\mathbf{Z}^n\|_2^2$, defined as in (4.259), is the energy of the finite difference scheme at time $t = nl$, and $\|\mathbf{Z}^n\|_\infty$ gives the modulus of the maximum error at time $t = nl$ but not the mesh point (mh, nl) at which this error occurs.

The solution and error norms at time $t = 4$ are given in Table 4.8 for $k = 2$, $h = 1/80$, $r = 2$ and $r = 4$, and in Table 4.9 for $k = 2$, $h = 1/640$, $r = 8$ and $r = 16$. The numerical results are in keeping with the theory; in particular, it is clear that all three methods, given in vector form by equations (4.237), (4.246) and (4.253) are unconditionally stable in that the norms all remain bounded. The initial function gives $\|\mathbf{U}^0\|_2 = 0.707$ to three significant figures; all entries in Tables 4.8 and 4.9 are to three significant figures.

It is easy to show that, in Table 4.8, the number of mesh points at which each scheme is applied is 12 800 for $r = 2$ and 6400 for $r = 4$; the CPU time is approximately halved for the two cases. In Table 4.9, the number of mesh points for $r = 8$ is 25 600 and for $r = 16$ is 12 800; again, the CPU time is

Table 4.8

r	Method (formula)	$\|U\|_2$	$\|Z\|_2$	$\|Z\|_\infty$	CPU time (s)
2	(4.287)	0.690	0.171(−1)	0.283(−1)	0.042
	(4.296)	0.685	0.752(−1)	0.142(−1)	0.047
	(4.303)	0.688	0.439(−1)	0.944(−1)	0.059
4	(4.287)	0.682	0.137	0.312	0.023
	(4.296)	0.655	0.111	0.107	0.025
	(4.303)	0.671	0.522(−1)	0.943(−1)	0.031

Table 4.9

r	Method (formula)	$\|U\|_2$	$\|Z\|_2$	$\|Z\|_\infty$	CPU time (s)
8	(4.287)	0.700	0.730(−1)	0.127	0.689
	(4.296)	0.705	0.190(−1)	0.276(−1)	0.791
	(4.303)	0.705	0.175(−2)	0.271(−2)	0.877
16	(4.287)	0.704	0.425(−1)	0.103	0.305
	(4.296)	0.698	0.373(−1)	0.538(−1)	0.347
	(4.303)	0.701	0.597(−2)	0.856(−2)	0.445

approximately halved for the two cases. The CPU times given include the calculations of the norms as well as the evaluations of the numerical experiments; this explains the vastly different CPU times for the two experiments having 12 800 mesh points.

PROBLEMS

1. The function $u(x, t)$ satisfies the equation

$$\frac{\partial^2 u}{\partial x^2} - x^2 \frac{\partial^2 u}{\partial t^2} = 0$$

and the initial conditions

$$u = x^2 \text{ and } \frac{\partial u}{\partial t} = 0 \text{ on } t = 0, \ -\infty < x < \infty.$$

Show that the characteristics through any point (x, t) in the domain of dependence have slopes $+x$ and $-x$. Show from first principles, with the notation used in the text, that $dp - x dq = 0$, $x \neq 0$, along the characteristic of slope x, and that $dp + x dq = 0$, $x \neq 0$, along the characteristic of slope $-x$.

Calculate to two decimal places a first approximation to the solution at the point where the characteristic with positive slope through the point $(0.2, 0)$ intersects the characteristic with negative slope through the point $(0.3, 0)$. Find a second approximation to x, t, p, q, u at this point.

2. Verify that the equation

$$(x^2 + 1)\frac{\partial^2 u}{\partial x^2} - \lambda^2 (t^2 + 1)\frac{\partial^2 u}{\partial t^2} = 6(x - x^3 - \lambda^2 t - \lambda^2 t^3)$$

is hyperbolic for all real x, t and all real non-zero values of the parameter λ.

Values of u, $p \equiv \partial u/\partial x$ and $q \equiv \partial u/\partial t$ are specified at all points on part of some initial curve L in the (x, t) plane which is not a characteristic. Two points T_1 and T_2 lie on L; the values of x, t, p, q, u at these points are given in the following table:

	x	t	u	p	q
T_1	0.2	0.1	0.009	0.12	0.03
T_2	0.3	0.1	0.028	0.27	0.03

Determine first approximations to x, t, p, q, u at the point T_3, where the characteristic of positive slope through T_1 intersects the characteristic of negative slope through T_2, for the case $\lambda = 1$.
Determine second approximations to the coordinates of T_3.

3. (a) Verify that the equation

$$\frac{\partial^2 u}{\partial x^2} + \left(\frac{\partial u}{\partial x} - u\right)\frac{\partial^2 u}{\partial x \partial t} - u\frac{\partial u}{\partial x}\frac{\partial^2 u}{\partial t^2} + x = 0$$

is hyperbolic provided $u \neq -\partial u/\partial x$ and find the slopes of its characteristics in the (x, t) plane.

(b) Verify that the equation

$$\frac{\partial^2 u}{\partial x^2} + (1 - 3x^2)\frac{\partial^2 u}{\partial x \partial t} - 3x^2\frac{\partial^2 u}{\partial t^2} + x = 0$$

is always hyperbolic and find the slopes of its characteristics.

4. Verify that using the $(0, 2)$ Padé approximant in equation (4.49) leads to the vector equation (4.54).

5. Use the second order central difference replacement of $\partial^2 u/\partial t^2$ and $\partial^2 u/\partial x^2$ to show that equation (4.60) becomes the nine-point, three level, implicit scheme given by (4.61). Show further that putting $a = \frac{1}{4}$ in (4.61) leads to a consistent finite difference scheme. Use the Fourier method to show that the scheme is also unconditionally stable.

6. Use the $(0, 4)$ Padé approximant in (4.49) to obtain the explicit scheme given by (4.94). Verify that this scheme is stable for $0 < cr \leq \frac{1}{2}\sqrt{6}$.

7. Given the wave equation $\partial^2 u/\partial t^2 = c^2 \partial^2 u/\partial x^2$ in a region $R = [0 < x < 1] \times t > 0$ with boundary conditions $u(0, t) = u(1, t) = 0$, $t > 0$, and initial conditions $u(x, 0) = \sin \pi x$, $0 \leq x \leq 1$. Use the equations (4.64), (4.65) to obtain $u(x, l)$ at the points $x = 0.1(0.1)0.9$, taking $h = 0.1$, for $l = 0.05(0.05)5.00$.

Hyperbolic Equations

8. Verify that three level difference schemes with stability equation of the form (4.75), namely $a\xi^2 - 2b\xi + a = 0$, are stable whenever $|b| \leq |a|$.

9. Apply the three level scheme given in vector form by (4.58), with the matrix A given by (4.113), to the mesh point (kh, mh, nl) to derive the difference scheme (4.122).

10. The differential equation $\partial^2 u/\partial t^2 = \partial^2 u/\partial x^2$ has initial conditions $u(x, 0) = x$, $\partial u(x, 0)/\partial t = \frac{1}{2} x$ for $0 \leq x \leq 1$ and boundary conditions $u(0, t) = 0$, $u(1, t) = 1$ for $t > 0$.

    Use (4.64) with $h = 0.25, l = 0.25$ to find $U(x, 0.25)$ for $x = 0.25(0.25)1.0$ and (4.55) with $h = 0.25, l = 0.25$ to find $U(x, 0.5)$ for $x = 0.25(0.25)1.0$.

11. The operators $\delta_x^2 u_m^n$, $\delta_t^2 u_m^n$ are defined by $\delta_x^2 = u_{m-1}^n - 2u_m^n + u_{m+1}^n$, $\delta_t^2 = u_m^{n-1} - 2u_m^n + u_m^{n+1}$, respectively. The wave eqution $\partial^2 u/\partial t^2 = \partial^2 u/\partial x^2$ $(0 < x < 1, t > 0)$ is approximated at the mesh point (mh, nl) by the difference equation

$$\delta_t^2 u_m^n = \frac{r^2}{3} \delta_x^2 (u_m^{n-1} + u_m^n + u_m^{n+1}); \ r = l/h.$$

Given that u, $\partial u/\partial t$ are known for $0 \leq x \leq 1, t = 0$ and that u is known for $x = 0, t > 0$ and $x = 1, t > 0$ show that the difference equation may be written in vector form as

$$A\mathbf{U}^{n+1} + B\mathbf{U}^n + A\mathbf{U}^{n-1} = \mathbf{c}^n,$$

where A, B, C are matrices of order N, $(N+1)h = 1$, given by $A = -(3 + 2r^2)I + r^2 D$, $B = (6 - 2r^2)I + r^2 D$ and $\mathbf{c}^n$ is a column vector of boundary conditions. The matrix I is the identity matrix of order N and D is the matrix of order N given by

$$D = \begin{bmatrix} 0 & 1 & & & \\ 1 & 0 & 1 & & 0 \\ & 1 & 0 & 1 & \\ & & & & \\ 0 & & & 1 & 0 \end{bmatrix}$$

12. Show that the theoretical solution of (4.153) with $\mathbf{U}(0) = \mathbf{g}$ is given by (4.155) and that (4.155) satisfies the recurrence relation (4.156) for a time step l. Verify that the difference scheme which arises by using the (0, 1) Padé approximant to the matrix exponential function in (4.156), is unconditionally unstable.

13. Verify that the finite difference schemes based on the (2, 0), (2, 1) (2, 2) Padé approximants to $\exp(-\frac{1}{2}alB)$ in (4.156) are unconditionally stable.

14. Wendroff's implicit formula is given by

$$(1+ar)U_m^{n+1} + (1-ar)U_{m-1}^{n+1} = (1-ar)U_m^n + (1+ar)U_{m-1}^n.$$

Discuss the application of the method to the problem given by (4.182), (4.183), (4.184). Show that the method is unconditionally stable and that its local truncation error is

$$-\frac{1}{6}a h^2 \frac{\partial^3 u}{\partial x^3} - \frac{1}{6}l^2 \frac{\partial^3 u}{\partial t^3} + \cdots.$$

15. Show that the characteristic of the hyperbolic equation

$$x^2 e^t \frac{\partial u}{\partial x} + \frac{\partial u}{\partial t} = -e^t u,$$

with initial condition $u(x, 0) = 1$, $0 < x < \infty$, through the point $(X, 0)$ in the (x, t) plane, is given by

$$t = \frac{1}{X} - \frac{1}{x}.$$

5

Parabolic Equations

5.1 INTRODUCTION

Second order parabolic equations are one of the three classes of the partial differential equation

$$a\frac{\partial^2 u}{\partial x^2} + b\frac{\partial^2 u}{\partial x \partial t} + c\frac{\partial^2 u}{\partial t^2} = e, \tag{5.1}$$

in which a, b, c, e are functions of x, t, u, $p \equiv \partial u/\partial x$, $q \equiv \partial u/\partial t$ but not of the second order derivatives; for parabolic equations $b^2 - 4ac = 0$.

The most widely used diffusion equation is given by

$$\frac{\partial u}{\partial t} = \frac{\partial^2 u}{\partial x^2} \tag{5.2}$$

in some region R; (5.2) arises from (5.1) by putting $a = 1$, $b = 0$, $c = 0$, $e = \partial u/\partial t$. Equation (5.2) is known as the one (space) dimensional heat equation, and is the model equation generally used to test the accuracy and reliability of a given numerical method. Writing $a = 1/\text{Re}$, $b = c = 0$, $e = \partial u/\partial t + \mu \partial u/\partial x$ ($\mu > 0$) gives

$$\frac{\partial u}{\partial t} + \mu \frac{\partial u}{\partial x} = \frac{1}{\text{Re}} \frac{\partial^2 u}{\partial x^2} \tag{5.3}$$

which is the one-dimensional diffusion-convection equation, Re being the Reynolds number. The left-hand side of (5.3), the hyperbolic part, contains a convection term which relates, for example, to the flow of heat by moving fluid.

The heat flow equation (5.2) is also an example of the equation

$$\beta(x, t)\frac{\partial u}{\partial t} = \frac{\partial}{\partial x}\left(\alpha(x, t)\frac{\partial u}{\partial x}\right) - \gamma(x, t)u \tag{5.4}$$

which is itself an example of (5.1) with $a = \alpha(x, t)$, $b = c = 0$ and $e = \beta(x, t)$

$\partial u/\partial t + \gamma(x, t)u - (\partial \alpha/\partial x)(\partial u/\partial x)$. In (5.4), which is known as the general linear parabolic partial differential equation, the conditions $\alpha(x, t) > 0$, $\beta(x, t) > 0$, $\gamma(x, t) \geq 0$ must hold at every point of the region R.

The region R is usually one of the following three types with the associated initial conditions and boundary conditions as specified:

(i) the open rectangle $X_0 \leq x \leq X_1$, $t > 0$. Here the heat flow problem consists of equation (5.4) together with the initial distribution

$$u(x, 0) = g(x), \quad X_0 \leq x \leq X_1$$

and boundary conditions of the form

$$a_0(X_0, t)u + b_0(X_0, t)\frac{\partial u}{\partial x} = c_0(X_0, t), \quad t > 0$$

$$a_1(X_1, t)u + b_1(X_1, t)\frac{\partial u}{\partial x} = c_1(X_1, t), \quad t > 0$$

with $a_0(x, t) \geq 0$, $a_1(x, t) \geq 0$, $b_0(x, t) \leq 0$, $b_1(x, t) \geq 0$ and with $a_0(x, t) - b_0(x, t) > 0$, $a_1(x, t) - b_1(x, t) > 0$.

The open rectangle is the most frequently occurring region in the numerical solution of the simple heat equation (5.2); the associated boundary conditions usually have $a_0 = a_1 = 1$ and $b_0 = b_1 = 0$. This formulation of the heat flow problem will be given most attention in this chapter;

(ii) the semi-infinite plane $-\infty < x < \infty$, $t > 0$. Here the heat flow problem is known as the *Cauchy initial value problem*; it consists of an equation of the form (5.4) together with the initial condition

$$u(x, 0) = g(x), \quad -\infty < x < \infty;$$

(iii) the quarter-plane $x > 0$, $t > 0$. In this final case the heat flow problem consists of a parabolic equation of the form (5.4) together with the initial condition

$$u(x, 0) = g(x), \quad 0 \leq x < \infty$$

and the boundary condition

$$a_0(0, t)u + b_0(0, t)\frac{\partial u}{\partial x} = c_0(0, t), \quad t > 0,$$

where, as in (i), $a_0(x, t) \geq 0$, $b_0(x, t) \leq 0$ and $a_0(x, t) - b_0(x, t) > 0$.

It will be clear to the reader that, without loss of generality, X_0 may be taken to be zero so that, in case (i), one edge of the boundary of R coincides with the t-axis. To that end, take $X_0 = 0$ and $X_1 = X$, thus facilitating the development and analysis of the family of numerical methods to be discussed in the following sections.

5.2 LOW ORDER FINITE DIFFERENCE METHODS

In this section three well known finite difference methods for solving second order parabolic equations will be developed.

Consider the constant coefficient heat equation in one space variable

$$\frac{\partial u}{\partial t} = \frac{\partial^2 u}{\partial x^2}; \quad 0 < x < X, \, t > 0 \tag{5.5}$$

with boundary conditions

$$u(0, t) = u(X, t) = 0; \quad t > 0 \tag{5.6}$$

and initial conditions

$$u(x, 0) = g(x); \quad 0 \leqslant x \leqslant X, \tag{5.7}$$

where $g(x)$ is a given continuous function of x. It is not specified that $g(0) = 0$ or $g(X) = 0$, so that discontinuities between initial conditions and boundary conditions may exist.

The interval $0 \leqslant x \leqslant X$ is divided into $N + 1$ subintervals each of width h, so that $(N+1)h = X$, and the independent variable t is discretized in steps of length l. The open region $R = [0 < x < X] \times [t > 0]$ and its boundary ∂R, consisting of the axes $t = 0$, $x = 0$ and the line $x = X$, have thus been covered by a rectangular mesh, the mesh points having coordinates (mh, nl), where $m = 0, 1, \ldots, N+1$ and $n = 0, 1, 2, \ldots$ As in previous chapters, $u(mh, nl)$, the solution of (5.5) at the mesh point (mh, nl), will be denoted by u_m^n, while the theoretical solution of an approximating difference scheme will be denoted by U_m^n. The computed solution of the difference scheme (the solution actually obtained, which may be subject to the errors discussed in Chapter 4) at this mesh point, will be denoted by $\widetilde{U}_m^n$.

The space derivative in (5.5) is now replaced by

$$\frac{\partial^2 u}{\partial x^2} = \{u(x-h, t) - 2u(x, t) + u(x+h, t)\}/h^2 + O(h^2) \tag{5.8}$$

and (5.5) with (5.8) are applied to all N interior mesh points at time level $t = nl$ ($n = 0, 1, \ldots$) to produce a system of N first order ordinary differential equations. Recalling from (5.6) that $U_0^n = U_{N+1}^n = 0$, this system is of the form

$$\frac{d\mathbf{U}(t)}{dt} = A\mathbf{U}(t), \tag{5.9}$$

where $\mathbf{U}(t) = [U_1(t), U_2(t), \ldots, U_N(t)]^T$ is the vector of approximations to the solution of (5.5) at time $t = nl$, and A is the matrix given by

$$A = h^{-2} \begin{bmatrix} -2 & 1 & & & 0 \\ 1 & -2 & 1 & & \\ & \ddots & \ddots & \ddots & \\ & & & & 1 \\ 0 & & & 1 & -2 \end{bmatrix} \qquad (5.10)$$

of which the eigenvalues are $\lambda_s = -4\sin^2[s\pi/\{2(N+1)\}]/h^2$ for $s = 1, 2, \ldots, N$.

A practical difficulty with (5.9) is that the system is stiff (see Chapter 1) because a large value of N must be used in some problems, leading to a large range of magnitudes of eigenvalues. The stiffness ratio has the value

$$\sin^2[N\pi/\{2(N+1)\}]/\sin^2[\pi/\{2(N+1)\}]$$

or $4(N+1)^2/\pi^2$ for large values of N.

Solving (5.9), with the initial vector $\mathbf{U}(0) = \mathbf{g}$ from (5.7) gives

$$\mathbf{U}(t) = \exp(tA)\mathbf{g} \qquad (5.11)$$

which may be written in the form of a recurrence relation as

$$\mathbf{U}(t+l) = \exp(lA)\mathbf{U}(t); \quad t = 0, l, 2l, \ldots \qquad (5.12)$$

To obtain numerical solutions $\mathbf{U}(t+l)$, Padé approximants to the matrix exponential function $\exp(lA)$ may be used in (5.12). In this way, a family of algorithms for the numerical solution of (5.5) subject to (5.6), (5.7) may be obtained. The $(0, 1), (1, 0), (1, 1)$ approximants give three well known methods; methods yielding higher order accuracy in time may be found by using higher order Padé approximants, as in section 5.6.

Replacing the exponential matrix function in (5.12) with its $(0, 1)$ Padé approximant (Table 4.3), gives

$$\mathbf{U}(t+l) = (I+lA)\mathbf{U}(t) \qquad (5.13)$$

in which there is an error in time of $0(l^2)$. Applying (5.13) to the mesh point (mh, nl) gives the four point explicit scheme

$$U_m^{n+1} = (1-2p)U_m^n + p(U_{m-1}^n + U_{m+1}^n), \qquad (5.14)$$

where $p = l/h^2$.

Replacing, now, the exponential matrix function in (5.12) by its $(1, 0)$ Padé approximant gives

$$\mathbf{U}(t+l) = (I-lA)^{-1}\mathbf{U}(t),$$

which, when written implicitly, becomes

$$(I-lA)\mathbf{U}(t+l) = \mathbf{U}(t); \tag{5.15}$$

the error in time in (5.15) is $O(l^2)$. Applying this fully implicit scheme to the mesh point (mh, nl) gives

$$-pU_{m-1}^{n+1} + (1+2p)U_m^{n+1} - pU_{m+1}^{n+1} = U_m^n. \tag{5.16}$$

The computed solution of this scheme is easily obtained from (5.15) using one of the methods outlined in Chapter 1 (subsection 1.2.3) for solving a linear system.

Finally in this section, the exponential matrix function in (5.12) is replaced by its $(1, 1)$ Padé approximant leading, after premultiplication by $(I-\frac{1}{2}lA)$, to

$$(I-\frac{1}{2}lA)\mathbf{U}(t+l) = (I + \frac{1}{2}lA)\mathbf{U}(t), \tag{5.17}$$

in which the error in time is $O(l^3)$. Applying (5.17) to the mesh point (mh, nl) gives the six point, two level, implicit formula

$$-\frac{1}{2}pU_{m-1}^{n+1} + (1+p)U_m^{n+1} - \frac{1}{2}pU_{m+1}^{n+1}$$

$$= \frac{1}{2}pU_{m-1}^n + (1-p)U_m^n + \frac{1}{2}pU_{m+1}^n \tag{5.18}$$

which is the well known Crank-Nicolson Scheme (1947). This method has enjoyed great popularity since its publication. It will be seen in section 5.3 that it is unconditionally stable but, as shown by Lawson and Morris (1978), it is nevertheless, necessary to impose a restriction on the ratio $r = l/h$. The computed solution of the Crank-Nicolson scheme may be obtained efficiently from the vector form (5.17) by rewriting it as

$$(I - \frac{1}{2}lA)\mathbf{U}^* = 2\mathbf{U}(t),$$

$$\mathbf{U}(t+l) = \mathbf{U}^* - \mathbf{U}(t).$$

This approach requires the storage of the vector $\mathbf{U}^*$ but needs $3N-2$ multiplication fewer than using a tridiagonal solver to find $\mathbf{U}(t+l)$ directly from (5.17).

Applying the explicit scheme (5.14) or the Crank-Nicolson scheme (5.18) to the mesh points $(h, 0)$ or $(Nh, 0)$ involves using the points $(0, 0)$ or $(X, 0)$ where there may be discontinuities between initial and boundary conditions, giving poor numerical results near the boundaries for all values of t. This phenomenon does not arise when using the fully implicit scheme (5.16) which does not use the points of discontinuity.

To illustrate the behaviour of each of these three schemes, the model problem

$$\frac{\partial u}{\partial t} = \frac{\partial^2 u}{\partial x^2}; \ 0 < x < 2; \ t > 0$$

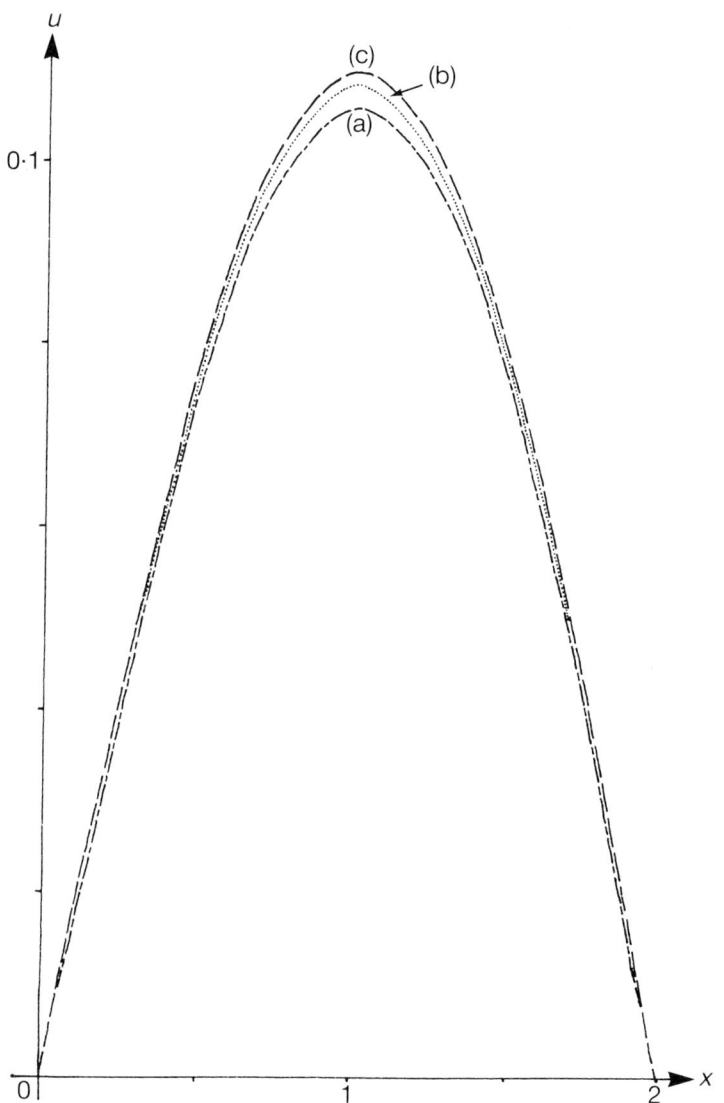

Fig. 5.1 – Solution of model problem at time $t = 1.0$ with $p = \frac{1}{2}$: (a) (0, 1) Padé approximant, (b) (1, 1) Padé approximant, (c) (1, 0) Padé approximant.

subject to $u(x, 0) = 1$ for $0 \leqslant x \leqslant 2$ and $u(0, t) = u(2, t) = 0$ for $t > 0$ is considered. This problem, which has theoretical solution

$$u(x, t) = \sum_{k=1}^{\infty} [1 - (-1)^k] \frac{2}{k\pi} \sin(\frac{1}{2} k\pi x) \exp(-\frac{1}{4} k^2 \pi^2 t),$$

is due to Lawson and Morris (1978). It is solved for $h = 0.1, l = 0.005$ ($p = \frac{1}{2}$) using schemes (5.14), (5.16), (5.18) and for $h = 0.05, l = 0.1$ ($p = 40$) using

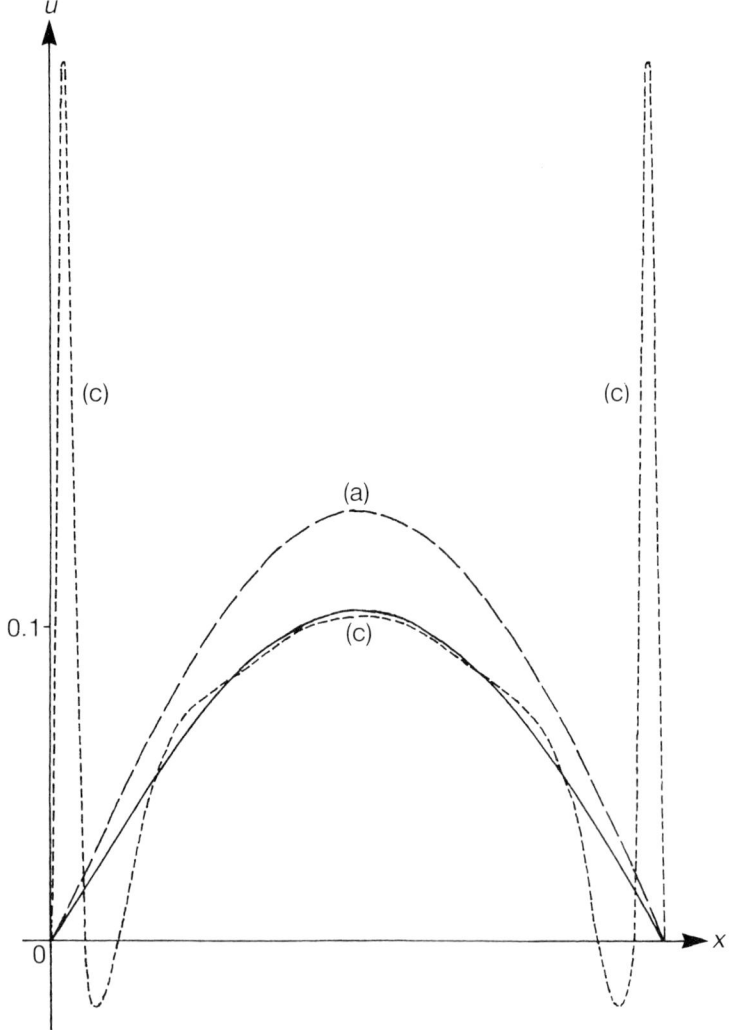

Fig. 5.2 — Solution of model problem at time $t = 1.0$ with $p = 40$: (a) theoretical solution, (b) (1, 0) Padé approximant, (c) (1, 1) Padé approximant.

schemes (5.16), (5.18); the computed solutions at time $t = 1.0$ are depicted in Figs. 5.1 and 5.2 respectively.

In the first of these experiments, the ratios p and r both satisfy the restrictions placed on them in section 5.3 and the computed solutions, depicted in Fig. 5.1, are satisfactory. The graph of the theoretical solution lies very close to the graph relating to the Crank-Nicolson method and is not shown in Fig. 5.1.

In the second experiment, the ratio p is much too large for the explicit method based on the (0, 1) Padé approximant to give a meaningful solution; the computed solution oscillates wildly near both boundaries and at $x = 1.0$, $t = 1.0$ it has the value 1.0, the theoretical solution being 0.108 approximately. As can be seen from Fig. 5.2, the Crank-Nicolson scheme is accurate in the interior of the region but has a large residual error near the boundary. This arises because the ratio $r = l/h$ violates a bound on it to be discussed in section 5.3; the magnitude of this residual error will be increased by refining h. The method based on the (1, 0) Padé approximant behaves smoothly over the whole interval $0 \leqslant x \leqslant 2$ but is not as accurate near the centre of the region as the Crank-Nicolson method.

5.3 ANALYSES OF THE METHODS

In this section, the tests which finite difference schemes for solving parabolic equations must pass before being used, are discussed. As a result of the analyses, the behaviour of some of the methods developed in section 5.1, and some of those to be developed in later sections, will be clarified.

5.3.1 Truncation errors and consistency

Suppose the parabolic equation is written in the form $G(u) = 0$. Each of the difference schemes (5.14), (5.16), (5.18) may be written in the form $F_{m,n}(U) = 0$. Replacing U by u at each mesh point occurring in the finite difference scheme, and carrying out the Taylor expansions about $u(mh, nl)$, the value of $l^{-1}F_{m,n}(u) - G(u_m^n)$ is the local truncation error at the mesh point (mh, nl). As with hyperbolic equations, the local truncation error is the difference between the finite difference scheme and the differential equation it is replacing; the local discretization error is, as before, the difference between the theoretical solutions of the differential and difference equations and is represented at the mesh point (mh, nl) by $z_m^n = u_m^n - U_m^n$.

If the local truncation error tends to zero as the mesh is refined, that is as $h, l \to 0$, the difference equation is said to be consistent with the differential equation.

Consider first of all the explicit scheme (5.14) for which

$$F_{m,n}(u) = u_m^{n+1} - (1-2p)u_m^n - p(u_{m-1}^n + u_{m+1}^n).$$

208   **Parabolic Equations**   [Ch. 5

Expanding u_m^{n+1}, u_{m-1}^n, u_{m+1}^n about u_m^n gives

$$F_{m,n} = l\left(\frac{\partial u}{\partial t} - \frac{\partial^2 u}{\partial x^2}\right)_m^n + \left(\frac{1}{2}l^2\frac{\partial^2 u}{\partial t^2} - \frac{1}{12}h^2 l\frac{\partial^4 u}{\partial x^4}\right)_m^n + \ldots,$$

so that the local truncation error of the explicit scheme (5.14) is

$$\left(\frac{1}{2}l\frac{\partial^2 u}{\partial t^2} - \frac{1}{12}h^2\frac{\partial^4 u}{\partial x^4} + \ldots\right)_m^n \tag{5.19}$$

and the scheme is therefore consistent.

In (5.19), the component $-\frac{1}{12}h^2 \partial^4 u/\partial x^4$ is due to the space discretization employed in section 5.2; this term will appear in the local truncation error of every finite difference scheme arising from the use of Padé approximants to the matrix exponential function in the recurrence relation (5.12). The term $\frac{1}{2}l\partial^2 u/\partial t^2$ in (5.19) relates to the use of the (0, 1) Padé approximant in (5.12).

The principal part of the local truncation error may thus be written down for every finite difference scheme yielded by (5.12). Each of the local truncation errors will be of the form

$$\left(C_q l^{q-1}\frac{\partial^q u}{\partial t^q} - \frac{1}{12}h^2\frac{\partial^4 u}{\partial x^4} + \ldots\right)_m^n \tag{5.20}$$

where, for the (M, K) Padé approximant, $q = M+K+1$, and C_q is contained in Table 4.5 of Chapter 4.

The local truncation error associated with the application of the fully implicit, backward difference scheme (5.16) to the mesh point (mh, nl) is thus

$$\left(-\frac{1}{2}l\frac{\partial^2 u}{\partial t^2} - \frac{1}{12}h^2\frac{\partial^4 u}{\partial x^4} + \ldots\right)_m^n, \tag{5.21}$$

and that of the Crank-Nicolson scheme (5.18) is

$$\left(-\frac{1}{12}l^2\frac{\partial^3 u}{\partial t^3} - \frac{1}{12}h^2\frac{\partial^4 u}{\partial x^4} + \ldots\right)_m^n. \tag{5.22}$$

Comparing (5.19), (5.21), (5.22) it is evident that the Crank-Nicolson scheme is more accurate than the explicit and fully implicit methods, in view of the $0(l^2)$ term in (5.22) and the $0(l)$ terms in (5.19) and (5.21). This explains the closeness of the Crank-Nicolson curve to the graph of the theoretical solution in Fig. 5.1; the behaviour of the Crank-Nicolson curve near the boundary in Fig. 5.2 will be explained in subsection 5.3.2. Comparing (5.19) with (5.21) it is seen that the coefficients C_q are opposite in sign for the explicit and fully implicit methods. This explains why the curve of the explicit solution is below

5.3.2 Conventional stability analyses

Description of the energy, Fourier, and matrix methods of analyzing the numerical stability of finite difference schemes, were given in subsection 4.6.2 of Chapter 4. These methods are, of course, applicable to parabolic partial differential equations and will be used in the present chapter.

The energy method was mentioned only briefly in Chapter 4 and was not used to analyze the stability of any of the finite difference methods, though it was used in the numerical example at the end of subsection 4.13.2.

Analyzing numerical stability using the energy method is a skill which the user can acquire with practice. Probably the easiest finite difference scheme to analyze by the energy method is the explicit scheme (5.14); this analysis will now be carried out in full.

The aim of the analysis is to prove that the energy, represented by the quantity $h \sum_{m=1}^{N} (z_m^n)^2$, remains bounded as $n \to \infty$, where z_m^n is the local discretization error. Before this can be done it must be shown that the theoretical solution of the differential equation itself, subject to its initial and boundary conditions, remains bounded.

In the case of the heat equation (5.5), with boundary conditions (5.6) and initial conditions (5.7), this is easily established by multiplying (5.5) by u and integrating with respect to x to obtain

$$\int_0^X u \frac{\partial u}{\partial t} dx = \int_0^X u \frac{\partial^2 u}{\partial x^2} dx. \qquad (5.23)$$

Integrating the right-hand side of (5.23) by parts gives

$$\frac{\partial}{\partial t} \int_0^X u^2 dx = \left[u \frac{\partial u}{\partial x} \right]_0^X - \int_0^X \left(\frac{\partial u}{\partial x} \right)^2 dx$$

$$= - \int_0^X \left(\frac{\partial u}{\partial x} \right)^2 dx \leq 0.$$

It follows from this that

$$\int_0^X u^2(x, t) dx \leq \int_0^X u^2(x, 0) dx = \int_0^X g^2(x) dx,$$

so that the quantity $\int_0^X u^2(x, t)dx$, and consequently the energy, remain bounded as $n \to \infty$.

It was shown in Chapter 4, and may easily be shown for parabolic equations also, that the growth of an initial perturbation represented by $Z_m^n = U_m^n - \tilde{U}_m^n$, where $\tilde{U}_m^n$ is the actual solution obtained at the mesh point (mh, nl), satisfies the difference scheme, so that for (5.14)

$$Z_m^{n+1} = pZ_{m-1}^n + (1-2p)Z_m^n + pZ_{m+1}^n, \tag{5.24}$$

where $m = 1, \ldots, N$ and it is assumed that $Z_0^n = Z_{N+1}^n = 0$ for all n. Rewriting (5.24) as

$$Z_m^{n+1} - Z_m^n = p(Z_{m-1}^n - 2Z_m^n + Z_{m+1}^n) \tag{5.25}$$

and multiplying both sides of (5.25) by $Z_m^{n+1} + Z_m^n$, then summing over all values of $m = 1, 2, \ldots, N$, gives

$$\|Z^{n+1}\|_2^2 - \|Z^n\|_2^2 = p \sum_{m=1}^{N} (Z_m^{n+1} + Z_m^n)(Z_{m-1}^n - 2Z_m^n + Z_{m+1}^n). \tag{5.26}$$

Recalling that $Z_0^n = Z_{N+1}^n = Z_0^{n+1} = Z_{N+1}^{n+1} = 0$, the reader can verify that (5.26) becomes

$$\|Z^{n+1}\|_2^2 - \|Z^n\|_2^2 = -p \left[\sum_{m=0}^{N} \{(Z_{m+1}^n - Z_m^n)^2 + (Z_{m+1}^{n+1} - Z_m^{n+1})(Z_{m+1}^n - Z_m^n)\} \right] \tag{5.27}$$

and, by defining

$$E_n = \|Z^n\|_2^2 - \frac{1}{2}p \sum_{m=0}^{N} (Z_{m+1}^n - Z_m^n)^2 \tag{5.28}$$

for all $n = 0, 1, 2, \ldots$, that

$$E_n \leqslant \|Z^n\|_2^2 \tag{5.29}$$

and

$$E_{n+1} - E_n = \|Z^{n+1}\|_2^2 - \|Z^n\|_2^2 - \frac{1}{2}p \sum_{m=0}^{N} \{(Z_{m+1}^{n+1} - Z_m^{n+1})^2 - (Z_{m+1}^n - Z_m^n)^2\}. \tag{5.30}$$

Using (5.27), it follows from (5.30) that

$$E_{n+1} - E_n = -\frac{1}{2}p \sum_{m=1}^{N} (Z_{m+1}^{n+1} - Z_m^{n+1} + Z_{m+1}^n - Z_m^n)^2 \leqslant 0,$$

from which the reader will conclude that E_n is monotone decreasing as $n \to \infty$.

The elementary inequality $(a-b)^2 \leq (a-b)^2 + (a+b)^2 = 2a^2 + 2b^2$ for any real a, b shows that

$$\sum_{m=0}^{N} (Z_{m+1}^n - Z_m^n)^2 \leq 2\sum_{m=0}^{N} (Z_m^n)^2 + 2\sum_{m=0}^{N} (Z_{m+1}^n)^2$$

$$= 4\sum_{m=0}^{N} (Z_m^n)^2 = 4\|\mathbf{Z}^n\|_2^2. \tag{5.31}$$

Substituting (5.31) in (5.28), rewritten in the form

$$\sum_{m=0}^{N} (Z_{m+1}^n - Z_m^n)^2 = \frac{2}{p}(\|\mathbf{Z}^n\|_2^2 - E_n),$$

leads to

$$(1-2p)\|\mathbf{Z}^n\|_2^2 \leq E_n$$

$$\leq E_{n-1} \leq \ldots \leq E_0 \leq \|\mathbf{Z}^0\|_2^2$$

since E_n is monotone decreasing and in view of (5.29). Thus

$$\|\mathbf{Z}^n\|_2^2 \leq \frac{1}{1-2p}\|\mathbf{Z}^0\|_2^2$$

and it may be concluded from the full analysis that the error vectors $\mathbf{Z}^n$ are bounded provided $0 < p < \frac{1}{2}$. This restriction on p is a sufficient condition for the stability of the explicit difference scheme (5.14).

The reader will have observed that the energy method for analyzing the stability of a finite difference scheme involves a much longer exercise than either of the Fourier or matrix methods. It must be remembered, however, that the energy method is applicable to a much wider class of problems than the Fourier and matrix methods and is therefore a very powerful tool.

Turning now to the fully implicit method (5.16) and using the Fourier method to analyze stability, the difference between the theoretical and computed solutions of the difference scheme at the mesh point (mh, nl) is considered to have typical term $e^{\alpha nl} e^{i\beta mh}$ where α is complex and β is real. This error satisfies the difference scheme and so, after cancelling by $e^{\alpha nl} e^{i\beta mh}$, the error equation becomes

$$[-p(\cos\beta h - i\sin\beta h) + 1 + 2p - p(\cos\beta h + i\sin\beta h)]\xi = 1$$

from which the amplification factor is found to be

$$\xi = \frac{1}{1 + 4p\sin^2\frac{1}{2}\beta h}. \tag{5.32}$$

This factor governs the growth of a perturbation of the initial conditions and must therefore satisfy the von Neumann necessary condition for stability, namely $|\xi| \leq 1$. Obviously ξ, as defined by (5.32), does satisfy the von Neumann condition for all $p \geq 0$ and the fully implicit method (5.16) is unconditionally stable (the possibility $p = 0$ can be discounted).

In the case of the Crank-Nicolson method (5.18), a conventional stability analysis shows that a perturbation $\mathbf{Z}^0$ of the initial data satisfies the equation

$$\mathbf{Z}^{n+1} = (I - \frac{1}{2} lA)^{-1} (I + \frac{1}{2} lA) \mathbf{Z}^n; \quad n = 1, 2, \ldots. \tag{5.33}$$

Since the eigenvalues λ_s ($s = 1, 2, \ldots, N$) of the matrix A are all negative, it is easy to show that all the eigenvalues of the amplification matrix $(I - \frac{1}{2} lA)^{-1} (I + \frac{1}{2} lA)$ are less than or equal to unity in modulus, so that the perturbation errors $\mathbf{Z}^n$ ($n = 0, 1, \ldots$) in (5.33) do not grow as $n \to \infty$.

It would seem therefore that the Crank-Nicolson scheme is unconditionally stable but, as can be readily seen from Figs. 5.1 and 5.2, some restriction must be placed on l if the Crank-Nicolson scheme is to be used to solve problems with discontinuities between initial conditions and boundary conditions. This restriction is not highlighted by the matrix method and further analysis must be carried out.

The eigenvalues of the amplification matrix in (5.33) are given by $\mu_s = (1 + \frac{1}{2} l\lambda_s)/(1 - \frac{1}{2} l\lambda_s)$, where λ_s are the eigenvalues of the matrix A. It is then easy to see that μ_N is close to -1 for small h (recall: $h = X/(N+1)$). Suppose now that the initial vector $\mathbf{g}$ can be written as a linear combination of the normalized eigenvectors $\mathbf{v}_s$ ($s = 1, 2, \ldots, N$) of the matrix A, that is

$$\mathbf{g} = \sum_{s=1}^{N} c_s \mathbf{v}_s, \tag{5.34}$$

where the c_s ($s = 1, 2, \ldots, N$) are constants. Then from (5.11) and (5.12)

$$\mathbf{U}(l) = \exp(lA) \mathbf{g} \cong \sum_{s=1}^{N} c_s (I - \frac{1}{2} lA)^{-1} (I + \frac{1}{2} lA) \mathbf{v}_s = \sum_{s=1}^{N} c_s \mu_s \mathbf{v}_s,$$

$$\mathbf{U}(2l) = \exp(lA) \mathbf{U}(l) \cong \sum_{s=1}^{N} c_s \mu_s (I - \frac{1}{2} lA)^{-1} (I + \frac{1}{2} lA) \mathbf{v}_s = \sum_{s=1}^{N} c_s \mu_s^2 \mathbf{v}_s,$$

and so on, so that, for any $n = 0, 1, \ldots$,

$$\mathbf{U}^n = \sum_{s=1}^{N} c_s \mu_s^n \mathbf{v}_s. \tag{5.35}$$

Analyses of the Methods

It follows that the components $c_N \mathbf{v}_N, c_{N-1}\mathbf{v}_{N-1}, \ldots$ of the initial conditions are preserved at subsequent time steps. In view of μ_N being close to -1 for large N, these components will alternate in sign, and for problems with discontinuities between initial conditions and boundary conditions, the effect will be to propagate the discontinuities and distort $\mathbf{U}$, if too large a value of l is used, by producing oscillations.

Not all components are oscillatory; those corresponding to the eigenvalues of A for which the numerator of the amplification factor $(1+\tfrac{1}{2}l\lambda_s)/(1-\tfrac{1}{2}l\lambda_s)$ are non-negative are non-oscillatory. Oscillatory components do not have a grave effect on the solution of the difference scheme provided they decay faster than the others.

Referring to equation (5.35) it is clear that the component $c_s \mathbf{v}_s (s = 1, 2, \ldots, N)$ is damped at each time step by the corresponding eigenvalue μ_s of the amplification matrix. Lawson and Morris (1978) noted that, to ensure that the higher frequency component $c_N \mathbf{v}_N$ is damped to zero faster than the lowest frequency component $c_1 \mathbf{v}_1$, the inequality $|\mu_N| < |\mu_1|$ is required. (This same argument could also be applied to the fully implicit method based on the (1, 0) Padé approximant, for which none of the components are oscillatory). This leads to

$$\frac{-1-\tfrac{1}{2}l\lambda_1}{1-\tfrac{1}{2}l\lambda_1} < \frac{1+\tfrac{1}{2}l\lambda_N}{1-\tfrac{1}{2}l\lambda_N} < \frac{1+\tfrac{1}{2}l\lambda_1}{1-\tfrac{1}{2}l\lambda_1}. \tag{5.36}$$

The right-hand inequality in (5.36) is trivially satisfied; the left-hand side gives $l^2 \lambda_1 \lambda_N < 4$.

For large values of N, $\lambda_1 \cong -4[\pi^2/\{4(N+1)^2 h^2\}] = -\pi^2/X^2$ (since, for any small angle θ, $\sin\theta \cong \theta$), and $\lambda_N \cong -4/h^2$. Thus

$$r = \frac{l}{h} < \frac{X}{\pi} \text{ (approximately)}, \tag{5.37}$$

which was clearly violated by the values $l = 0.1$ and $h = 0.05$ in the second experiment of section 5.2, and explains the erratic behaviour of U near the boundaries $x = 0$, $x = X$. The restriction (5.37) was not violated in the first experiment which accounts for the smooth behaviour of U calculated using the Crank-Nicolson method over the whole of the range $0 \leqslant x \leqslant X = 2$ in Fig. 5.1.

5.3.3 Alternative definitions of stability

It has been remarked already that for a large number, N, of points at each time level, the range of eigenvalues λ_s of the matrix A is large and, for large time steps, the magnitude of $l\lambda_s$ can also be large. The von Neumann necessary condition for stability, which requires $|(Q_M(l\lambda_s))^{-1} P_K(l\lambda_s)| \leqslant 1$, where, in the notation of subsection 4.5.2 of Chapter 4, $R_{M,K}(\theta) = P_K(\theta)/Q_M(\theta) + 0(\theta^{M+K+1})$ is the (M, K) Padé approximant to e^θ, is therefore a minimum requirement for stability. Otherwise, some components of **U** might increase in magnitude while corresponding components of **u** decrease in magnitude. Gourlay and Morris (1980) refer to $R_{m,k}(l\lambda)$ as the *symbol* of the method.

The finite difference method arising from the use of the (M, K) Padé approximant is said to be A_0-*stable* if $\|R_{M,K}(lA)\| \leq 1$ for all values of $p > 0$ in some norm. It follows that A_0-stability guarantees unconditional stability in the conventional sense of perturbations in the initial data not being magnified for the same range of values of p as $t \to \infty$. Lambert (1973) has noted that, using the (M, K) Padé approximant to the exponential matrix function in (5.12) gives a method which is A_0-stable provided $M \geq K$. Twizell and Khaliq (1981) noted that the family of multiderivative methods arising from the use of such Padé approximants to solve a system of first order ordinary differential equations of the form (5.9) is *absolutely stable* for $M \geq K$. This concept of absolute stability corresponds to the conventional concept of stability of finite difference methods for partial differential equations, and the family of multiderivative methods gives rise to some of the methods to be discussed in section 5.6.

An A_0-stable method for which, additionally, $\lim_{\theta \to -\infty} R_{M,K}(\theta) = 0$ for any real θ, gives rise to a finite difference method which is L_0-*stable*. This property ensures that all components of the approximate solution **U** are damped as t increases. Lambert (1973) has noted that the Padé approximant $R_{M,K}$ gives a finite difference method which is L_0-stable provided $M > K$. The finite difference method (5.16), which is based on the $(1, 0)$ Padé approximant, is therefore L_0-stable. This property of the method explains why it was able to cope well with the discontinuities between initial and boundary conditions, even for a large value of p, as in Fig. 5.2. The Crank–Nicolson method (5.18) is not L_0-stable and the failure of the method to damp offending components of **U** is depicted well in Fig. 5.2.

The reader will, hopefully, have observed that the properties of A_0-stability and L_0-stability are assigned to a finite difference method, written in a matrix form such as (5.12) in which the eigenvalues of the matrix are real, depending on its treatment of the less important components of the solution.

When the eigenvalues $\lambda_s (s = 1, 2, \ldots, N)$ are complex, as was the case for first order hyperbolic equations in section 4.11 of Chapter 4, stability conditions become less lenient. The requirement on the method based on the (M, K) Padé approximant is that the symbol $R_{M,K}(\theta)$ satisfies $|R_{M,K}(l\lambda_s)| < 1$ for all values of l and all eigenvalues $\lambda_s (s = 1, \ldots, N)$. If $|\arg(\lambda_s)| < \alpha$ for all s, it is sufficient to require that the method be $A(\alpha)$-*stable* which means that the method is stable for those values of l for which $|R_{M,K}(l\lambda_s)| < 1$ and $|\arg(l\lambda_s)| < \alpha$ for all eigenvalues λ_s.

In particular, if $\alpha = \frac{1}{2}\pi$ the method is said to be *A-stable*; that is, it is stable for any problem in which $\text{Re}(\lambda_s) \leq 0$. This corresponds, in the theory of ordinary differential equations, to the region of stability of a linear multistep method, or a multiderivative method, covering the whole of the left-half plane. Lambert (1973) notes that a method is A-stable if $M = K$.

The reader is warned that, in Chapter 4, the matrix A is replaced by one of the matrices $-\frac{1}{2} aB$, $-aC$, $-\frac{1}{2} aD$, $a > 0$, so that the description of A-stability leads to the requirement $\text{Re}(\lambda_s^*) \geq 0$, where λ_s^* refers to the eigenvalues of B, C or D.

If, in addition to the requirements for $A(\alpha)$-stability, the requirement $|R_{M,K}(l\lambda_s)| \to 0$ as $\text{Re}(l\lambda_s) \to -\infty$ for any eigenvalue λ_s is met, the method is said to be $L(\alpha)$-stable. In particular, the method is L-stable for $\alpha = \frac{1}{2}\pi$. Lambert (1973) notes that a method based on the (M, K) Padé approximant is L-stable whenever $M = K + 1$ or $M = K + 2$.

If $\alpha = 0$, the eigenvalues $\lambda_s (s = 1, \ldots, N)$ are real and the properties $A(0)$-stability and $L(0)$-stability are obviously the properties A_0-stability and L_0-stability, respectively. In the theory of ordinary differential equations, the corresponding region of stability in the complex plane degenerates to an interval of stability of the negative real axis.

5.3.4 Convergence

A convergence analysis for parabolic equations is carried out in precisely the same way as for elliptic or hyperbolic equations, by determining the restriction, if any, on the ratio p for which the local discretization error $z_m^n = u_m^n - U_m^n$, at the *fixed* mesh point (mh, nl), tends to zero as the mesh is refined in such a way that both $h, l \to 0$.

For the explicit method based on the $(0, 1)$ Padé approximant, for instance, the local discretization error satisfies the equation

$$z_m^{n+1} = (1-2p)z_m^n + p(z_{m-1}^n + z_{m+1}^n) + lT_m^n, \tag{5.38}$$

where $T_m^n = 0(l + h^2)$ is the local truncation error (5.19).

If $0 < p \leq \frac{1}{2}$, none of the coefficients on the right-hand side of (5.38) is negative and it follows that

$$|z_m^{n+1}| \leq (1-2p)|z_m^n| + p|z_{m+1}^n| + p|z_{m-1}^n| + k^{(n)}(l^2 + lh^2)M$$

where $k^{(n)} = \max \{|[\partial^2 u/\partial t^2]_m^n|, |[\partial^4 u/\partial x^4]_m^n|\}$ and M is some constant. Thus

$$|z_m^{n+1}| \leq \hat{z}^n + k^{(n)} (l^2 + lh^2)M,$$

where $\hat{z}^n = \max |z_m^n|$ for $m = 1, 2, \ldots, N$, and hence

$$\hat{z}^{n+1} \leq \hat{z}^n + k^{(n)}(l^2 + lh^2)M. \tag{5.39}$$

Assuming that the same initial data are used to solve the differential equation and the difference equation, $\hat{z}^0 = 0$ and it is easy to see that

$$\hat{z}^n \leq nK_n(l^2 + lh^2)M = (nl)K_n(l + h)M,$$

where $K_n = k^{(1)} + k^{(2)} + \ldots + k^{(n)}$. It follows that $\hat{z}^n \to 0$ as $h, l \to 0$ since nl is fixed, thus establishing convergence provided $0 < p \leq \frac{1}{2}$.

The convergence of the fully implicit method (5.16) and the Crank-Nicolson method (5.18) can be established in a similar way.

5.4 EXTRAPOLATIONS

The component of the principal part of the local truncation error due to the approximation of the space derivative in (5.5) by its central difference replacement will not change. As with first order hyperbolic equations, however, the component arising from the (M, K) Padé approximant can be improved by one or two powers of l by the process of extrapolation.

The extrapolating procedure is carried out as in Chapter 4; first of all $\mathbf{U}^{(1)} \equiv \mathbf{U}^{(1)} (t+2l)$ is computed by applying (5.12) over two single time steps with the matrix exponential function replaced by its (M, K) Padé approximant; secondly, $\mathbf{U}^{(2)} \equiv \mathbf{U}^{(2)} (t+2l)$ is computed by applying the approximated form of (5.12) over a double time step. The extrapolated value $\mathbf{U}^{(E)} \equiv \mathbf{U}^{(E)} (t+2l)$ is determined from one of the formulas

$$\mathbf{U}^{(E)} = (2^{M+K}\mathbf{U}^{(1)} - \mathbf{U}^{(2)})/(2^{M+K} - 1) + 0(l^{M+K+2}) \tag{5.40}$$

for $M \neq K$, or

$$\mathbf{U}^{(E)} = (2^{2M}\mathbf{U}^{(1)} - \mathbf{U}^{(2)})/(2^{2M} - 1) + 0(l^{2M+3}) \tag{5.41}$$

for $M = K$.

The extrapolation formulas for the three finite difference methods derived in section 5.2 are given in Table 4.5 of Chapter 4. The principal part of the local truncation error of the extrapolated form of each finite difference method, will be of the form

$$\left(E_q l^{q-1} \frac{\partial^q u}{\partial t^q} - \frac{1}{12} h^2 \frac{\partial^4 u}{\partial x^4}\right)_m^n, \tag{5.42}$$

where now, $q = M+K+2$ for $M \neq K$ and $q = 2M+3$ for $M=K$. The constants E_q (Twizell and Khaliq, 1981) are also contained in Table 4.5 of Chapter 4.

Using the (M, K) Padé approximants, $M \neq K$, equation (5.40) may be written in the form

$$U^{(E)} = \frac{2^{M+K}}{2^{M+K}-1}[(Q_M(lA)^{-1}P_K(lA)]^2$$
$$-\frac{1}{2^{M+K}-1}[(Q_M(2lA))^{-1}P_K(2lA)] \; U(t) \quad (5.43)$$

from which it follows that a perturbation of the initial conditions satisfies

$$Z^{n+2} = \frac{2^{M+K}}{2^{M+K}-1}[(Q_M(lA)^{-1}P_K(lA)]^2$$
$$-\frac{1}{2^{M+K}-1}[(Q_M(2lA))^{-1}P_K(2lA)] \; Z^n. \quad (5.44)$$

The condition to be satisfied for A_0-stability is $|R_{M,K}(l\lambda_s)| \leq 1$, where $\lambda_s < 0$ is any eigenvalue of A, and

$$R_{M,K}(l\lambda) = \frac{2^{M+K}}{2^{M+K}-1}\left(\frac{P_K(l\lambda)}{Q_M(l\lambda)}\right)^2 - \frac{1}{2^{M+K}-1}\left(\frac{P_K(2l\lambda)}{Q_M(2l\lambda)}\right). \quad (5.45)$$

For L_0-stability, the additional condition is that $R_{M,K}(l\lambda)$ tends asymptotically to zero as $l\lambda \to -\infty$. For $M=K$, the amplification factor of the method is given by

$$R_{M,M}(l\lambda) = \frac{2^{2M}}{2^{2M}-1}\left(\frac{P_M(l\lambda)}{P_M(-l\lambda)}\right)^2 - \frac{1}{2^{2M}-1}\left(\frac{P_M(2l\lambda)}{P_M(-2l\lambda)}\right). \quad (5.46)$$

Writing $z = -l\lambda$, z is positive since all eigenvalues of A are negative; then (5.45), (5.46) can be written as

$$S_{M,K}(z) = \frac{2^{M+K}}{2^{M+K}-1}\left(\frac{P_K(-z)}{Q_M(-z)}\right)^2 - \frac{1}{2^{M+K}-1}\left(\frac{P_K(-2z)}{Q_M(-2z)}\right) \quad (5.47)$$

and

$$S_{M,M}(z) = \frac{2^{2M}}{2^{2M}-1}\left(\frac{P_M(-z)}{P_M(-z)}\right)^2 - \frac{1}{2^{2M}-1}\left(\frac{P_M(-2z)}{P_M(-2z)}\right) \quad (5.48)$$

respectively. The expression $S_{M,K}(z)$, or $S_{M,M}(z)$ when $M = K$, is known as the symbol of the extrapolated form of the method based on the (M, K) Padé approximant, as in Gourlay and Morris (1980); the notations $R_{M,K}(l\lambda)$ and $S_{M,K}(z)$ are clearly equivalent, as are $R_{M,M}(l\lambda)$ and $S_{M,M}(z)$. Henceforth, $R_{M,K}$ will be reserved for the amplification of the raw form of a method and $S_{M,K}$ will be used for its extrapolated form.

Examining the symbols of the extrapolated forms of the finite difference methods derived in section 5.2, gives first of all

$$S_{0,1}(z) = 2(1-z)^2 - (1-2z) = 1 - 2z + 2z^2 \qquad (5.49)$$

for the (0, 1) Padé approximant. It is easy to show, remembering that $z = -\lambda_s = 4p \sin^2[s\pi/\{2(N+1)\}]$ for $s = 1, 2, \ldots, N$, that $|S_{0,1}(z)| \leqslant 1$ provided $0 < p \leqslant \frac{1}{2}$. It may therefore be concluded that, in view of the more severe stability restriction, the (0, 1) method should not be extrapolated.

The symbol for the extrapolated form of the fully implicit scheme is given by

$$S_{1,0}(z) = 2\left(\frac{1}{1+z}\right)^2 - \left(\frac{1}{1+2z}\right)$$

$$= \frac{1+2z-z^2}{1+4z+5z^2+2z^3}. \qquad (5.50)$$

It may easily be verified that $|S_{1,0}(z)| \leqslant 1$ for all $z \geqslant 0$, from which it follows that the extrapolated form of the (1, 0) method is stable for all $p > 0$ (neglecting the possibility $p=0$); that is, the method is at least A_0-stable.

Dividing the numerator and the denominator of (5.50) by z^3 gives

$$S_{1,0}(z) = \left(\frac{1}{z^3} + \frac{2}{z^2} - \frac{1}{z}\right) \bigg/ \left(\frac{1}{z^3} + \frac{4}{z^2} + \frac{5}{z} + 2\right)$$

so that

$$\lim_{z \to \infty} S_{1,0}(z) = 0$$

and the extrapolated form of the fully implicit scheme (5.16) is L_0-stable. Its symbols in raw and extrapolated forms are graphed in Fig. 5.3.

Finally, the symbol for the extrapolated form of the Crank-Nicolson scheme (5.18), contained in Fig. 5.4, is given by

$$S_{1,1}(z) = \frac{4}{3}\left(\frac{1-\frac{1}{2}z}{1+\frac{1}{2}z}\right)^2 - \frac{1}{3}\left(\frac{1-z}{1+z}\right)$$

$$= \left(3 - \frac{9}{4}z^2 + \frac{5}{4}z^3\right) \bigg/ \left(3 + 6z + \frac{15}{4}z^2 + \frac{3}{4}z^3\right). \qquad (5.51)$$

from which it may be shown that $|S_{1,1}(z)| \leqslant 1$ only for $0 \leqslant z \leqslant 6 + 4\sqrt{3}$. Then, since $z = -\lambda_s = 4p \sin^2[s\pi/\{2(N+1)\}]$ for $s = 1, 2, \ldots, N$, it is easy to see that the extrapolated form of the Crank-Nicolson scheme is stable only for $0 < p \leqslant 3.23205$ (approximately).

Extrapolations

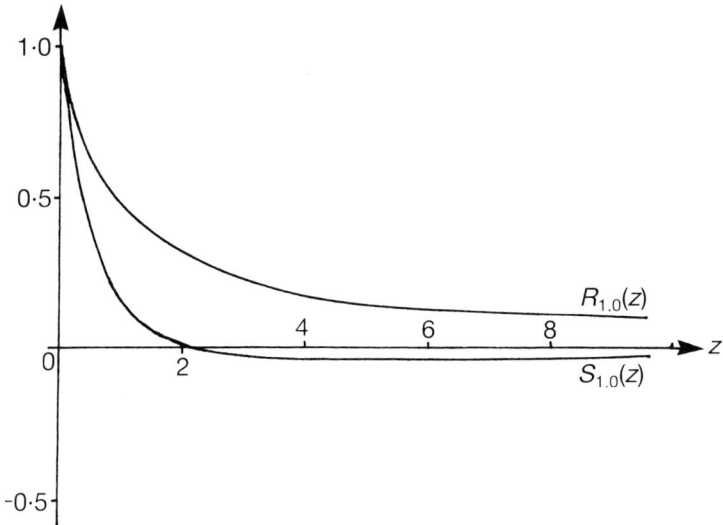

Fig. 5.3 – Amplification symbols $R_{1,0}(z)$, $S_{1,0}(z)$.

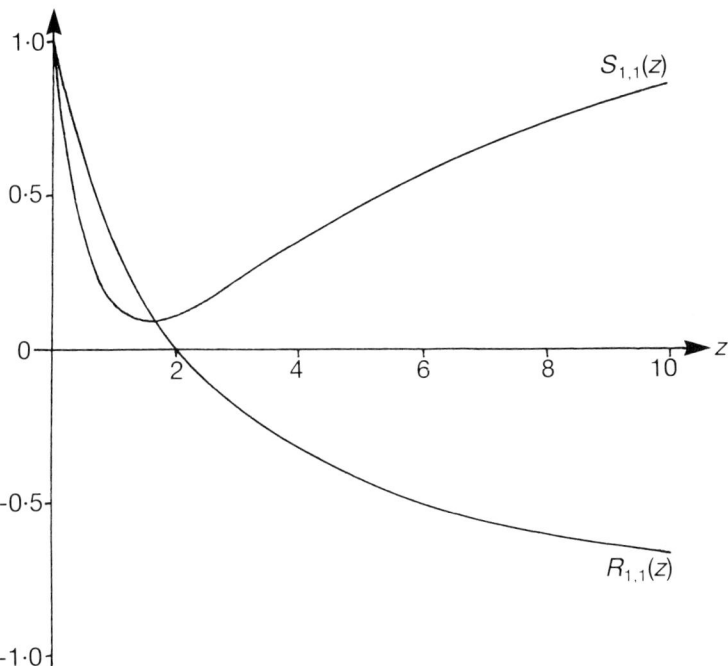

Fig. 5.4 – Amplification symbols $R_{1,1}(z)$, $S_{1,1}(z)$.

220   **Parabolic Equations**   [Ch. 5

However, this range of values of p may lead to a value of l and, subsequently, a value of r which exceeds that permitted by (5.37), and the reader must be cautious about extrapolating the Crank–Nicolson method.

Summarizing the preceding analyses, it appears that, of the low order methods, the extrapolated form of the fully implicit scheme is the most attractive to the user. Generally speaking, a method which is not L_0-stable should not be extrapolated.

5.5 EXTENSION TO TWO-SPACE DIMENSIONS

Consider now the constant coefficient heat equation in two-space variables given by

$$\frac{\partial u}{\partial t} = \frac{\partial^2 u}{\partial x^2} + \frac{\partial^2 u}{\partial y^2}; \quad 0 < x, y < X, t > 0 \tag{5.52}$$

with homogeneous Dirichlet boundary conditions ($u \equiv 0$) on the boundary $\partial\Omega$ of the square Ω defined by the lines $x = 0, y = 0, x = X, y = X$, and initial conditions

$$u(x, y, 0) = g(x, y); \quad (x, y) \in \Omega \times \partial\Omega. \tag{5.53}$$

It may be assumed that $g(x, y)$, which is a given continuous function of x, y, does not necessarily have the value zero for $(x, y) \in \partial\Omega$, so that discontinuities between initial conditions and boundary conditions may exist.

Both intervals $0 \leq x \leq X$ and $0 \leq y \leq X$ are divided into $N+1$ subintervals each of width h, so that $(N+1)h = X$ as before, and the independent variable t is incremented in steps of l. At each time level $t = nl$ ($n = 0, 1, 2, \ldots,$) the square Ω together with its boundary $\partial\Omega$ have been superimposed by a square grid of N^2 points within Ω and $N+2$ equally spaced points along each side of $\partial\Omega$.

The solution $u(x, y, t)$ of the two-dimensional heat equation is sought at each point (kh, mh, nl) in the region $R = \Omega \times [t > 0]$, where $k, m = 1, 2, \ldots, N$ and $n = 1, 2, \ldots$; for $k, m = 0$ and $N+1$ the boundary condition $u \equiv 0$ applies and for $n = 0$ the initial function g applies. As in Chapter 4, the solution $u(kh, mh, nl)$ of the differential equation at the mesh point (kh, mh, nl) will be denoted by $u_{k,m}^n$, while the theoretical solution of an approximating difference scheme will be denoted by $U_{k,m}^n$. The elements $U_{k,m}^n$ will be ordered in rows parallel to the x-axis and written in vector form as

$$\mathbf{U}^n = (U_{1,1}^n, U_{2,1}^n, \ldots, U_{N,1}^n; U_{1,2}^n, U_{2,2}^n, \ldots, U_{N,2}^n; \ldots; U_{1,N}^n, U_{2,N}^n, \ldots,$$

$$U_{N,N}^n)^T. \tag{5.54}$$

Extension to Two-Space Dimensions

The space derivatives in (5.52) may be approximated by the central difference replacements

$$\frac{\partial^2 u}{\partial x^2} = \{u(x-h, y, t) - 2u(x, y, t) + u(x+h, y, t)\}/h^2 + O(h^2) \quad (5.55)$$

and

$$\frac{\partial^2 u}{\partial y^2} = \{u(x, y-h, t) - 2u(x, y, t) + u(x, y+h, t)\}/h^2 + O(h^2). \quad (5.56)$$

Considering the time level $t = nl$, the differential equation (5.52) is applied to all N^2 interior mesh points of the square Ω with the space derivatives approximated by (5.55) and (5.56).

The N^2 applications result in a system of N^2 first order ordinary differential equations of the form (5.9), in which the square matrix A is now of order N^2 and may be split into the constituent matrices B, C such that $A = B+C$. The matrix B arises from the use of (5.55) in (5.52); it is block diagonal with tridiagonal blocks and has the form

$$B = h^{-2} \begin{bmatrix} B_1 & & & \\ & B_1 & & O \\ & & B_1 & \\ & O & & B_1 \end{bmatrix}, \quad (5.57)$$

where B_1 is the tridiagonal matrix of order N given by

$$B_1 = \begin{bmatrix} -2 & 1 & & 0 \\ 1 & -2 & 1 & \\ & \ddots & \ddots & \ddots \\ 0 & & 1 & -2 \end{bmatrix}. \quad (5.58)$$

The matrix C arises from the use of (5.56) in (5.52); it is block tridiagonal with diagonal blocks and has the form

$$C = h^{-2} \begin{bmatrix} -2I & I & & & 0 \\ I & -2I & I & & \\ & \ddots & \ddots & \ddots & \\ & & I & -2I & I \\ 0 & & & I & -2I \end{bmatrix} \quad (5.59)$$

where I is the identity matrix of order N. The N^2 eigenvalues of the matrix A are real and negative and are given by

$$\lambda_{i,j} = -4h^{-2}\left(\sin^2 \frac{i\pi}{2(N+1)} + \sin^2 \frac{j\pi}{2(N+1)}\right); \; i, j = 1, \ldots, N. \quad (5.60)$$

Solving the system of ordinary differential equations, subject to the initial condition $\mathbf{U}(0) = \mathbf{g}$, gives

$$\mathbf{U}(t) = \exp\{t(B+C)\}\mathbf{g}$$

which satisfies the recurrence relation

$$\mathbf{U}(t+l) = \exp\{l(B+C)\}\mathbf{U}(t); \quad t = 0, l, 2l, \ldots . \tag{5.61}$$

It is this recurrence relation, which is the two-dimensional analog of (5.12), which will yield a family of finite difference methods for solving the two-dimensional heat equation. In this section of the chapter low order methods (in time) will be developed and analyzed; higher order methods will be considered in section 5.6.

An explicit scheme may be determined by using the (0, 1) Padé approximant to the matrix exponential function in (5.61); this gives

$$\mathbf{U}(t+l) = [I+l(B+C)]\mathbf{U}(t) \tag{5.62}$$

which, when applied to the mesh point (kh, mh, nl) in R, gives

$$U_{k,m}^{n+1} = U_m^n + \frac{l}{h^2}(U_{k-1,m}^n - 2U_{k,m}^n + U_{k+1,m}^n)$$

$$+ \frac{l}{h^2}(U_{k,m-1}^n - 2U_m^n + U_{k,m+1}^n).$$

Gathering terms leads to the six-point, two level explicit scheme

$$U_{k,m}^{n+1} = (1-4p)U_{k,m}^n + p(U_{k,m-1}^n + U_{k-1,m}^n + U_{k+1,m}^n + U_{k,m+1}^n). \tag{5.63}$$

The local truncation error of (5.63) is found by examining

$$F_{k,m,n}(u) = u_{k,m}^{n+1} - (1-4p)u_m^n - p(u_{k,m-1}^n + u_{k-1,m}^n + u_{k+1,m}^n + u_{k,m+1}^n);$$

expanding all terms about $u_{k,m}^n$ gives

$$F_{k,m,n}(u) = l\left(\frac{\partial u}{\partial t} - \frac{\partial^2 u}{\partial x^2} - \frac{\partial^2 u}{\partial y^2}\right) + \frac{1}{2}l^2\frac{\partial^2 u}{\partial t^2} - \frac{1}{2}lh^2\left(\frac{\partial^4 u}{\partial x^4} + \frac{\partial^4 u}{\partial y^4}\right) + \ldots,$$

the right-hand side of which is evaluated at the mesh point (kh, mh, nl). The principal part of the local truncation error of (5.63) is therefore

$$\left[\frac{1}{2}l\frac{\partial^2 u}{\partial t^2} - \frac{1}{2}\left(\frac{\partial^4 u}{\partial x^4} + \frac{\partial^4 u}{\partial y^4}\right)\right]_{k,m}^n. \tag{5.64}$$

The first term of (5.64) is due to the replacement of the matrix exponential

Sec. 5.5]   Extension to Two-Space Dimensions   223

function in (5.61) with the (0, 1) Padé approximant; this first component of the principal part of the local truncation error for other Padé approximants may be found by referring to Table 4.5 of Chapter 4. The second term in (5.64) is due to the space discretization and will be present in every finite difference method, in raw or extrapolated form, originating from (5.61). This component of the local truncation error is a simple extension of the one-dimensional equivalent determined in section 5.3. The principal part of the local truncation error of any method derived from (5.61) may thus be written down, as in sections 5.3 and 5.4, by referring to Table 4.5 of Chapter 4.

Like the one-dimensional explicit method given by (5.14), equation (5.63) enjoys only conditional stability. The amplification factor $R_{0,1}(l\lambda)$ must not exceed unity in magnitude for stability in the conventional sense. This leads to $|1 + l\lambda_{i,j}| \leqslant 1$ for all $i, j = 1, \ldots, N$ from which it is easy to show that the finite difference scheme is stable provided $-l\lambda_{i,j} \leqslant 2$. Referring to (5.60), from which it may be deduced that $|\lambda_{i,j}| < 8h^{-2}$, it follows that the von Neumann necessary condition for the stability of (5.63) is $p \leqslant \frac{1}{4}$.

This result, together with the analogous result $p \leqslant \frac{1}{2}$ for the one-dimensional equation, may be extended to the heat equation

$$\frac{\partial u}{\partial t} = \frac{\partial^2 u}{\partial x_1^2} + \frac{\partial^2 u}{\partial x_2^2} + \ldots + \frac{\partial^2 u}{\partial x_s^2} \tag{5.65}$$

in s dimensions $x_1, x_2, \ldots, x_s$ and time t for which appropriate initial and boundary conditions are given.

Replacing the space derivatives in (5.65) by central difference approximations leads to a system of ordinary differential equations of the form (5.9) where the vector $\mathbf{U}$ has N^s components, and the square matrix A, which may be split into s constituent matrices such that $A = A_1 + A_2 + \ldots + A_s$, is of order N^s with real, negative eigenvalues given by

$$\lambda_{i_1, i_2, \ldots, i_s} = -4h^{-2} \sum_{j=1}^{s} [i_j\pi/\{2(N+1)\}] \tag{5.66}$$

with each $i_j = 1, 2, \ldots, N$ ($j = 1, 2, \ldots, s$). It is clear from (5.66) that $-4sh^{-2} < \lambda_{i_1, i_2, \ldots, i_s} < 0$, so that the condition $|R_{0,1}(l\lambda)| \leqslant 1$ leads to the von Neumann criterion $p \leqslant \frac{1}{2s}$ for the stability of the explicit method based on the (0, 1) Padé approximant. None of the explicit schemes, for any $s \geqslant 1$, should be extrapolated.

Returning to the two-dimensional heat flow problem and to the recurrence relation (5.61), a fully implicit method may be developed by first writing (5.61) in the form

$$\mathbf{U}(t+l) = \exp(lC) \exp(lB) \mathbf{U}(t) \tag{5.67}$$

which has error $O(l^2)$ in time. Equation (5.67) may be approximated further, also with error $O(l^2)$, by introducing the $(1,0)$ Padé approximants to the matrix exponential functions $\exp(lC)$, $\exp(lB)$ and writing (5.67) in the split form

$$(I-lB)\mathbf{U}^* = \mathbf{U}(t), \qquad (5.68)$$
$$(I-lC)\mathbf{U}(t+l) = \mathbf{U}^*. \qquad (5.69)$$

This splitting has the disadvantage of introducing an additonal, intermediate vector $\mathbf{U}^*$ with N^2 components, which must be stored. On the other hand it is easily found from (5.68) using a tridiagonal solver, the solution $\mathbf{U}(t+l)$ then being determined from (5.69) after a consistent re-ordering of the components of $\mathbf{U}^*$ and $\mathbf{U}(t+l)$. The alternative method of solution is to write $\exp[l(B+C)] = [I-l(B+C)]^{-1} + O(l^2)$ in (5.69) and to solve implicitly for $\mathbf{U}(t+l)$ from the resulting linear system which has coefficient matrix of band width $2N+1$.

Writing (5.68), (5.69) in terms of functional values at the mesh points (kh, mh, nl), as in (5.63), for instance, leads to a very complicated expression which serves little, if any, purpose; certainly, such an expression would not be used to solve for the elements of $\mathbf{U}(t+l)$.

The fully implicit algorithm defined by (5.68), (5.69) is first order accurate in time and is L_0-stable. The accuracy in time may be improved by extrapolating, as in Lawson and Morris (1978), the algorithm being written over two single time steps as

$$\begin{aligned}(I-lB)\mathbf{U}^* &= \mathbf{U}(t), \\ (I-lC)\mathbf{U}(t+l) &= \mathbf{U}^*, \\ (I-lC)\mathbf{U}^+ &= \mathbf{U}(t+l), \\ (I-lB)\mathbf{U}^{(0)}(t+2l) &= \mathbf{U}^+. \end{aligned} \qquad (5.70)$$

Eliminating the intermediate vectors $\mathbf{U}^*$, $\mathbf{U}(t+l)$, $\mathbf{U}^+$ in (5.70) gives

$$\mathbf{U}^{(0)}(t+2l) = (I-lB)^{-1}(I-lC)^{-1}(I-lC)^{-1}(I-lB)^{-1}\mathbf{U}(t) \qquad (5.71)$$

and expanding the matrix inverses leads to

$$\mathbf{U}^{(0)}(t+2l) = [I+2l(B+C) + l^2(3B^2+3C^2+2BC+2CB)]\mathbf{U}(t) + O(l^3). \qquad (5.72)$$

The Maclaurin expansion of $\exp[2l(B+C)]$ in (5.61) written over a double time step $2l$ gives

$$\mathbf{U}^{(M)}(t+2l) = [I+2l(B+C) + 2l^2(B^2+C^2+BC+CB)]\mathbf{U}(t) + O(l^3), \qquad (5.73)$$

so that (5.70) is verified to be first order accurate.

Lawson and Morris (1978) proceed by writing (5.61) as

$$\mathbf{U}(t+l) = \exp(lB)\exp(lC)\mathbf{U}(t) \qquad (5.74)$$

in addition to (5.67). Equations (5.74), (5.67) are then written over a double time step $2l$ and the matrix exponential functions are all replaced by their $(1, 0)$ Padé approximants, giving

$$\mathbf{U}^{(1)}(t+2l) = (I-2lB)^{-1}(I-2lC)^{-1}\mathbf{U}(t)$$
$$= [I+2l(B+C) + 4l^2(B^2+BC+C^2)]\mathbf{U}(t) + 0(l^3) \qquad (5.75)$$

and

$$\mathbf{U}^{(2)}(t+2l) = (I-2lC)^{-1}(I-2lB)^{-1}\mathbf{U}(t)$$
$$= [I+2l(B+C) + 4l^2(B^2+CB+C^2)]\mathbf{U}(t) + 0(l^3). \qquad (5.76)$$

Comparing (5.75), (5.76) with (5.73) it is seen that, in addition to $\mathbf{U}^{(0)}(t+2l)$, the approximations $\mathbf{U}^{(1)}(t+2l)$ and $\mathbf{U}^{(2)}(t+2l)$ are first order accurate in time, their errors being $0(l^2)$.

Lawson and Morris then point out that the extrapolation

$$\mathbf{U}^{(E)}(t+2l) = 2\mathbf{U}^{(0)} - \frac{1}{2}(\mathbf{U}^{(1)} + \mathbf{U}^{(2)}) \qquad (5.77)$$

is second order accurate in time, the first order approximants (5.72), (5.75), (5.76) having been combined to give higher order accuracy. Lawson and Morris state that (5.77) is L_0-stable.

Taking the fully implicit method in the form

$$[I-l(B+C)]\mathbf{U}(t+l) = \mathbf{U}(t) \qquad (5.78)$$

gives, at the mesh point (kh, mh, nl), the equivalent six-point scheme

$$(1+4p)U_{k,m}^{n+1} - p(U_{k,m-1}^{n+1} + U_{k-1,m}^{n+1} + U_{k+1,m}^{n+1} + U_{k,m+1}^{n+1}) = U_{k,m}^n. \qquad (5.79)$$

The principal part of the local truncation error of (5.79) is

$$\left[-\frac{1}{2}l\frac{\partial^2 u}{\partial t^2} - \frac{1}{2}h^2\left(\frac{\partial^4 u}{\partial x^4} + \frac{\partial^4 u}{\partial y^4}\right)\right]_{k,m}^n \qquad (5.80)$$

which, after extrapolation, becomes

$$\left[\frac{4}{3}l^2\frac{\partial^3 u}{\partial t^3} - \frac{1}{2}h^2\left(\frac{\partial^4 u}{\partial x^4} + \frac{\partial^4 u}{\partial y^4}\right)\right]_{k,m}^n \qquad (5.81)$$

Using the $(1, 1)$ Padé approximant to the matrix exponential function in (5.61) gives

$$[I - \frac{1}{2}l(B+C)]\mathbf{U}(t+l) = [I + \frac{1}{2}l(B+C)]\mathbf{U}(t) \qquad (5.82)$$

which, when applied to the mesh point (kh, mh, nl), becomes

$$(1+2p)U_{k,m}^{n+1} - \frac{1}{2}p(U_{k,m-1}^{n+1} + U_{k-1,m}^{n+1} + U_{k+1,m}^{n+1} + U_{k,m+1}^{n+1})$$

$$= (1-2p)U_{k,m}^{n} + \frac{1}{2}p\,(U_{k,m-1}^{n} + U_{k-1,m}^{n} + U_{k+1,m}^{n} + U_{k,m+1}^{n}). \tag{5.83}$$

The principal part of the local truncation error of (5.83) is

$$\left[-\frac{1}{12}l^2 \frac{\partial^3 u}{\partial t^3} - \frac{1}{2}h^2\left(\frac{\partial^4 u}{\partial x^4} + \frac{\partial^4 u}{\partial y^4}\right)\right]_{k,m}^{n} \tag{5.84}$$

and the scheme is A_0-stable. It does, however, impose a similar restriction on the time step as did the Crank-Nicolson method (5.18) for the one-dimensional heat equation.

Lawson and Morris (1977, 1978) discuss the implementation of this implicit method, noting that (5.61) may be approximated to $0(l^2)$ accuracy by

$$\mathbf{U}(t+l) = \exp(\tfrac{1}{2}lB)\exp(lC)\exp(\tfrac{1}{2}lB)\,\mathbf{U}(t). \tag{5.85}$$

Then, by approximating the matrix exponential functions in (5.85) by the $(1, 0)$, $(1, 1)$, $(0, 1)$ Padé approximants, (5.85) becomes

$$(I-\tfrac{1}{2}lC)\mathbf{U}^* = (I+\tfrac{1}{2}lB)\mathbf{U}(t), \tag{5.86}$$

$$(I-\tfrac{1}{2}lB)\mathbf{U}(t+l) = (I+\tfrac{1}{2}lC)\mathbf{U}^*, \tag{5.87}$$

where $\mathbf{U}^*$ is an intermediate vector. Written in this form the implicit method is known as the Peaceman-Rachford scheme (1955).

Lawson and Morris (1977) claim a saving of 30-40 per cent in computational arithmetic by implementing the Peaceman-Rachford scheme in the form

$$\mathbf{U}^{(1)}(t+l) = (I+\tfrac{1}{2}lB)\mathbf{U}(t), \tag{5.88}$$

$$(I-\tfrac{1}{2}lC)\mathbf{U}^* = \mathbf{U}^{(1)}(t+l), \tag{5.89}$$

$$(I-\tfrac{1}{2}lB)\mathbf{U}(t+l) = 2\mathbf{U}^* \tag{5.90}$$

and continuing with

$$\mathbf{U}^{(1)}(t+2l) = 2\mathbf{U}(t+l) - 2\mathbf{U}^* + \mathbf{U}^{(1)}(t+l), \tag{5.91}$$

$$(I-\tfrac{1}{2}lC)\mathbf{U}^+ = \mathbf{U}^{(1)}(t+l), \tag{5.92}$$

$$(I-\tfrac{1}{2}lB)\mathbf{U}(t+2l) = 2\mathbf{U}^+ - \mathbf{U}^{(1)}(t+2l). \tag{5.93}$$

The additonal storage requirement of this implementation is one vector of order N^2.

5.6 HIGHER ORDER APPROXIMATIONS IN TIME

5.6.1 Introduction

It is evident from the theoretical findings of earlier sections of this chapter, that there is much to be gained from using higher order Padé approximants to the matrix exponential functions in (5.12) or (5.61) provided that, in problems with discontinuities between initial conditions and boundary conditions, an A_0-stable method is not used in conjunction with a time step which is too large relative to the space discretization.

It is, therefore, constructive to examine the algorithms resulting from the use of the (0, 2), (1, 2), (2, 0), (2, 1), (2, 2) Padé approximants. All five of these algorithms require the matrix A given by (5.10) to be squared, and may lead to the numerical solution of a linear system which has a quindiagonal coefficient matrix; in the case of the (0, 2) approximant the finite difference scheme is explicit.

In addition to the improvements to be gained using higher order Padé approximants, even higher order accuracy may be achieved, in certain cases, by extrapolation. It was noted in section 5.4 of this chapter that methods which are not L_0-stable should not be extrapolated; this generalization should be adhered to when using the higher order methods of this section. One-space and two-space dimensional problems will be examined separately.

5.6.2 One-space dimension

Using first of all the (0, 2) Padé approximant in (5.12) gives

$$\mathbf{U}(t+l) = (I + lA + \tfrac{1}{2} l^2 A^2) \mathbf{U}(t). \tag{5.94}$$

The matrix A^2 is given by

$$h^4 A^2 = \begin{bmatrix} 5 & -4 & 1 & & & \\ -4 & 6 & -4 & 1 & & 0 \\ 1 & -4 & 6 & -4 & 1 & \\ & \ddots & \ddots & \ddots & \ddots & \ddots \\ 0 & 1 & -4 & 6 & -4 \\ & & & 1 & -4 & 5 \end{bmatrix} \tag{5.95}$$

and has eigenvalues $16 \sin^4[s\pi/\{2(N+1)\}]/h^4$ for $s = 1, 2, \ldots, N$.

Applying (5.94) to each mesh point (mh, nl) with $m = 2, \ldots, N-1$ and $n = 0, 1, \ldots$ gives the six-point explicit scheme

$$U_m^{n+1} = (1-2p+3p^2)U_m^n + p(1-2p)(U_{m-1}^n + U_{m+1}^n) + \frac{1}{2}p^2(U_{m-2}^n + U_{m+2}^n) \tag{5.96}$$

which has local truncation error with principal part given by

$$\left(-\frac{1}{12}h^2\frac{\partial^4 u}{\partial x^4}+\frac{1}{6}l^2\frac{\partial^3 u}{\partial t^3}\right)^n_m. \tag{5.97}$$

For $m = 1$, (5.96) must be modified to

$$U_1^{n+1} = (1-2p+\frac{5}{2}p^2)U_1^n + p(1-2p)U_2^n + \frac{1}{2}p^2 U_3^n, \tag{5.98}$$

and for $m = N$ to

$$U_N^{n+1} = \frac{1}{2}p^2 U_{N-2}^n + p(1-2p)U_{N-1}^n + (1-2p+\frac{5}{2}p^2)U_N^n, \tag{5.99}$$

neither of which achieves the accuracy of (5.96). In a paper on hyperbolic equations, Oliger (1974) showed that such lower order difference equations near the boundary do not affect the stability or convergence properties of the scheme as a whole, and the existing numerical evidence suggests that this is true for second order parabolic equations also.

A stability analysis shows that (5.94) is stable only for $0 < p \leq \frac{1}{2}$, which is the same as the stability restriction for the scheme based on the (0, 1) Padé approximant. In view of the loss of error of (5.94) when applied to the points adjacent to the boundaries $x = 0$ and $x = X$, there is no advantage to be gained by using (5.94) in preference to (5.13).

The scheme based on the (1, 2) Padé approximant must also be dismissed as being of limited usefulness. This scheme may be written in vector form as

$$(I - \frac{1}{3}lA)\mathbf{U}(t+l) = (I + \frac{2}{3}lA + \frac{1}{6}l^2A^2)\mathbf{U}(t) \tag{5.100}$$

and, when applied to the general mesh point (mh, nl) with $m = 2, \ldots, N-1$ and $n = 0, 1, 2, \ldots$, leads to an eight-point implicit scheme requiring the use of a tridiagonal solver. In spite of being implicit the scheme is stable only for $0 < p \leq 1.5$ and, if an implicit scheme is to be used at all, clearly one which is at least A_0-stable should be chosen.

Using the (2, 0) Padé approximant to the matrix exponential function in (5.12) gives

$$\mathbf{U}(t+l) = (I-lA+\tfrac{1}{2}l^2A^2)^{-1}\,\mathbf{U}(t) \tag{5.101}$$

suggesting the fully implicit scheme

$$(I-lA+\tfrac{1}{2}l^2A^2)\mathbf{U}(t+l) = \mathbf{U}(t). \tag{5.102}$$

Higher Order Approximations in Time

A stability analysis shows that (5.102) is L_0-stable and that the principal part of the local truncation error at the mesh point (mh, nl), for $m = 2, \ldots, N-1$ and $n = 0, 1, \ldots$, is

$$\left(-\frac{1}{12}h^2\frac{\partial^4 u}{\partial x^4} + \frac{1}{6}l^2\frac{\partial^3 u}{\partial t^3}\right)_m^n. \tag{5.103}$$

Like the methods based on the (0, 2) and (1, 2) Padé approximants, however, this accuracy is not achieved at points adjacent to the boundary. Unlike first order hyperbolic equations, where the use of higher order Padé approximants led to loss of accuracy at three points adjacent to the boundary $x = 0$, accuracy is lost at only one point adjacent to each boundary line $x = 0$ and $x = X$.

Applying (5.102) to the mesh point (mh, nl) at time $t = nl$ ($m = 1, \ldots, N$; $n = 0, 1, \ldots$) leads to a linear system, the unknowns of which are the components of the N-vector $\mathbf{U}(t+l)$. This linear system is of the form

$$E\mathbf{U}(t+l) = \boldsymbol{\phi}^n \tag{5.104}$$

where E is a constant matrix of order N; it has the form

$$E = \begin{bmatrix} e_4 & e_2 & e_3 & & & & & \\ e_2 & e_1 & e_2 & e_3 & & & 0 & \\ e_3 & e_2 & e_1 & e_2 & e_3 & & & \\ & \ddots & \ddots & \ddots & \ddots & \ddots & & \\ & & e_3 & e_2 & e_1 & e_2 & e_3 & \\ & & & e_3 & e_2 & e_1 & e_2 & \\ & 0 & & & e_3 & e_2 & e_4 & \end{bmatrix} \tag{5.105}$$

where

$$e_1 = 1+2p+3p^2,\ e_2 = -p-2p^2,\ e_3 = \tfrac{1}{2}p^2,$$

$$e_4 = 1+2p + \tfrac{5}{2}p^2. \tag{5.106}$$

The vector $\boldsymbol{\phi}^n = (\phi_1^n, \phi_2^n, \ldots, \phi_N^n)^T$ is easily seen from (5.102) to be $\mathbf{U}(t)$.

The solution $\mathbf{U}(t+l)$ of (5.102) is found from (5.104) by applying a quindiagonal solver. An alternative approach for finding $\mathbf{U}(t+l)$ is to write $E = I - lA + \tfrac{1}{2}l^2A^2$ in its complex factor form, namely

$$E = \tfrac{1}{2}[(1+i)I - lA][(1-i)I - lA],\ i = +\sqrt{-1}. \tag{5.107}$$

This suggests the complex splitting

$$[(1-i)I - lA]\mathbf{U}^* = \mathbf{U}(t)$$
$$\tfrac{1}{2}[(1+i)I - lA]\mathbf{U}(t+l) = \mathbf{U}^*, \tag{5.108}$$

where **U*** is an intermediate vector. In (5.108) the vectors **U**(t), **U***, **U**(t+l) must be treated as complex; the imaginary part of each element of **U**(t) is, of course, zero and the imaginary part of each element of **U**(t+l) is found to vanish. The order of accuracy of **U**(t+l) is unchanged by the complex splitting.

The vector **U**(t+l) is obtained from (5.108) by the application of two tridiagonal solvers using complex arithmetic. This is less efficient, in that it uses more CPU time, than using one quindiagonal solver with real arithmetic and (5.104) is therefore to be preferred to (5.108).

It was noted earlier that (5.102) is L_0-stable; it may therefore be extrapolated, to improve the accuracy in time, using equation (5.40), following which the principal part of its local truncation error at points not adjacent to either boundary $x = 0, x = X$ is

$$\left(-\frac{1}{12}h^2 \frac{\partial^4 u}{\partial x^4} - \frac{1}{3}l^3 \frac{\partial^4 u}{\partial t^4}\right)_m^n, \quad m = 2, \ldots, N-1. \tag{5.109}$$

The amplification symbol of the extrapolated form of (5.102), shown in Fig. 5.5, is

$$S_{2,0}(z) = \frac{4}{3}\left(\frac{1}{1+z+\frac{1}{2}z^2}\right)^2 - \frac{1}{3}\left(\frac{1}{1+2z+2z^2}\right). \tag{5.110}$$

Clearly $\lim_{z \to \infty} S_{2,0}(z) = 0$, the convergence to zero being monotonic, and the extrapolated form of (5.102) is L_0-stable.

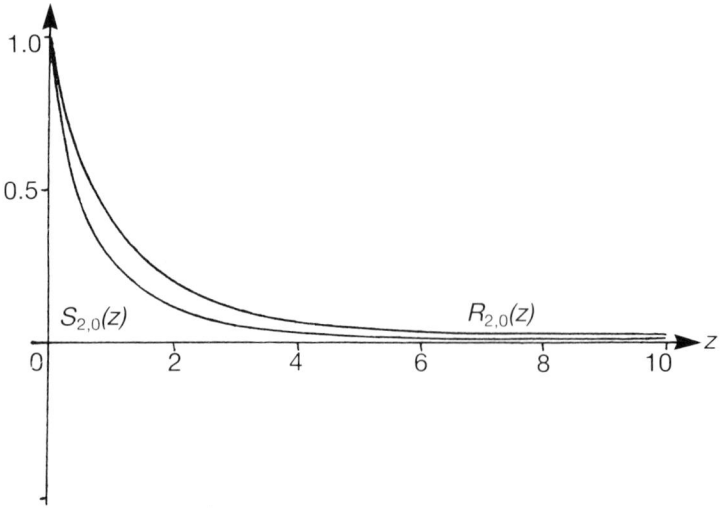

Fig. 5.5 – Amplification symbols $R_{2,0}(z), S_{2,0}(z)$.

Sec. 5.6] **Higher Order Approximations in Time** 231

The extension of (5.102) to problems with two space variables will be considered in subsection 5.6.2.

The extrapolation of (5.102) produces a scheme which is third order accurate in time. The same accuracy in time can be achieved when the $(2,1)$ Padé approximant is used in (5.12), giving

$$\mathbf{U}(t+l) = (I - \frac{2}{3} lA + \frac{1}{6} l^2 A^2)^{-1} (I + \frac{1}{3} lA) \mathbf{U}(t). \tag{5.111}$$

Written implicitly, the vector equation (5.111) becomes

$$(I - \frac{2}{3} lA + \frac{1}{6} l^2 A^2) \mathbf{U}(t+l) = (I + \frac{1}{3} lA) \mathbf{U}(t) \tag{5.112}$$

which, when applied to the mesh point (mh, nl) for $m = 1, 2, \ldots, N$ and $n = 0, 1, \ldots$, leads to a linear system of the form (5.104). The elements of the matrix E in (5.104) are now given by

$$e_1 = 1 + \frac{4}{3} p + p^2, \quad e_2 = -\frac{2}{3} p - \frac{2}{3} p^2, \quad e_3 = \frac{1}{6} p^2,$$

$$e_4 = 1 + \frac{4}{3} p + \frac{5}{6} p^2 \tag{5.113}$$

and the elements of the vector $\boldsymbol{\phi}^n$ by

$$\phi_1^n = (1 - \frac{2}{3} p) U_1^n + \frac{1}{3} p U_2^n,$$

$$\phi_m^n = \frac{1}{3} p U_{m-1}^n + (1 - \frac{2}{3} p) U_m^n + \frac{1}{3} p U_{m+1}^n; \quad m = 2, \ldots, N-1,$$

$$\phi_N^n = \frac{1}{3} p U_{N-1}^n + (1 - \frac{2}{3} p) U_N^n. \tag{5.114}$$

The solution $\mathbf{U}(t+l)$ of (5.112) is determined by applying a quindiagonal solver, described in subsection 1.2.3 of Chapter 1, to (5.104) with (5.113), (5.114). In view of the discussion of (5.108), it is not worthwhile to consider a complex splitting of (5.112).

The principal part of the local truncation error of (5.112) at the interior mesh points (mh, nl), $(m = 2, \ldots, N-1; n = 0, 1, \ldots)$, is given by

$$\left(-\frac{1}{12} h^2 \frac{\partial^4 u}{\partial x^4} + \frac{1}{72} l^3 \frac{\partial^4 u}{\partial t^4} \right)_m^n, \tag{5.115}$$

this accuracy not being achieved for $m = 1, N$. Following extrapolation of the algorithm (5.112) using (5.40), the principal part of the local truncation error becomes

$$\left(-\frac{1}{12}h^2\frac{\partial^4 u}{\partial x^4} - \frac{8}{945}l^4\frac{\partial^5 u}{\partial t^5}\right)_m^n ; \quad m \neq 1, N. \tag{5.116}$$

The amplification factor of (5.112) is

$$R_{2,1}(z) = (1 - \frac{1}{3}z)/(1 + \frac{2}{3}z + \frac{1}{6}z^2) \tag{5.117}$$

from which it follows that (5.112) is L_0-stable. The function $R_{2,1}(z)$ is negative for $z > 3$, as can be seen from Fig. 5.6, and does, in fact, tend to zero more slowly than the fully implicit method (5.15) based on the $(1,0)$ Padé approximant. The amplification symbol of the extrapolated from of (5.112) is

$$S_{2,1}(z) = \frac{8}{7}\left(\frac{1 - \frac{1}{3}z}{1 + \frac{2}{3}z + \frac{1}{6}z^2}\right)^2 - \frac{1}{7}\left(\frac{1 - \frac{2}{3}z}{1 + \frac{4}{3}z + \frac{2}{3}z^2}\right), \tag{5.118}$$

the graph of which is contained in Fig. 5.6. The extrapolated form of (5.112) is easily shown to be L_0stable also.

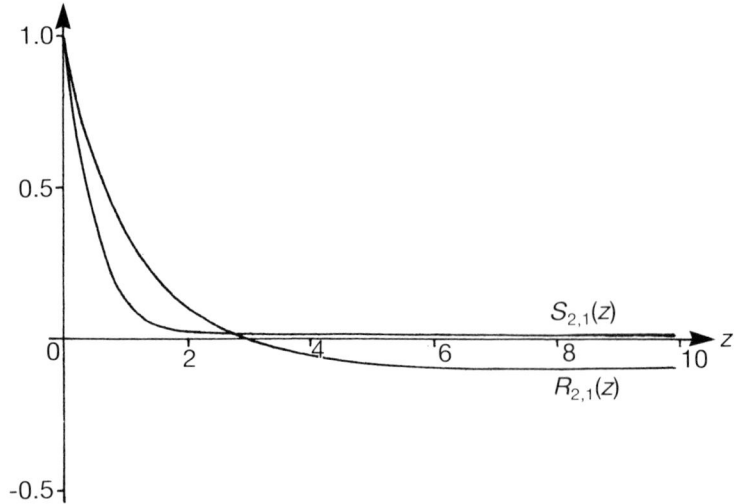

Fig. 5.6 — Amplification symbols $R_{2,1}(z)$, $S_{2,1}(z)$.

Higher Order Approximations in Time

The final algorithm to be discussed in this subsection of the chapter is that obtained by replacing the matrix exponential function in (5.12) by its (2, 2) Padé approximant giving

$$\mathbf{U}(t+l) = (I - \tfrac{1}{2}lA + \tfrac{1}{12}l^2A^2)^{-1}(I + \tfrac{1}{2}lA + \tfrac{1}{12}l^2A^2)\mathbf{U}(t). \qquad (5.119)$$

Written implicitly (5.119) becomes

$$(I - \tfrac{1}{2}lA + \tfrac{1}{12}l^2A^2)\,\mathbf{U}(t+l) = (I + \tfrac{1}{2}lA + \tfrac{1}{12}l^2A^2)\,\mathbf{U}(t) \qquad (5.120)$$

and, applying (5.120) to each of the N mesh points at time level $t = nl$, again leads to $\mathbf{U}(t+l)$ being determined from a linear system of the form (5.104). The elements of the matrix E in (5.104) are now given by

$$e_1 = 1 + p + \tfrac{1}{2}p^2,\ e_2 = -\tfrac{1}{2}p - \tfrac{1}{3}p^2,\ e_3 = \tfrac{1}{12}p^2,$$

$$e_4 = 1 + p + \tfrac{5}{12}p^2, \qquad (5.121)$$

while the elements of $\boldsymbol{\phi}^n$ become

$$\phi_1^n = (1 - p + \tfrac{5}{12}p^2)\,U_1^n + p(\tfrac{1}{2} - \tfrac{1}{3}p)\,U_2^n + \tfrac{1}{12}p^2 U_3^n,$$

$$\phi_m^n = \tfrac{1}{12}p^2 U_{m-2}^n + p(\tfrac{1}{2} - \tfrac{1}{3}p)\,U_{m-1}^n + (1 - p + \tfrac{1}{2}p^2)U_m^n$$

$$+ p(\tfrac{1}{2} - \tfrac{1}{3}p)\,U_{m-1}^n + \tfrac{1}{12}p^2 U_{m+2}^n;\ m = 2,\ldots,N-1,$$

$$\phi_N^n = \tfrac{1}{12}p^2 U_{N-2}^n + p(\tfrac{1}{2} - \tfrac{1}{3}p)\,U_{N-1}^n + (1 - p + \tfrac{5}{12}p^2)U_N^n. \qquad (5.122)$$

The solution of (5.120) is determined from (5.104) with (5.121), (5.122) using a quindiagonal solver. As with the methods based on the (2, 0) and (2, 1) Padé approximants, the reader is not advised to determine the solution of (5.120) by considering a complex splitting of its quadratic factors.

The amplification factor of (5.120) is given by

$$R_{2,2}(z) = (1 - \tfrac{1}{2}z + \tfrac{1}{12}z^2)/(1 + \tfrac{1}{2}z + \tfrac{1}{12}z^2) \qquad (5.123)$$

which is seen never to exceed unity in modulus but to approach $+1$ as $z \to \infty$.

The method (5.120) is thus A_0-stable and its amplification factor is depicted in Fig. 5.7. The symbol $R_{2,2}(z)$ is never negative for positive values of z and the difficulties regarding oscillations encountered with the (1, 1) Padé approximant (the Crank-Nicolson method) in subsection 5.3.2, do not arise with the (2, 2) method.

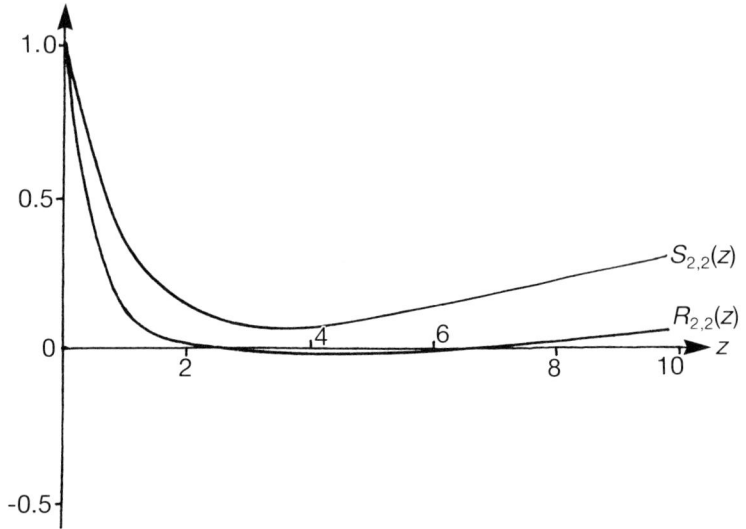

Fig. 5.7 – Amplifcation symbols $R_{2,2}(z)$, $S_{2,2}(z)$.

It was noted earlier that the extrapolation of an A_0-stable method will magnify any difficulties apparent in the raw form of the method. The implicit method (5.120) may be extrapolated using (5.41). The amplification symbol of the extrapolated form of the method is graphed in Fig. 5.7 and is given by

$$S_{2,2}(z) = \frac{16}{15}\left(\frac{1 - \frac{1}{2}z + \frac{1}{12}z^2}{1 + \frac{1}{2}z + \frac{1}{12}z^2}\right)^2 - \frac{1}{15}\left(\frac{1 - z + \frac{1}{3}z^2}{1 + z + \frac{1}{3}z^2}\right). \tag{5.124}$$

A mechanical calulation shows that $S_{2,2}(z)$, like $R_{2,2}(z)$ given by (5.123), tends asymptotically to $+1$ as $z \to \infty$. However, $S_{2,2}(z)$ is negative for $2.7 < z < 6.3$ (approximately), from which it follows that for p in the approximate range $0.675 < p < 1.575$, the oscillations encountered with the Crank-Nicolson method (5.17), will arise during the use of the extrapolated form of (5.120).

For $m \neq 1, N$ the local truncation error of (5.120) is

$$\left(-\frac{1}{12}h^2 \frac{\partial^4 u}{\partial x^4} + \frac{1}{720}l^4 \frac{\partial^5 u}{\partial t^5}\right)_m^n, \tag{5.125}$$

which, following extrapolation, becomes

$$\left(-\frac{1}{12}h^2 \frac{\partial^4 u}{\partial x^4} - \frac{1}{1890}l^6 \frac{\partial^7 u}{\partial t^7}\right)_m^n. \tag{5.126}$$

In the event of a higher order approximant to the space derivative being used in (5.5), instead of (5.8), further improvements to the accuracy in time can be attained using even higher order Padé approximants in (5.12).

An alternative way of devising a method which is third order accurate in time is, following Gourlay and Morris (1980), to define an operator $L(l, \psi)$ by

$$L(l, \psi) = [I - l(1-\psi)A]^{-1} [I + l\psi A], \tag{5.127}$$

so that $L(2l, \psi) = [I - 2l(1-\psi)A]^{-1} [I + 2l\psi A]$ and $L(3l, \psi) = [I - 3l(1-\psi)A]^{-1} [I + 3l\psi A]$, and to consider the family of approximations

$$\mathbf{U}^{(1)}(t+3l) = [L(l, \psi)]^3 \mathbf{U}(t), \tag{5.128}$$

$$\mathbf{U}^{(2)}(t+3l) = L(2l, \psi)L(l, \psi)\mathbf{U}(t), \tag{5.129}$$

$$\mathbf{U}^{(3)}(t+3l) = L(3l, \psi)\mathbf{U}(t). \tag{5.130}$$

The three estimates (5.128), (5.129), (5.130) are combined to give a relation of the form

$$\mathbf{U}(t+3l) = \alpha \mathbf{U}^{(1)} + \beta \mathbf{U}^{(2)} + (1-\alpha-\beta)\mathbf{U}^{(3)} \tag{5.131}$$

which agrees with the Taylor expansion of $\mathbf{U}(t+3l)$ about $\mathbf{U}(t)$ up to and including the term in l^3; this expansion is

$$\mathbf{U}(t+3l) = (I + 3lA + \frac{9}{2}l^2 A^2 + \frac{9}{2}l^3 A^3 + \frac{27}{8}l^4 A^4 \ldots)\mathbf{U}(t). \tag{5.132}$$

It is tiresome, though not difficult, to check that third order accuracy is achieved by (5.131), for any value of ψ, if $\alpha = \frac{9}{2}$ and $\beta = -\frac{9}{2}$. Like the three methods based on the (2, 0), (2, 1), (2, 2) Padé approximants discussed above, any method derived from (5.128), (5.129), (5.130), (5.131) will also be subject to the error associated with (5.8) and the space discretizations giving rise to the matrix A.

The stability of any scheme arising from (5.128), (5.129), (5.130), (5.131) must also be investigated. The reader will have concluded that L_0-stability is the property to aim for, and the stability analysis will look for the values of ψ which give L_0-stability together with third order accuracy.

The amplification factor of the method (5.131) is

$$S(z) = \alpha\left(\frac{1-\psi z}{1+(1-\psi)z}\right)^3 + \beta\left(\frac{1-\psi z}{1+(1-\psi)z}\right)\left(\frac{1-2\psi z}{1+2(1-\psi)z}\right) + (1-\alpha-\beta)\left(\frac{1-3\psi z}{1+3(1-\psi)z}\right) \tag{5.133}$$

and the reader can easily show that, for $\alpha = -\beta = \frac{9}{2}$, $|S(z)| \leq 1$ and

$$\lim_{z \to \infty} S(z) = \psi(\psi+\tfrac{1}{2})(\psi+2)/(\psi-1), \tag{5.134}$$

so that for L_0-stability $\psi = 0, -\tfrac{1}{2}$ or -2.

In their discussion of these three values of ψ, Gourlay and Morris (1980) show quite conclusively that the choice $\psi = 0$ gives the best numerical result when applied to the problem discussed in section 5.2. This choice of ψ is, of course, exactly equivalent to choosing the $(1, 0)$ Padé approximant.

The implementation of the third order algorithm defined by (5.128), (5.129), (5.130), (5.131) is not as straightforward as the elegant look of these four equations would suggest. The strategy to be followed is as follows:

(i) intermediate vectors $\mathbf{V}^{(1)}$ and $\mathbf{V}^{(2)}$ are computed and used to determine $\mathbf{U}^{(1)}$ as follows:

$$[I - l(1-\psi)A]\mathbf{V}^{(1)} = [I + l\psi A]\mathbf{U}(t), \tag{5.135}$$

$$[I - l(1-\psi)A]\mathbf{V}^{(2)} = [I + l\psi A]\mathbf{V}^{(1)}, \tag{5.136}$$

$$[I - l(1-\psi)A]\mathbf{U}^{(1)} = [I + l\psi A]\mathbf{V}^{(2)}; \tag{5.137}$$

(ii) the intermediate vector $\mathbf{V}^{(1)}$ is used to determine $\mathbf{U}^{(2)}$ from

$$[I - 2l(1-\psi)A]\mathbf{U}^{(2)} = [I + 2l\psi A]\mathbf{V}^{(1)}; \tag{5.138}$$

(iii) the vector $\mathbf{U}^{(3)}$ is determined from

$$[I - 3l(1-\psi)A]\mathbf{U}^{(3)} = [I + 3l\psi A]\mathbf{U}(t); \tag{5.139}$$

(iv) $U(t+3l)$ is determined from

$$\mathbf{U}(t+3l) = \frac{9}{2}\mathbf{U}^{(1)} - \frac{9}{2}\mathbf{U}^{(2)} + \mathbf{U}^{(3)}. \tag{5.140}$$

It is clear that this strategy requires five applications of a tridiagonal solver to determine $\mathbf{U}(t+3l)$, compared with one application of a quindiagonal solver using, say, the equivalent third order method based on the $(2, 1)$ Padé approximant.

Gourlay and Morris (1980) use the same multistage concept to derive an L_0-stable method which is fourth order accurate in time. The method is defined by the sequence of vectors

$$\mathbf{U}^{(1)}(t+4l) = [L(l, \psi)]^4 \mathbf{U}(t), \tag{5.141}$$
$$\mathbf{U}^{(2)}(t+4l) = L(3l, \psi)L(l, \psi)\mathbf{U}(t), \tag{5.142}$$
$$\mathbf{U}^{(3)}(t+4l) = [L(2l, \psi)]^2 \mathbf{U}(t), \tag{5.143}$$
$$\mathbf{U}^{(4)}(t+4l) = L(2l, \psi)[L(l, \psi)]^2 \mathbf{U}(t), \tag{5.144}$$
$$\mathbf{U}^{(5)}(t+4l) = L(4l, \psi)\mathbf{U}(t). \tag{5.145}$$

These five estimates (5.141) through (5.145), are combined to give a relation of the form

$$\mathbf{U}(t+4l) = \alpha\mathbf{U}^{(1)} + \beta\mathbf{U}^{(2)} + \gamma\mathbf{U}^{(3)} + \delta\mathbf{U}^{(4)} + (1-\alpha-\beta-\gamma-\delta)\mathbf{U}^{(5)} \tag{5.146}$$

in which the parameters α, β, γ, δ are chosen so that (5.146) agrees with the Taylor expansion of $\mathbf{U}(t+4l)$ up to and including the term in l^4; this expansion is

$$\mathbf{U}(t+4l) = (I + 4lA + 8l^2A^2 + \frac{32}{3}l^3A^3 + \frac{32}{3}l^4A^4 + \frac{128}{15}l^5A^5 + \ldots)\mathbf{U}(t). \tag{5.147}$$

Gourlay and Morris show that fourth order accuracy is attained if

$$\theta = \frac{1}{2} \text{ and } 10\alpha + 6\beta + 8\gamma + 9\delta = \frac{32}{3} \tag{5.148}$$

or

$$8 - 6\alpha - 3\beta - 4\gamma - 5\delta = 0, \quad 2\alpha + \delta = \frac{16}{3}, \text{ and}$$

$$\left(\frac{16}{3} + \alpha - 3\beta\right) - 4\left(\frac{16}{3} + \alpha - 3\beta\right)\theta + \left(\frac{112}{3} + 4\alpha - 15\beta\right)\theta^2 - \left(32 - 6\beta\right)\theta^3 = 0. \tag{5.149}$$

To these conditions must be added the requirements to guarantee L_0-stability. The amplification symbol of (5.146) is given by

$$S(z) = \alpha\left(\frac{1-\psi z}{1+(1-\psi)z}\right)^4 + \beta\left(\frac{1-3\psi z}{1+3(1-\psi)z}\right)\left(\frac{1-\psi z}{1+(1-\psi)z}\right)$$

$$+ \gamma\left(\frac{1-2\psi z}{1+2(1-\psi)z}\right)^2 + \delta\left(\frac{1-2\psi z}{1+2(1-\psi)z}\right)\left(\frac{1-\psi z}{1+(1-\psi)z}\right)^2$$

$$+ (1-\alpha-\beta-\gamma-\delta)\left(\frac{1-4\psi z}{1+4(1-\psi)z}\right), \tag{5.150}$$

and for L_0-stability, it is necessary that $\lim_{z \to \infty} S(z) = 0$. This leads to

$$\frac{\alpha\psi^4}{(1-\psi)^4} + \frac{\beta\psi^2}{(1-\psi)^2} + \frac{\gamma\psi^2}{(1-\psi)^2} - \frac{\delta\psi^3}{(1-\psi)^3} - (1-\alpha-\beta-\gamma-\delta)\frac{\psi}{1-\psi} = 0,$$

(5.151)

so that $\alpha, \beta, \gamma, \delta, \psi$ must be chosen so that either (5.148), (5.151) or (5.149), (5.151) are satisfied.

Gourlay and Morris (1980) list five sets of values of the parameters $\alpha, \beta, \gamma, \delta, \psi$ for each of these two pairs of conditions to be satisfied. They conclude that the sets

$$\alpha = 8, \beta = \frac{40}{9}, \gamma = 0, \delta = \frac{-32}{3}, \psi = 0 \qquad (5.152)$$

$$\alpha = 0, \beta = \frac{16}{9}, \gamma = -6, \delta = \frac{16}{3}, \psi = 0 \qquad (5.153)$$

are equally efficient and produce similar accuracies, with (5.152) being marginally superior for larger values of p.

The five sets of values of the parameters relating to (5.148), (5.151) in which $\psi = \frac{1}{2}$ are noted to be more efficient in that they require fewer applications of a tridiagonal solver. However, Gourlay and Morris report substantially larger errors for each of these five sets of parameters, compared with the five sets relating to $\psi = 0$, for the same value of $p = l/h^2$; their observations are based on numerical experiments for solving the problem introduced in section 5.2. They explored the possibility, suggested by the increased efficiency of the methods for which $\psi = \frac{1}{2}$, of decreasing p and using more time steps, but they found the ensuing computed results still inferior to the methods for which $\psi = 0$.

The implementation of the general five-stage formulation given by (5.141) through (5.146) is carried out by the following strategy:

(i) intermediate vectors $\mathbf{V}^{(1)}, \mathbf{V}^{(2)}, \mathbf{V}^{(3)}$ are computed and used to determine $\mathbf{U}^{(1)}(t+4l)$ as follows:

$$[I-l(1-\psi)A]\mathbf{V}^{(1)} = [I+l\psi A]\mathbf{U}(t), \qquad (5.154)$$
$$[I-l(1-\psi)A]\mathbf{V}^{(2)} = [I+l\psi A]\mathbf{V}^{(1)}, \qquad (5.155)$$
$$[I-l(1-\psi)A]\mathbf{V}^{(3)} = [I+l\psi A]\mathbf{V}^{(2)}, \qquad (5.156)$$
$$[I-l(1-\psi)A]\mathbf{U}^{(1)} = [I+l\psi A]\mathbf{V}^{(3)}, \qquad (5.157)$$

(ii) the intermediate vector $\mathbf{V}^{(1)}$ is used to determine $\mathbf{U}^{(2)}(t+4l)$ from

$$[I-3l(1-\psi)A]\mathbf{U}^{(2)} = [I+3l\psi A]\mathbf{V}^{(1)}; \qquad (5.158)$$

Sec. 5.6] Higher Order Approximations in Time 239

(iii) an intermediate vector $\mathbf{V}^{(4)}$ is computed and used to determine $\mathbf{U}^{(3)}(t+4l)$ as follows:

$$[I-2l(1-\psi)A]\mathbf{V}^{(4)} = [I+2l\psi A]\mathbf{U}(t), \qquad (5.159)$$

$$[I-2l(1-\psi)A]\mathbf{U}^{(3)} = [I+2l\psi A]\mathbf{V}^{(4)}; \qquad (5.160)$$

(iv) the intermediate vector $\mathbf{V}^{(2)}$ is used to determine $\mathbf{U}^{(4)}(t+4l)$ from

$$[I-2l(1-\psi)A]\mathbf{U}^{(4)} = [I+2l\psi A]\mathbf{V}^{(2)}; \qquad (5.161)$$

(v) the vector $\mathbf{U}^{(5)}(t+4l)$ is determined from

$$[I-4l(1-\psi)A]\mathbf{U}^{(5)} = [I+4l\psi A]\mathbf{U}(t); \qquad (5.162)$$

(vi) $\mathbf{U}(t+4l)$ is determined from (5.146).

In any set of parameter values in which none of $\alpha, \beta, \gamma, \delta$ are equal to zero, the strategy for obtaining the solution requires nine applications of a tridiagonal solver, compared with three applications of a quindiagonal solver using, say, the extrapolated form of the method based on the (2, 1) Padé approximant which is L_0-stable and fourth order accurate in time. Fourth order accuracy may also be achieved using the method based on the (2, 2) Padé approximant. The implementation of this method requires only one application of a quindiagonal solver and is therefore more economical than the extrapolated form of the (2, 1) method. It is, however, only A_0-stable and is likely to produce less satisfactory results for certain problems.

The two sets of parameters given by (5.152), (5.153) each have one of $\alpha, \beta, \gamma, \delta$ equal to zero, leading in each case to a decrease in the count of tridiagonal solver applications. The reader can easily check that in each of these cases, the count is reduced to 7: relating to (5.152), in which $\gamma = 0$, the steps described by equations (5.159) and (5.160) are not required, and relating to (5.153), in which $\alpha = 0$, the steps described by (5.156) and (5.157) are not required.

To illustrate the behaviour of some of the schemes discussed in this subsection, the model problem (5.5) with the conditions introduced in section 5.2, namely $X = 2$ and $g(x) = 1$ for $0 \leq x \leq 2$, is solved using the (2, 0), (2, 1) L_0-stable methods and their extrapolated forms, the (2, 2) A_0-stable method, and the best third- and fourth- order accurate methods of Gourlay and Morris. All methods were tested using $l = 0.025$, $h = 0.05$ giving $p = 10$, and $l = 0.1$, $h = 0.05$ giving $p = 40$. The maximum errors at time $t = 1.2$ are given in Table 5.1 where GM refers to Gourlay and Morris.

The reader will note that for $p = 10$ the second order method based on the (2, 0) approximant gives better results than the best third order method of Gourlay and Morris ($\alpha = -\beta = \frac{9}{2}$, $\psi = 0$) and that the third order method

240   **Parabolic Equations**   [Ch. 5

Table 5.1

Method	Order	Maximum errors	
		$p = 10$	$p = 40$
(2, 0) method (5.102)	2	0.18(−3)	0.17(−2)
Extrapolated (2, 0)	3	0.74(−4)	0.41(−3)
(2, 1) method (5.112)	3	0.69(−4)	0.28(−4)
GM (5.140)	3	0.47(−3)	0.17(−2)
Extrapolated (2, 1)	4	0.67(−4)	0.87(−4)
(2, 2) method (5.120)	4	0.66(−4)	0.68(−1)
GM (5.146), (5.152)	4	0.17(−4)	0.38(−3)
GM (5.146), (5.153)	4	0.13(−5)	0.83(−3)

based on the (2, 1) approximant gives results which are almost as accurate as some of the fourth order methods. For $p = 40$, the second order method based on the (2, 0) approximant is seen to be as accurate as the best third order method of Gourlay and Morris, and the third order method based on the (2, 1) approximant is seen to give better numerical results than any of the fourth order methods.

The reader will observe from Table 5.1 that extrapolation does not, unfortunately, always produce the improvement in accuracy predicted by the theory. Table 5.1 also shows that the A_0-stable method (5.120) which is based on the diagonal (2, 2) Padé approximant gives poor results for large values of p.

5.6.3 Two-space dimensions

Some of the difficulties encountered in implementing the methods developed in subsection 5.6.2, are magnified in the case of two space dimensions. In particular, the square matrix A is now of order N^2 and is split into the form $A = B+C$ where B, C are given by (5.57), (5.58), (5.59), so that when the second power of A is required the matrices B^2, BC, CB, C^2 must be determined. Another difficulty is with the poor stability properties of explicit methods and with the poor results given by A_0-stable methods when used to solve certain problems.

The method which will be developed in this subsection is that based on the (2, 0) Padé approximant to the exponential function. This method is second order accurate in time, the same as the Peaceman-Rachford method discussed in section 5.5; it is L_0-stable, unlike the Peaceman-Rachford method which is A_0-stable. In its extrapolated form the (2, 0) method is third order accurate and retains the property of L_0-stability.

The uniform space discretization described in section 5.5 will be retained and the theoretical solution of the scheme is given by the vector **U** of (5.54).

Sec. 5.6]   **Higher Order Approximations in Time**   241

The solution of the problem satisfies (5.61) as in section 5.5, and this recurrence relation will be used in the two alternate forms (5.74), (5.67) both of which have $O(l^2)$ error; a simple combination of (5.74), (5.67) after the (2, 0) Padé approximants to $\exp(lB)$ and $\exp(lC)$ have been introduced, improves the error to $O(l^3)$. Later, this error will be improved further to $O(l^4)$, following extrapolation of the combination.

Using the (2, 0) Padé approximant, equations (5.74), (5.67) may be written

$$\mathbf{U}^*(t+l) = (I-lB+\tfrac{1}{2}l^2B^2)^{-1}(I-lC+\tfrac{1}{2}l^2C^2)^{-1}\mathbf{U}(t) \qquad (5.163)$$

$$\mathbf{U}^+(t+l) = (I-lC+\tfrac{1}{2}l^2C^2)^{-1}(I-lB+\tfrac{1}{2}l^2B^2)^{-1}\mathbf{U}(t), \qquad (5.164)$$

respectively. Expanding the matrix inverses in (5.163), (5.164) confirms that each is only first order accurate in time when compared with the Maclaurin expansion of $\exp[l(B+C)]$ given by

$$\exp[l(B+C)] = I + l(B+C) + \tfrac{1}{2}l^2(B^2+C^2+BC+CB)$$
$$+ \tfrac{1}{6}l^3(B^3+C^3+B^2C+BC^2+CB^2+C^2B+BCB+CBC) + \ldots \quad (5.165)$$

Combining $\mathbf{U}^*$ and $\mathbf{U}^+$ by the linear relation

$$\mathbf{U}(t+l) = \tfrac{1}{2}(\mathbf{U}^*+\mathbf{U}^+) \qquad (5.166)$$

gives

$$\mathbf{U}(t+l) = [I + l(B+C) + \tfrac{1}{2}l^2(B^2+C^2+BC+CB) + O(l^3)]\mathbf{U}(t) \qquad (5.167)$$

which is second order accurate in time.

The splittings (5.163), (5.164) and the relation (5.166) are formalized by the following algorithm which requires four applications of a quindiagonal solver:

$$(I-lC+\tfrac{1}{2}l^2C^2)\mathbf{V}^{(1)} = \mathbf{U}(t),$$
$$(I-lB+\tfrac{1}{2}l^2B^2)\mathbf{U}^* = \mathbf{V}^{(1)};$$
$$(I-lB+\tfrac{1}{2}l^2B^2)\mathbf{V}^{(2)} = \mathbf{U}(t),$$
$$(I-lC+\tfrac{1}{2}l^2C^2)\mathbf{U}^+ = \mathbf{V}^{(2)};$$
$$\mathbf{U}(t+l) = \tfrac{1}{2}(\mathbf{U}^*+\mathbf{U}^+). \qquad (5.168)$$

It is a very easy task to verify that the algorithm is L_0-stable and it is therefore left as an exercise for the reader.

The second order accuracy of the method may be extrapolated to third order by, first of all, considering (5.163), (5.164) over two single time steps to give

$$\mathbf{U}^{**}(t+2l) = [(I-lB+\tfrac{1}{2}l^2B^2)^{-1}(I-lC+\tfrac{1}{2}l^2C^2)^{-1}]^2\mathbf{U}(t), \qquad (5.169)$$

$$\mathbf{U}^{++}(t+2l) = [(I-lC+\tfrac{1}{2}l^2C^2)^{-1}(I-lB+\tfrac{1}{2}l^2B^2)^{-1}]^2\mathbf{U}(t). \qquad (5.170)$$

Expanding the matrix inverses in (5.169), (5.170) verifies that each is only first order accurate when compared with the expansion of $\exp[2l(B+C)]$ given by

$$\exp[2l(B+C)] = I + 2l(B+C) + 2l^2(B^2+C^2+BC+CB)$$
$$+ \tfrac{4}{3} l^3 (B^3+C^3+B^2C+BC^2+C^2B+CB^2+BCB+CBC) + \ldots \quad (5.171)$$

Substituting the expansions of (5.169), (5.170) in

$$\mathbf{U}^{(0)}(t+2l) = \tfrac{1}{2}(\mathbf{U}^{**}+\mathbf{U}^{++}),$$

however, gives

$$\mathbf{U}^{(0)}(t+2l) = [I+2l(B+C)+2l^2(B^2+C^2+BC+CB)$$
$$+ l^3\{B^3+C^3+\tfrac{3}{2}(BC^2+B^2C+CB^2+C^2B)+BCB+CBC\}+O(l^4)]\mathbf{U}(t), \quad (5.172)$$

showing that $\mathbf{U}^{(0)}$ is second order accurate in time.

Writing (5.163), (5.164) over a double time step $2l$ gives

$$\mathbf{U}^{(1)}(t+2l) = (I-2lB+2l^2B^2)^{-1}(I-2lC+2l^2C^2)^{-1}\mathbf{U}(t), \quad (5.173)$$
$$\mathbf{U}^{(2)}(t+2l) = (I-2lC+2l^2C^2)^{-1}(I-2lB+2l^2B^2)^{-1}\mathbf{U}(t). \quad (5.174)$$

Expanding the matrix inverses in (5.173), (5.174) gives

$$\mathbf{U}^{(1)}(t+2l) = [I+2l(B+C)+2l^2(B^2+C^2+2BC)$$
$$+4l^3(B^2C+BC^2)+O(l^4)]\mathbf{U}(t), \quad (5.175)$$

$$\mathbf{U}^{(2)}(t+2l) = [I+2l(B+C)+2l^2(B^2+C^2+2CB)$$
$$+4l^3(C^2B+CB^2)+O(l^4)]\mathbf{U}(t), \quad (5.176)$$

respectively, showing that each is first order accurate in time.

The linear combination of (5.172), (5.175), (5.176), defined by

$$\mathbf{U}^{(E)}(t+2l) = \frac{4}{3}\mathbf{U}^{(0)} - \frac{1}{6}(\mathbf{U}^{(1)}+\mathbf{U}^{(2)}),$$

is easily verified to be third order accurate in time when compared with the Maclaurin expansion (5.171).

The principal part of the local truncation error of (5.166) when applied to the mesh point (kh, mh, nl), with $k, m = 2, \ldots, N-1$ and $n = 0, 1, 2, \ldots$, is easily written down from (5.64) and Table 4.5 of Chapter 4. It is

$$\left[\frac{1}{6}l^2 \frac{\partial^3 u}{\partial t^3} - \frac{1}{12}h^2\left(\frac{\partial^4 u}{\partial x^4}+\frac{\partial^4 u}{\partial y^2}\right)\right]_{k,m}^n \quad ; k, m \neq 1, N \quad (5.177)$$

which, following extrapolation using (5.178), becomes

$$\left[-\frac{1}{3}l^3\frac{\partial^4 u}{\partial t^4} - \frac{1}{12}h^2\left(\frac{\partial^4 u}{\partial x^4} + \frac{\partial^4 u}{\partial y^4}\right)\right]_{k,m}^n \quad ; k, m \neq 1, N. \quad (5.178)$$

It is equally easy to verify that the extrapolated form of the method based on the (2, 0) Padé approximant is L_0-stable also, and this, too, is left as an exercise to the reader.

The implementation of the algorithm based on the (2, 0) Padé approximant may be carried out by means of the following strategy:

(i) intermediate vectors $\mathbf{V}^{(1)}$, $\mathbf{V}^{(2)}$ are introduced and used to find the estimates $\mathbf{U}^*(t+l)$, $\mathbf{U}^+(t+l)$, as follows:

$$(I-lC+\tfrac{1}{2}l^2C^2)\mathbf{V}^{(1)} = \mathbf{U}(t), \quad (5.179)$$

$$(I-lB+\tfrac{1}{2}l^2B^2)\mathbf{U}^*(t+l) = \mathbf{V}^{(1)}, \quad (5.180)$$

$$(I-lB+\tfrac{1}{2}l^2B^2)\mathbf{V}^{(2)} = \mathbf{U}(t), \quad (5.181)$$

$$(I-lC+\tfrac{1}{2}l^2C^2)\mathbf{U}^+(t+l) = \mathbf{V}^{(2)}; \quad (5.182)$$

(ii) intermediate vectors $\mathbf{V}^{(3)}$, $\mathbf{V}^{(4)}$ are introduced to extend the estimates $\mathbf{U}^*(t+l)$, $\mathbf{U}^+(t+l)$ over a second single time step as follows:

$$(I-lC+\tfrac{1}{2}l^2C^2)\mathbf{V}^{(3)} = \mathbf{U}^*(t+l), \quad (5.183)$$

$$(I-lB+\tfrac{1}{2}l^2B^2)\mathbf{U}^{**}(t+2l) = \mathbf{V}^{(3)}; \quad (5.184)$$

$$(I-lB+\tfrac{1}{2}l^2B^2)\mathbf{V}^{(4)} = \mathbf{U}^+(t+l), \quad (5.185)$$

$$(I-lC+\tfrac{1}{2}l^2C^2)\mathbf{U}^{++}(t+2l) = \mathbf{V}^{(4)}; \quad (5.186)$$

(iii) the second order estimate $\mathbf{U}^{(0)}$ is now calculated from

$$\mathbf{U}^{(0)} = \tfrac{1}{2}(\mathbf{U}^{**}+\mathbf{U}^{++}); \quad (5.187)$$

(iv) intermediate vectors $\mathbf{V}^{(5)}$, $\mathbf{V}^{(6)}$ are introduced and used with a double time step to find the estimates $\mathbf{U}^{(1)}(t+2l)$ and $\mathbf{U}^{(2)}(t+2l)$ as follows:

$$(I-2lC+2l^2C^2)\mathbf{V}^{(5)} = \mathbf{U}(t), \quad (5.188)$$

$$(I-2lB+2l^2B^2)\mathbf{U}^{(1)}(t+2l) = \mathbf{V}^{(5)}; \quad (5.189)$$

$$(I-2lB+2l^2B^2)\mathbf{V}^{(6)} = \mathbf{U}(t), \quad (5.190)$$

$$(I-2lC+2l^2C^2)\mathbf{U}^{(2)}(t+2l) = \mathbf{V}^{(6)}; \quad (5.191)$$

(v) the third order accurate estimate $\mathbf{U}^{(E)}(t+2l)$, given by

$$\mathbf{U}^{(E)}(t+2l) = \frac{4}{3}\mathbf{U}^{(0)} - \frac{1}{6}(\mathbf{U}^{(1)}+\mathbf{U}^{(2)}), \quad (5.192)$$

is now calculated.

In order to illustrate the different behaviours of A_0-stable and L_0-stable methods in two space variables, the following model problem, which was introduced in the paper by Lawson and Morris (1978), is solved using the A_0-stable Peaceman-Rachford method and the L_0-stable method based on the (2, 0) Padé approximant described by the equations included in (5.168); these methods are both second order accurate in time. The problem is

$$\frac{\partial u}{\partial t} = \frac{\partial^2 u}{\partial x^2} + \frac{\partial^2 u}{\partial y^2}; \quad 0 < x, y < 2, t > 0 \tag{5.193}$$

subject to the initial conditions

$$u(x, y, 0) = \sin(\tfrac{1}{2}\pi y); \quad 0 \leqslant x, y \leqslant 2 \tag{5.194}$$

and the boundary conditions

$$u(x, y, t) = 0; \quad x = 0, \ y = 0, \ x = 2, \ y = 2, \ t > 0. \tag{5.195}$$

The initial distribution is shown in Fig. 5.8 (note the discontinuity between the boundary conditions and initial conditions for $y = 0, y = 2$), and the theoretical solution

$$u(x, y, t) = \sin \tfrac{1}{2}\pi y \sum_{k=1}^{\infty} [1-(-1)^k] \frac{2}{k\pi} \sin(\tfrac{1}{2}k\pi x) \exp[-\tfrac{1}{4}\pi^2(k^2+1)t] \tag{5.196}$$

is depicted a time $t = 1.0$ in Fig. 5.9. The maximum value of u at time $t = 1.0$ occurs for $x = 1, y = 1$ and is approximately 0.00916.

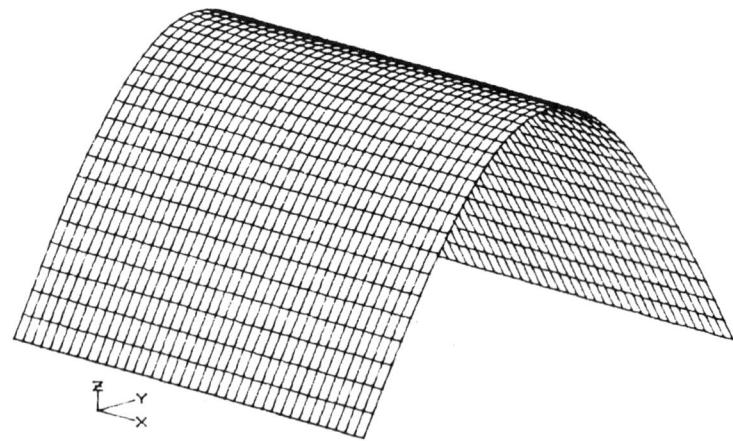

Z:0.000 TO 1.000

Fig. 5.8 – Initial distribution for model problem with two space dimensions.

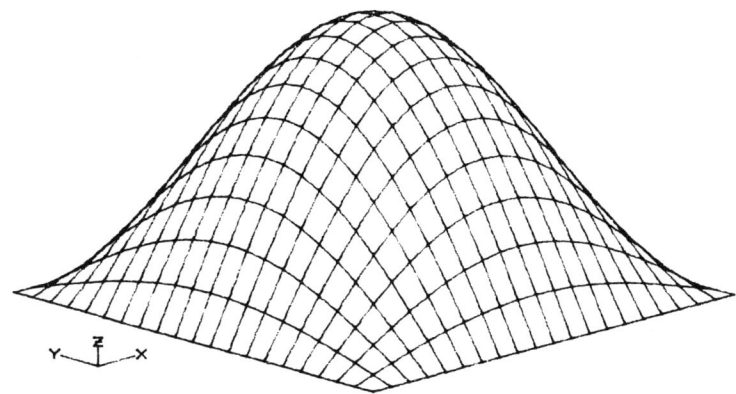

Z:0.00000 TO 0.00916

Fig. 5.9 – Theoretical solution at time $t = 1.0$ for the model problem with two dimensions.

Following Lawson and Morris (1978), the solution is computed at time $t = 1.0$ using $h = 0.05$ and $l = 0.1$; the ratios $p = l/h^2$, $r = l/h$ therefore have the values $p = 40$, $r = 2$. The distribution for $t = 1.0$ is shown in Fig. 5.10 for the A_0-stable Peaceman-Rachford scheme, in Fig. 5.11 for the L_0-stable scheme (5.168), and in Fig. 5.12 for its extrapolated form.

It has been necessary to scale down the interior peak in Fig. 5.10; in fact, the Peaceman-Rachford method, like the Crank-Nicolson method in one space variable, is very accurate away from the boundary of the region, and if Fig. 5.9 and 5.10 had been drawn to the same scale, the interior peaks would have been almost coincident. The peaks near the boundary of Fig. 5.10, however, would then have been almost twice their apparent height.

The presence of these peaks near the boundaries $y = 0$, $y = 2$ of Fig. 5.10 show that the time step $l = 0.1$ is too large for use with a space step of $h = 0.05$ when the Peaceman-Rachford method is used to solve the model problem.

On the other hand, it is evident from Fig. 5.11 that the L_0-stable method (5.168), like the one-dimensional equivalent (5.102), does not suffer from this restriction and the computed solution is a smooth representation of the theoretical solution (5.196).

The maximum errors for this experiment using the A_0-stable Peaceman-Rachford method and the L_0-stable method (5.168) are included in Table 5.2. The maximum errors for the Peaceman-Rachford method are those near the boundaries $y = 0$, $y = 2$ and, because of symmetry, these errors are the same; the maximum error for the method based on the (2, 0) Padé approximant occurs at the point where $x = 1, y = 1$.

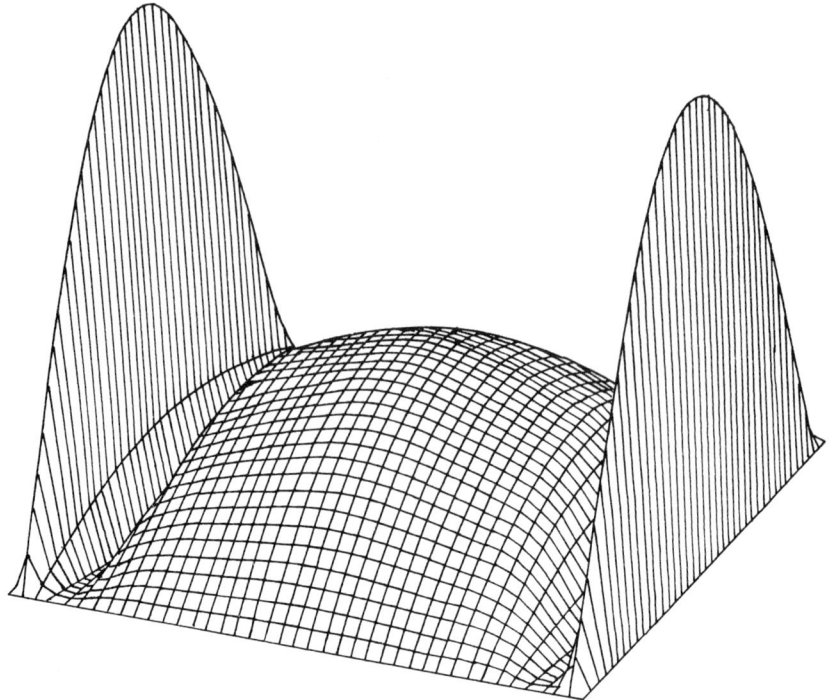

Fig. 5.10 — Computed solution at time $t = 1.0$ using the Peaceman–Rachford method with $h = 0.05, l = 0.1$.

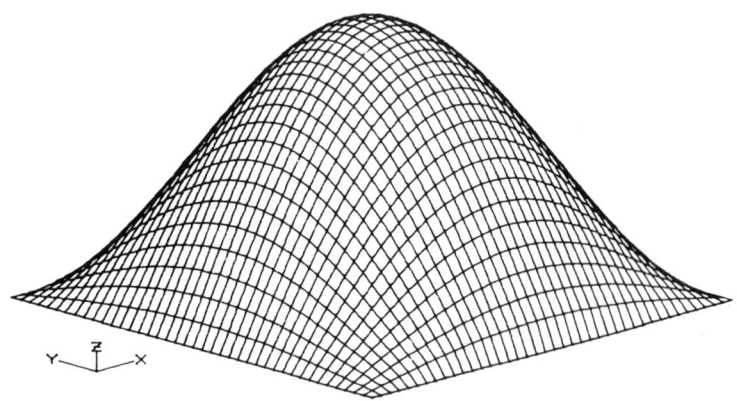

Z:0.00000 TO 0.00950

Fig. 5.11 — Computed solution at time $t = 1.0$ using the (2, 0) method with $h = 0.05, l = 0.1$.

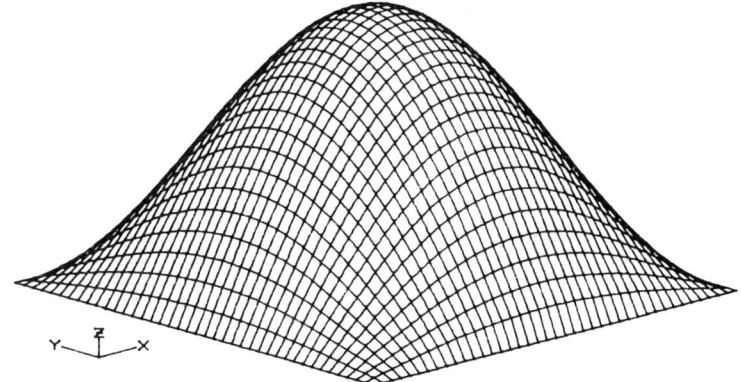

Z:0.00000 TO 0.00919

Fig. 5.12 – Computed solution at time $t = 1.0$ using the extrapolated form of the (2, 0) method with $h = 0.05$, $l = 0.1$.

Table 5.2

Method	Order	Maximum errors ($h = 0.05$)	
		$l = 0.1$	$l = 0.025$
Peaceman-Rachford	Second	0.23(−1)	0.33(−4)
(2, 0) method (5.168)	Second	0.34(−3)	0.46(−4)
Extrapolated (2, 0)	Third	0.35(−4)	0.80(−5)

The maximum error for the extrapolated form of the method based on the (2, 0) Padé approximant, which is L_0-stable and third order accurate in time, is also included in Table 5.2; this maximum error also occurs at the point where $x = 1, y = 1$.

The same model problem can be used to show that the Peaceman-Rachford method gives good results when the ratio $r = l/h$ is not too large. The maximum errors at time $t = 1.0$ for $h = 0.05$, as before, and l decreased to 0.025, are also given in Table 5.2. For this experiment, the maximum errors for all three methods occur at the point where $x = 1, y = 1$ showing that the discontinuities between initial conditions and boundary conditions have been smoothed out by all three methods. The superiority of the Peaceman-Rachford results over those of the (2, 0) method for this experiment is due to the more favourable coefficient of l^3 in (5.84) compared with (5.177).

5.7 NON-CONSTANT COEFFICIENTS

5.7.1 Coefficients depending on the space variable x

The parabolic differential equation in this type of problem takes the form

$$\frac{\partial u}{\partial t} = a(x)\frac{\partial^2 u}{\partial x^2} \; ; \; 0 < x < X, \; t > 0 \qquad (5.197)$$

with suitable initial conditions and boundary conditions specified; the function $a(x) \neq 0$ for all values of x. Equation (5.197) is of the form

$$\frac{\partial u}{\partial t} = L(x, t, D, D^2)u \qquad (5.198)$$

where $D \equiv \partial/\partial x$ and the operator L is linear.

Now, by Taylor's series

$$u(x, t+l) = (1 + l\frac{\partial}{\partial t} + \frac{1}{2}l^2 \frac{\partial^2}{\partial t^2} + \ldots)u = \exp(l\frac{\partial}{\partial t})u \qquad (5.199)$$

and writing $x = mh$, $t = nl$ in (5.199) and retaining the notation $u(mh, nl) \equiv u_m^n$, then, provided L is independent of t,

$$u_m^{n+1} = \exp(l\frac{\partial}{\partial t})u_m^n = \exp(lL)u_m^n. \qquad (5.200)$$

Writing $L \equiv aD^2$, (5.200) becomes

$$u_m^{n+1} = \exp(alD^2)u_m^n. \qquad (5.201)$$

Replacing the exponential term in (5.201) by its (0, 1) Padé approximant and replacing D^2u by the second order central difference approximant (5.8) gives the finite difference scheme

$$U_m^{n+1} = U_m^n + a(x)l[U_{m-1}^n - 2U_m^n + U_{m+1}^n]/h^2. \qquad (5.202)$$

Evaluating $a(x)$ at the mesh point (mh, nl) gives

$$U_m^{n+1} = (1 - 2a_m p)U_m^n + a_m p(U_{m-1}^n + U_{m+1}^n) \qquad (5.203)$$

where $a_m \equiv a(mh)$.

Equation (5.203) is obviously a more general case of the three point, two level explicit finite difference scheme (5.14); in fact (5.203) reduces to (5.14) upon writing $a(x) \equiv 1$.

Using the (1, 0) Padé approximant in (5.201) gives

$$(1 - alD^2)U_m^{n+1} = U_m^n \tag{5.204}$$

which gives

$$U_m^n - a_m l[U_{m-1}^{n+1} - 2U_m^n + U_{m+1}^{n+1}]/h^2 = U_m^n$$

or

$$(1 + 2a_m p)U_m^{n+1} - a_m p(U_{m-1}^{n+1} + U_{m+1}^{n+1}) = U_m^n \tag{5.205}$$

which is a more general case of (5.16) and, indeed, reduces to (5.16) for $a(x) \equiv 1$.
Using the (1, 1) Padé approximant in (5.201) leads in a similar way to

$$(1 + a_m p)U_m^{n+1} - \frac{1}{2} p(U_{m-1}^{n+1} + U_{m+1}^{n+1}) = (1 - a_m p)U_m^n + \frac{1}{2} p(U_{m-1}^n + U_{m+1}^n) \tag{5.206}$$

which reduces to (5.18) for $a(x) \equiv 1$ and is a more general form of the Crank-Nicolson method. Using higher order Padé approximants to the exponential term in (5.201) enables methods analogous to those of section 5.6 to be written down.
The matrix A given by (5.10) must be replaced by

$$A(x) = h^{-2} \begin{bmatrix} -2a_1 & a_2 & & & & & 0 \\ a_1 & -2a_2 & a_3 & & & & \\ & a_2 & -2a_3 & a_4 & & & \\ & & \ddots & \ddots & \ddots & & \\ 0 & & & & a_{N-2} & -2a_{N-1} & a_N \\ & & & & & a_{N-1} & -2a_N \end{bmatrix} \tag{5.207}$$

in equations (5.13), (5.15), (5.17), (5.94), (5.100), (5.102), (5.112), (5.120) following which the extrapolated forms of (5.15), (5.17), (5.102), (5.112), (5.120) may be used as discussed in sections 5.4 and 5.6.

5.7.2 The self-adjoint case
Here, the parabolic equation has the form

$$\frac{\partial u}{\partial t} = \frac{\partial}{\partial x}\left[a(x)\frac{\partial u}{\partial x}\right]; \quad 0 < x < X, \quad t > 0 \tag{5.208}$$

with suitable initial conditions and boundary conditions specified and, as before, $a(x) \neq 0$ for all values of x. The operator L in (5.198) is now $L \equiv aD^2 + a'D$, where a' denotes da/dx, so that (5.200) becomes

$$u_m^{n+1} = \exp(alD^2 + a'lD)u_m^n. \tag{5.209}$$

250   **Parabolic Equations**   [Ch. 5

Proceeding from (5.209) Padé approximants are used to replace $\exp(aD^2 + a'D)$, then Du and D^2u are replaced by their second order central difference replacements $Du_m^n = (U_{m+1}^n - U_{m-1}^n)/2h$, $D^2u = (U_{m+1}^n - 2U_m^n + U_{m-1}^n)/h^2$. For the $(0, 1)$ Padé approximant, for instance, this gives

$$U_m^{n+1} = U_m^n + a_m l(U_{m+1}^n - 2U_m^n + U_{m-1}^n)/h^2 + a'_m l(U_{m+1}^n - U_{m-1}^n)/2h$$

which leads to the explicit formula

$$U_m^{n+1} = (a_m p - \tfrac{1}{2} a'_m r) U_{m-1}^n + (1 - 2a_m p) U_m^n + (a_m p + \tfrac{1}{2} a'_m r) U_{m+1}^n \tag{5.210}$$

in which, as before, $p = l/h^2$ and $r = l/h$.

Low order implicit formulas may be obtained from (5.209) by replacing $\exp(alD^2 + a'lD)$ by its $(1, 0)$ and $(1, 1)$ Padé approximants giving, respectively,

$$-(a_m p - \tfrac{1}{2} a'_m r) U_{m-1}^{n+1} + (1 + 2a_m p) U_m^{n+1} - (a_m p + \tfrac{1}{2} a'_m r) U_{m+1}^{n+1} = U_m^n \tag{5.211}$$

and

$$-\tfrac{1}{2}(a_m p - \tfrac{1}{2} a'_m r) U_{m-1}^{n+1} + (1 + a_m p) U_m^{n+1} - \tfrac{1}{2}(a_m p + \tfrac{1}{2} a'_m r) U_{m+1}^{n+1}$$

$$= \tfrac{1}{2}(a_m p - \tfrac{1}{2} a'_m r) U_{m-1}^n + (1 - a_m p) U_m^n + \tfrac{1}{2}(a_m p + \tfrac{1}{2} a'_m r). \tag{5.212}$$

Writing $a(x) \equiv 1$, equations (5.210), (5.211), (5.212) become (5.14), (5.16), (5.18) respectively.

5.7.3 Variable coefficients depending on x and t

Second order parabolic equations of this type have the form

$$\frac{\partial u}{\partial t} = a(x, t) \frac{\partial^2 u}{\partial x^2} \tag{5.213}$$

with appropriate boundary conditions and initial conditions specified. One approach to solving (5.213) requires evaluation of some terms at the intermediate time $t = (n + \tfrac{1}{2})l$. The numerical value of $a(x, t)$ at the point where $x = mh$, $t = (n + \tfrac{1}{2})l$ for instance, will be denoted by $a_m^{n+\tfrac{1}{2}}$; it is clear that $(mh, (n + \tfrac{1}{2})l)$ is not a mesh point, but this does not prohibit its use.

One of the problems at the end of this chapter requires verification of the formula

$$\frac{1}{2h^2} \delta_x^2 (u_m^{n+1} + u_m^n) = \left(\frac{\partial^2 u}{\partial x^2}\right)_m^{n+\tfrac{1}{2}} + \frac{h^2}{12} \left(\frac{\partial^4 u}{\partial x^4}\right)_m^{n+\tfrac{1}{2}} + 0(h^4 + l^2), \tag{5.214}$$

where $\delta_x^2(u_m^n) = u_{m-1}^n - 2u_m^n + u_{m+1}^n$, and using (5.213) it follows that

$$\frac{1}{2h^2}\delta_x^2(u_m^{n+1} + u_m^n) = \left(\frac{1}{a}\frac{\partial u}{\partial t}\right)_m^{n+\frac{1}{2}} + \frac{h^2}{12}\frac{\partial^2}{\partial x^2}\left(\frac{1}{a}\frac{\partial u}{\partial t}\right)_m^{n+\frac{1}{2}} + O(h^4 + l^2). \quad (5.215)$$

Another problem at the end of the chapter is to verifiy the formula

$$\frac{1}{h^2 l}\delta_x^2\left[(a_m^{n+\frac{1}{2}})^{-1}(u_m^{n+1} - u_m^n)\right] = \frac{\partial^2}{\partial x^2}\left(\frac{1}{a}\frac{\partial u}{\partial t}\right)_m^{n+\frac{1}{2}} + O(h^2 + l^2). \quad (5.216)$$

Substituting for (5.216) in (5.215) gives

$$\frac{1}{2h^2}\delta_x^2(u_m^{n+1} + u_m^n) = \frac{1}{a_m^{n+\frac{1}{2}} l}(u_m^{n+1} - u_m^n)$$

$$+ \frac{1}{12p}\frac{1}{h^2}\delta_x^2\left[\frac{1}{a_m^{n+\frac{1}{2}}}(u_m^{n+1} - u_m^n)\right] + O(h^4 + l^2)$$

which leads to the finite difference scheme

$$\frac{1}{pa_m^{n+\frac{1}{2}}}(U_m^{n+1} - U_m^n) = \frac{1}{2}\delta_x^2\left[1 - \frac{1}{6pa_m^{n+\frac{1}{2}}}\right]U_m^{n+1} + \frac{1}{2}\delta_x^2$$

$$\left[1 + \frac{1}{6pa_m^{n+\frac{1}{2}}}\right]U_m^n + O(h^6 + l^3). \quad (5.217)$$

This is usually written in the more convenient form

$$[1 + \frac{1}{12}a_m^{n+\frac{1}{2}}\delta_x^2(a_m^{n+\frac{1}{2}})^{-1} - \frac{1}{2}pa_m^{n+\frac{1}{2}}\delta_x^2]U_m^{n+1}$$

$$= [1 + \frac{1}{12}a_m^{n+\frac{1}{2}}\delta_x^2(a_m^{n+\frac{1}{2}})^{-1} + \frac{1}{2}pa_m^{n+\frac{1}{2}}\delta_x^2]U_m^n \quad (5.218)$$

which has the same error as (5.217).

5.7.4 Non-linear equations
In the more sophisticated mathematical models of heat flow, see, for example, Twizell and Smith (1981), the parabolic equation is non-linear. Fortunately, many finite difference methods developed for linear parabolic equations carry over to non-linear equations, though added difficulties arise in computing the solution.

An explicit method leads to the numerical solution of a single non-linear difference equation while an implicit method leads to the numerical solution of a system of non-linear equations. The literature on the solution of single non-linear equations, and of systems of non-linear equations, is large. Indeed the Newton-Raphson method, for which adequate computer software is available, and one-point functional iteration methods may be relied on to give good results provided certain criteria are met (see Chapter 1).

A more detailed discussion of non-linear parabolic equations, including solution by linearization techniques and three-time level methods are given in the texts by Smith (1978) and Mitchell and Griffiths (1980).

5.8 DIFFUSION-CONVECTION EQUATIONS

The linear form of the one-dimensional diffusion-convection equation, which has important applications in convective heat transfer (in the arms and legs, for instance, where flowing blood makes a strong contribution to heat transfer) and other viscous flow problems, is given by (5.3). Without loss of generality Re may be taken to be unity, so that the model equation becomes

$$\frac{\partial u}{\partial t} = \frac{\partial^2 u}{\partial x^2} - \mu \frac{\partial u}{\partial x}; \ 0 < x < X, t > 0. \tag{5.219}$$

The parameter $\mu > 0$ is the convection parameter and may be large.

The solution of (5.219) is usually discussed (Siemieniuch and Gladwell (1978), Morton (1980), Mitchell and Griffiths (1980)) with respect to the initial conditions.

$$u(x, 0) = g(x), 0 < x < X \tag{5.220}$$

and the boundary conditions

$$u(0, t) = v(t), t > 0, \tag{5.221}$$

$$\frac{\partial u}{\partial x}(X, t) = 0, \ t > 0. \tag{5.222}$$

The finite difference methods for the solution of (5.219) with (5.220), (5.221), (5.222) which have appeared in the literature to date have used two-time levels. Like the methods for the solution of the simple diffusion equation (5.2), the explicit methods derived for (5.219) have suffered stability restrictions on $p = l/h^2$ which can be severe for large values of μ. Implicit methods, on the other hand, have not had stability restrictions, but have required the solution to be obtained by solving a linear system.

In obtaining a numerical solution to the diffusion-convection problem, the interval $0 \leqslant x \leqslant X$ is divided into N subintervals each of width h, so that

Sec. 5.8] Diffusion-Convection Equations 253

$Nh = X$, and the time variable t is discretized in steps of length l. The open rectangular region $R = [0 < x < X] \times [t > 0]$ in the (x, t) plane which has boundary ∂R consisting of the lines $t = 0, x = 0, x = X$, have therefore been covered by a rectangular mesh, the mesh points having coordinates (mh, nl) with $m = 0, 1, \ldots, N$ and $n = 0, 1, 2, \ldots$. The notations u, U, $\widetilde{U}$ used in earlier chapters, and in previous sections of this chapter, will be retained for the diffusion-convection problem.

The space derivatives in the differential equation (5.219) and the second boundary condition (5.222) are replaced by the central difference approximants

$$\frac{\partial^2 u}{\partial x^2} = \{u(x-h, t) - 2u(x, t) + u(x+h, t)\}/h^2 + O(h^2) \quad (5.223)$$

$$\frac{\partial u}{\partial x} = \{u(x+h, t) - u(x-h, t)\}/2h + O(h^2). \quad (5.224)$$

Equation (5.219) with (5.223), (5.224) is now applied to the mesh points (mh, nl), $m = 1, 2, \ldots, N$ at time level $t = nl$ ($n = 0, 1, \ldots$). This produces a system of N first order ordinary differential equations which is of the form

$$\frac{d\mathbf{U}}{dt} = A\mathbf{U}(t) + \mathbf{b}_t. \quad (5.225)$$

Recalling that $U_0^n = v(nl)$ from the boundary condition (5.221), it may be shown that A, which is clearly a square matrix of order N, is given by

$$h^2 A = \begin{bmatrix} -2 & 1-\tfrac{1}{2}\mu h & & & 0 \\ 1+\tfrac{1}{2}\mu h & -2 & 1-\tfrac{1}{2}\mu h & & \\ & \ddots & \ddots & \ddots & \\ 0 & & 1+\tfrac{1}{2}\mu h & -2 & 1-\tfrac{1}{2}\mu h \\ & & & 2 & -2 \end{bmatrix} \quad (5.226)$$

The vector $\mathbf{b}_t = h^{-2}[(1 + \tfrac{1}{2}\mu h)v_t, 0, \ldots, 0]^T$, T denoting transpose, is also of order N; the term $v_t = v(nl)$ is the numerical (frozen) value of $v(t)$ at time $t = nl$.

Siemieniuch and Gladwell (1978) show that for $\mu h > 2$, the eigenvalues of A are given by

$$\lambda_j = -2h^{-2} + i\gamma_j;\ j = 1, \ldots, N \quad (5.227)$$

where $-2h^{-2}(\tfrac{1}{4}\mu^2 h^2 - 1)^{\tfrac{1}{2}} \leq \gamma_j \leq 2h^{-2}(\tfrac{1}{4}\mu^2 h^2 - 1)^{\tfrac{1}{2}}$; for $\mu h = 2$ the matrix A is bidiagonal and each eigenvalue is seen to have the value $-2h^{-2}$; for $0 \leq \mu h < 2$ Siemieniuch and Gladwell (1978) derive bounds for the eigenvalues given by

$$-2h^{-2} - 2h^{-2}(1 - \tfrac{1}{4}\mu^2 h^2)^{\tfrac{1}{2}} \leq \lambda_j \leq -2h^{-2} + 2h^{-2}(1 - \tfrac{1}{4}\mu^2 h^2)^{\tfrac{1}{2}} < 0. \quad (5.228)$$

It is easy to verify that the solution of (5.225) with the initial vector $\mathbf{U}(0)=\mathbf{g}$ satisfies

$$\mathbf{U}(t) = -A^{-1}\mathbf{b}_t + \exp(tA)\{\mathbf{g} + A^{-1}\mathbf{b}_t\} \qquad (5.229)$$

and that (5.229) satisfies the recurrence relation

$$\mathbf{U}(t+l) = -A^{-1}\mathbf{b}_t + \exp(lA)\{\mathbf{U}(t) + A^{-1}\mathbf{b}_t\}. \qquad (5.230)$$

To obtain an explicit solution to (5.219) with (5.220), (5.221), (5.222), the (0, 1) Padé approximant may be used to replace the matrix exponential function in (5.230), giving

$$\mathbf{U}(t+l) = -A^{-1}\mathbf{b}_t + (I + lA)\{\mathbf{U}(t) + A^{-1}\mathbf{b}_t\} + 0(l^2) \qquad (5.231)$$

which is first order accurate in time. Gathering terms, (5.231) becomes

$$\mathbf{U}(t+l) = (I + lA)\mathbf{U}(t) + l\mathbf{b}_t \qquad (5.232)$$

which, when applied to the mesh point (mh, nl), gives

$$U_m^{n+1} = (p + \tfrac{1}{2}\mu r)U_{m-1}^n + (1 - 2p)U_m^n + (p - \tfrac{1}{2}\mu r)U_{m+1}^n \qquad (5.233)$$

for $m \neq N$. For $m = N$, $U_{N+1}^n = U_{N-1}^n$ from (5.222) and (5.224) so that (5.233) takes the simplified form

$$U_N^{n+1} = 2p\, U_{N-1}^n + (1-2p)\, U_N^n. \qquad (5.234)$$

Writing the diffusion-convection equation in the form $G(u) = 0$ and the difference scheme (5.233) in the form $F_{m,n}(U) = 0$, the local truncation error of (5.233) at the mesh point (mh, nl) is given by $l^{-1}F_{m,n}(u) - G(u_m^n)$. Expanding u_m^{n+1}, u_{m-1}^n, u_{m+1}^n about u_m^n using Taylor's series, the principal part of the local truncation error is easily verified to be

$$\left(\frac{1}{6}\mu h^2 \frac{\partial^3 u}{\partial x^3} + \frac{1}{2}l\frac{\partial^2 u}{\partial t^2}\right)_m^n. \qquad (5.235)$$

The component $\tfrac{1}{6}\mu h^2 \partial^3 u/\partial x^3$ arises from the space discretization and the use of (5.223), (5.224); this component will be present in any finite difference scheme arising from the replacement of $\exp(lA)$ in (5.230) by a suitable Padé approximant. The component $\tfrac{1}{2}l\, \partial^2 u/\partial t^2$ arises from the use of the (0, 1) Padé approximant in (5.230) and is given in Table 4.5 of Chapter 4.

It is clear that the local truncation error tends to zero as the grid is refined, that is, as $h, l \to 0$. The finite difference scheme arising from the (0, 1) Padé approximant is therefore consistent with the differential equation.

An investigation of the stability of the finite difference scheme (5.233) using the matrix method illustrates once more the inadequacies of this method in relation to certain problems. The matrix method requires that no eigenvalue of the amplification matrix $I + lA$ should exceed unity in modulus if a perturbation in the initial conditions is not to grow as time increases. Siemieniuch and Gladwell (1978) show that, for stability, the matrix method thus requires

$$0 < p < \begin{cases} [1 + (1 - \tfrac{1}{4} \mu^2 h^2)]^{-1}, & 0 \leq \mu h \leq 2 \\ 4/\mu^2 h^2, & \mu h \geq 2. \end{cases} \qquad (5.236)$$

Morton (1980) shows that these intervals are too large. Employing the Fourier method for analyzing stability, the growth of $Z_m^n = U_m^n - \tilde{U}_m^n$ may be analyzed in the normal way be writing $Z_m^n = e^{\alpha nl} e^{i\beta mh}$ in (5.233) where β is real and α is, in general, complex. This gives $e^{\alpha(n+1)l} e^{i\beta mh} = (p + \tfrac{1}{2}\mu r) e^{\alpha nl} e^{i\beta(m-1)h} + (1 - 2p) e^{\alpha nl} e^{i\beta mh} + (p - \tfrac{1}{2}\mu r) e^{\alpha nl} e^{i\beta(m+1)h}$ which, after dividing by $e^{\alpha nl} e^{i\beta mh}$ and writing $\xi = e^{\alpha l}$, becomes

$$\xi = 1 - 4p \sin^2 \tfrac{1}{2} \beta h - i\mu r \sin \beta h. \qquad (5.237)$$

The von Neumann criterion for stability stipulates that the amplification factor ξ should not exceed unity in modulus. Now,

$$|\xi|^2 = 1 - 8p \sin^2 \tfrac{1}{2} \beta h + 16 p^2 \sin^4 \tfrac{1}{2} \beta h + \mu^2 r^2 \sin^2 \beta h$$

$$= 1 + 4(\mu^2 r^2 - 2p) \sin^2 \tfrac{1}{2} \beta h + 4(4 p^2 - \mu^2 r^2) \sin^4 \tfrac{1}{2} \beta h$$

$$\leq (1 - 4p)^2,$$

and it is easy to show from this that the necessary condition for the stability of (5.233) is $0 < p \leq \tfrac{1}{2}$. This condition is therefore necessary and sufficient for convergence. The reader is referred to the paper by Morton (1980) which gives a much more detailed analysis of the stability of (5.233).

In contrast to the (0, 1) Padé approximant, the (1, 0) Padé approximant may be used in (5.230) to give an implicit scheme for solving (5.219). Substituting this approximant in (5.230) and gathering terms gives

$$(I - lA)\mathbf{U}(t + l) - l\mathbf{b}_{t+l} = \mathbf{U}(t) \qquad (5.238)$$

where $\mathbf{b}_{t+l} = h^{-2}[(1 + \tfrac{1}{2}\mu h) v_{t+l}, 0, \ldots, 0]^T$, v_{t+l} being the numerical value of $v(t + l)$. Applying (5.238) to the mesh point (mh, nl) gives

$$-(p + \tfrac{1}{2} \mu r) U_{m-1}^{n+1} + (1 + 2p) U_m^{n+1} - (p - \tfrac{1}{2} \mu r) U_{m+1}^{n+1} = U_m^n \qquad (5.239)$$

for $m \neq N$ and

$$-2pU_{N-1}^{n+1} + (1+2p)U_N^{n+1} = U_N^n \qquad (5.240)$$

when $m = N$, since $U_{N-1}^{n+1} = U_{N+1}^{n+1}$ from (5.222) and (5.224).

The principal part of the local truncation error of the four-point, fully implicit scheme (5.239) is

$$\left(\frac{1}{6}\mu h^2 \frac{\partial^3 u}{\partial x^3} - \frac{1}{2}l\frac{\partial^2 u}{\partial t^2}\right)_m^n \qquad (5.241)$$

where the component $-\frac{1}{2}l\,\partial^2 u/\partial t^2$ is found in Table 4.5 of Chapter 4.

To investigate the stability of (5.239) by the Fourier method, substitution of $Z_m^n = e^{\alpha n l}e^{i\beta m h}$ into

$$-(p+\tfrac{1}{2}\mu r)Z_{m-1}^{n+1} + (1+2p)Z_m^{n+1} - (p-\tfrac{1}{2}\mu r)Z_{m+1}^{n+1} = Z_m^n$$

leads to

$$\xi = \frac{1 + 4p\sin^2\beta h + i\mu r \sin\beta h}{(1 + 4p\sin^2\tfrac{1}{2}\beta h)^2 + (\mu r \sin\beta h)^2},$$

from which it follows that

$$|\xi|^2 = \frac{1}{(1 + 4p\sin^2\tfrac{1}{2}\beta h)^2 + (\mu r \sin\beta h)^2}. \qquad (5.242)$$

Equation (5.242) gives $|\xi| \leq 1$ so that the scheme (5.239) is unconditionally stable in the conventional sense that a perturbation $\mathbf{Z}^0$ of the initial vector will not grow as time increases; the method is thus A-stable, at least.

To investigate the scheme with regard to L-stability, the three cases $\mu h > 2$, $\mu h = 2$, $0 \leq \mu h < 2$ must be examined separately. For $\mu h > 2$, the amplification symbol $R_{1,0}(z)$, where $z = -l\lambda_j$ with λ_j defined by (5.227), is given by

$$R_{1,0}(z) = \frac{1}{1+z} = \frac{1 + 2p + il\gamma_j}{(1+2p)^2 + l^2\gamma^2} \qquad (5.243)$$

for each eigenvalue λ_j of A. The method is L-stable provided $|R_{1,0}(z)| \to 0$ as $\mathrm{Re}(l\lambda_j) = -2p \to -\infty$ for any eigenvalue λ_j of A.

Using (5.243), $|R_{1,0}(z)|$ is given by

$$|R_{1,0}(z)| = \frac{1}{[(1+2p)^2 + l^2\gamma_j^2]^{\frac{1}{2}}}$$

which clearly tends to zero as $\mathrm{Re}(l\lambda_j) \to -\infty$ so that (5.238) is L-stable for $\mu h > 2$.

Sec. 5.8] **Diffusion-Convection Equations** 257

Turning next to the case $\mu h = 2$, all eigenvalues of A have the value $-2/h^2$ so that $z = 2p$ and

$$R_{1,0}(z) = \frac{1}{1+z} = \frac{1}{1+2p} \to 0 \text{ as } z = 2p \to \infty \qquad (5.244)$$

so that (5.238) is L_0-stable for $\mu h = 2$ and the solution is non-oscillatory.

Finally, for $0 \leq \mu h < 2$, all eigenvalues λ_j of A are real and negative with bounds given by (5.228) and $z = -l\lambda_j$ satisfies

$$0 < 2p - 2p(1 - \tfrac{1}{4}\mu^2 h^2)^{\frac{1}{2}} \leq z \leq 2p + 2p(1 - \tfrac{1}{4}\mu^2 h^2)^{\frac{1}{2}}. \qquad (5.245)$$

Clearly $z \to \infty$ only as $p \to \infty$ since $0 \leq \mu h < 2$. The symbol $R_{1,0}(z)$ is bounded by

$$\frac{1}{1 + 2p[1 + (1 - \tfrac{1}{4}\mu^2 h^2)^{\frac{1}{2}}]} \leq R_{1,0}(z) \leq \frac{1}{1 + 2p[1 - (1 - \tfrac{1}{4}\mu^2 h^2)^{\frac{1}{2}}]} \qquad (5.246)$$

and clearly $R_{1,0}(z) \to 0$ as $z = 2p \to \infty$ so that (5.238) is L_0-stable for $0 \leq \mu h < 2$ and the solution is non-oscillatory.

The finite difference scheme arising from the replacement of $\exp(lA)$ in (5.230) by its (1, 0) Padé approximant has been seen to be consistent with the differential equation (5.219) and to be stable for all positive values of p and all positive values of μr. It is therefore a convergent method.

In view of its satisfactory L-stability properties, the method can be usefully extrapolated to give second order accuracy in time using (5.40). The local truncation error of the extrapolated form of the method has principal part given by

$$\left(\frac{1}{6}\mu h^2 \frac{\partial^3 u}{\partial x^3} + \frac{4}{3}l^2 \frac{\partial^3 u}{\partial t^3}\right)_m^n. \qquad (5.247)$$

The property of L-stability must again be examined for the three cases $\mu h > 2$, $\mu h = 2$ and $0 \leq \mu h < 2$. For $\mu h > 2$, the amplification symbol is given by

$$S_{1,0}(z) = 2\left(\frac{1}{1+z}\right)^2 - \left(\frac{1}{1+2z}\right) = 2\left(\frac{1 + 2p + il\gamma_j}{(1+2p)^2 + l^2\gamma_j^2}\right)^2 - \frac{1 + 4p + 2il\gamma_j}{(1+4p)^2 + 4l^2\gamma_j^2} \qquad (5.248)$$

for each eigenvalue λ_j of A. The extrapolated form of (5.238) is L-stable when $\mu h > 2$ provided $|S_{1,0}| \to 0$ as $\text{Re}(l\lambda_j) = -2p \to -\infty$ for any eigenvalue λ_j of A. Separating real and imaginary parts of (5.248) gives

$$S_{1,0}(z) = \left[\frac{2\{(1+2p)^2 - l^2\gamma_j^2\}}{\{(1+2p)^2 + l^2\gamma_j^2\}^2} - \frac{1+4p}{(1+4p)^2 + 4l^2\gamma_j^2}\right]$$

$$+ 2i\left[\frac{2(1+2p)l\gamma_j}{\{(1+2p)^2 + l^2\gamma_j^2\}^2} - \frac{l\gamma_j}{(1+4p)^2 + 4l^2\gamma_j^2}\right] \qquad (5.249)$$

from which it follows that

$$|S_{1,0}(z)|^2 = \left[\frac{2\{(1+2p)^2 - l^2\gamma_j^2\}}{\{(1+2p)^2 + l^2\gamma_j^2\}^2} - \frac{1+4p}{(1+4p)^2 + 4l^2\gamma_j^2}\right]^2$$

$$+ 4\left[\frac{2l\gamma_j(1+2p)}{\{(1+2p)^2 + l^2\gamma_j^2\}^2} - \frac{l\gamma_j}{(1+4p)^2 + 4l^2\gamma_j^2}\right]^2$$

$$= \left[\frac{2\{(2p)^{-4} + \tfrac{1}{4}p^{-3} + \tfrac{1}{4}p^{-2} - l^2\gamma_j^2(2p)^{-4}\}}{\{\tfrac{1}{4}p^{-2} + p^{-1} + 1 + \tfrac{1}{4}l^2\gamma_j^2 p^{-2}\}^2}\right.$$

$$\left. - \frac{(2p)^{-4} + \tfrac{1}{4}p^{-3}}{(2p)^{-4} + \tfrac{1}{2}p^{-3} + 1 + \tfrac{1}{4}l^2\gamma_j^2 p^{-4}}\right]^2$$

$$+ 4\left[\frac{\tfrac{1}{8}l\gamma_j p^{-4} + \tfrac{1}{4}l\gamma_j p^{-3}}{\{\tfrac{1}{4}p^{-2} + p^{-1} + 1 + \tfrac{1}{4}l^2\gamma_j^2 p^{-2}\}^2}\right.$$

$$\left. - \frac{l\gamma_j(4p)^{-2}}{(4p)^{-2} + \tfrac{1}{2}p^{-1} + 1 + \tfrac{1}{4}l^2\gamma_j^2 p^{-2}}\right]^2 \quad (5.250)$$

It is seen from (5.250) that $|S_{1,0}(z)|^2 \to 0$ and, consequently, $|S_{1,0}(z)| \to 0$ as $\operatorname{Re}(l\lambda_j) = -2p \to -\infty$ so that the extrapolated form of (5.238) is L-stable for $\mu h > 2$. The real part of $S_{1,0}(z)$ is, however, negative for values of $p < 1.3$ and $l^2\gamma_j^2 < 0.7 + 0.2p + 0.8p^2$ (approximately).

Turning next to the case when $\mu h = 2$, the amplification symbol of the extrapolated form of the method is given by

$$S_{1,0}(z) = 2\left(\frac{1}{1+2p}\right)^2 - \frac{1}{1+4p} \quad (5.251)$$

which tends to zero as $z = 2p \to \infty$, but is negative for $p > \tfrac{1}{2}(1 + \sqrt{2})$ so that small oscillations are evident for $p > \tfrac{1}{2}(1 + \sqrt{2})$. The extrapolated form of (5.238) is nevertheless L_0-stable for $\mu h = 2$.

In the case $0 \leq \mu h < 2$, the amplification symbol $S_{1,0}(z)$ is bounded by

$$\frac{1}{2p^2[1+(1-\tfrac{1}{4}\mu^2 h^2)^{\tfrac{1}{2}}]^2} - \frac{1}{4p[1+(1-\tfrac{1}{4}\mu^2 h^2)^{\tfrac{1}{2}}]}$$

$$\leq S_{1,0}(z) \leq \frac{1}{2p^2[1-(1-\tfrac{1}{4}\mu^2 h^2)^{\tfrac{1}{2}}]^2} - \frac{1}{4p[1-(1-\tfrac{1}{4}\mu^2 h^2)^{\tfrac{1}{2}}]} \quad (5.252)$$

so that, here also, $S_{1,0}(z) \to 0$ as $z = 2p \to \infty$ and the extrapolated form of (5.238) is L_0-stable for $0 \leq \mu h < 2$, though for μh in this interval the solution oscillates for $p > 2$.

Sec. 5.8] **Diffusion-Convection Equations** 259

The conclusion to be reached regarding the fully implicit method (5.238) is that it, and its extrapolated form are consistent, L-stable and convergent for all values of the convection parameter μ.

The (1, 1) Padé approximant provides a third replacement for the matrix exponential function in (5.238). Substituting this replacement and gathering terms leads to

$$(I - \tfrac{1}{2}lA)\mathbf{U}(t+l) - \tfrac{1}{2}l\mathbf{b}_{t+l} = (I + \tfrac{1}{2}lA)\mathbf{U}(t) + \tfrac{1}{2}l\mathbf{b}_t. \tag{5.253}$$

Applying (5.253) to the mesh point (mh, nl) gives

$$-\tfrac{1}{2}(p + \tfrac{1}{2}\mu r)U_{m-1}^{n+1} + (1+p)U_m^{n+1} - \tfrac{1}{2}(p - \tfrac{1}{2}\mu r)U_{m+1}^{n+1}$$

$$= \tfrac{1}{2}(p + \tfrac{1}{2}\mu r)U_{m-1}^n + (1-p)U_m^n + \tfrac{1}{2}(p - \tfrac{1}{2}\mu r)U_{m+1}^n ; \tag{5.254}$$

for $m = N$, however, (5.222) and (5.224) imply $U_{N-1}^n = U_{N+1}^n$ and $U_{N-1}^{n+1} = U_{N+1}^{n+1}$ so that (5.254) takes the form

$$-pU_{N-1}^{n+1} + (1+p)U_N^{n+1} = pU_{N-1}^n + (1-p)U_N^n. \tag{5.255}$$

The principal part of the local truncation error of the general six-point, two-level, implicit scheme (5.254) is given by

$$\left(\tfrac{1}{6}\mu h^2 \frac{\partial^3 u}{\partial x^3} - \tfrac{1}{12}l^2 \frac{\partial^3 u}{\partial t^3}\right)_m^n. \tag{5.256}$$

The scheme would thus seem to be more accurate than the extrapolated form of the method based on the (1, 0) Padé approximant.

To investigate the scheme with regard to stability, the three cases $\mu h > 2$, $\mu h = 2$, $0 \leqslant \mu h < 2$ must again be examined separately. For $\mu h > 2$ the amplification factor $R_{1,1}(z)$, where $z = -l\lambda_j$ with λ_j defined by (5.227), is given by

$$R_{1,1}(z) = \frac{1 - \tfrac{1}{2}z}{1 + \tfrac{1}{2}z} = \frac{1 - p^2 - \tfrac{1}{4}l^2\gamma_j^2 + il\gamma_j}{(1+p)^2 + \tfrac{1}{4}l^2\gamma_j^2} \tag{5.257}$$

for each eigenvalue $\lambda_j = -2h^{-2} + i\gamma_j$ of A. It follows from (5.257) that

$$|R_{1,1}(z)|^2 = \frac{(1 - p^2 - \tfrac{1}{4}l^2\gamma_j^2) + l^2\gamma_j^2}{\{(1+p)^2 + \tfrac{1}{4}l^2\gamma_j^2\}^2}$$

$$= \frac{(p^{-2} - 1 - \tfrac{1}{4}l^2\gamma_j^2 p^{-2})^2 + l^2\gamma_j^2 p^{-4}}{\{(p^{-1} - 1)^2 + \tfrac{1}{4}l^2\gamma_j^2 p^{-2}\}^2}$$

$$\to 1 \text{ as } \operatorname{Re}(l\lambda_j) = -2p \to \infty$$

so that (5.253) is A-stable for $\mu h > 2$.

For $\mu h = 2$, all eigenvalues of A have the value $-2h^{-2}$ so that $z = 2p$ and

$$R_{1,1}(z) = \frac{1 - \tfrac{1}{2}z}{1 + \tfrac{1}{2}z} = \frac{1-p}{1+p} = \frac{p^{-1}-1}{p^{-1}+1}$$

$$\to -1 \text{ as } z = 2p \to -\infty$$

so that (5.253) is A_0-stable when $\mu h = 2$ and the oscillations which affect the analogous Crank–Nicolson method for the diffusion equation (5.2), discussed in subsection 5.3.2, are present. The procedure, outlined in subsection 5.3.2, for limiting the value of r thus removing the oscillations, cannot be followed as all the eigenvalues of A have the same value.

Turning finally to the case $0 \leq \mu h < 2$, all eigenvalues λ_j of A are real and negative wtih bounds by (5.228) and $z = -l\lambda_j$ satisfies (5.245) from which it was seen that $z \to \infty$ only as $p \to \infty$ since $0 \leq \mu h < 2$. The symbol $R_{1,1}(z)$ thus has the bounds $R_{1,1}^{(1)}(z)$, $R_{1,1}^{(2)}(z)$ given by

$$R_{1,1}^{(1)}(z) = \frac{1 - p[1 - (1 - \tfrac{1}{4}\mu^2 h^2)^{\tfrac{1}{2}}]}{1 + p[1 - (1 - \tfrac{1}{4}\mu^2 h^2)^{\tfrac{1}{2}}]} = \frac{p^{-1} - [1 - (1 - \tfrac{1}{4}\mu^2 h^2)^{\tfrac{1}{2}}]}{p^{-1} + [1 - (1 - \tfrac{1}{4}\mu^2 h^2)^{\tfrac{1}{2}}]}, \quad (5.258)$$

$$R_{1,1}^{(2)}(z) = \frac{1 - p[1 + (1 - \tfrac{1}{4}\mu^2 h^2)^{\tfrac{1}{2}}]}{1 + p[1 + (1 - \tfrac{1}{4}\mu^2 h^2)^{\tfrac{1}{2}}]} = \frac{p^{-1} - [1 + (1 - \tfrac{1}{4}\mu^2 h^2)^{\tfrac{1}{2}}]}{p^{-1} + [1 + (1 - \tfrac{1}{4}\mu^2 h^2)^{\tfrac{1}{2}}]}. \quad (5.259)$$

Clearly both of these bounds tend to -1 as $p \to \infty$ and the method is A_0-stable for $0 \leq \mu h < 2$ also. In view of its overall A-stability property the method is not suitable for extrapolation.

In conclusion, it is noted that the finite difference method developed by replacing the matrix exponential function in (5.230) by its (1, 1) Padé approximant, is consistent and unconditionally stable. It is, therefore, convergent.

PROBLEMS

1. Write a computer program to compute the solution of the equation

$$\frac{\partial u}{\partial t} = \frac{\partial^2 u}{\partial x^2}; \quad 0 \leq x \leq 1$$

with boundary conditions $u(0, t) = u(1, t) = 0$ for all $t > 0$ and initial conditions $u(x, 0) = x(1 - x)$ for $0 \leq x \leq 1$. Obtain the solution at $x = 0.1(0.1)0.9$, $t = 1.0$ using $h = l = 0.1$.

Use now the initial condition $u(x, 0) = \tfrac{1}{2}$ and repeat the computation using (a) $h = l = 0.1$, (b) $h = 0.1$, $l = 0.05$. Explain the behaviour of the computed results.

2. Given the simple heat equation $\partial u/\partial t = \partial^2 u/\partial x^2$ with appropriate initial and boundary conditions specified,
   (i) show that the finite difference scheme of Du Font and Frankel:

   $$U_m^{n+1} = U_m^{n-1} + 2p(U_{m+1}^n - U_m^{n+1} - U_m^{n-1} + U_{m-1}^n); \ p = l/h^2$$

   is consistent with the differential equation only if $r = l/h \to 0$ as $l \to 0$, where h, l are the increments in space and time, respectively. If $l/h \to C$, some constant, as $l \to 0$, what differential equation does the difference scheme approximate? Is this second differential equation parabolic?
   (ii) show that the difference scheme

   $$U_m^{n+1} = U_m^{n-1} + 2p(U_{m-1}^n - 2U_m^n + U_{m+1}^n)$$

   is unstable for all positive values of r.

3. The heat equation $\partial u/\partial t = \partial^2 u/\partial x^2$ is approximated by the difference scheme

   $$(U_m^{n+1} - U_m^{n-1})/2l - [U_{m+1}^n - 2\{\theta U_m^{n+1} + (1-\theta)U_m^{n-1}\} + U_{m-1}^n]/h^2 = 0$$

   where θ is a variable parameter. Show that the local truncation error is

   $$\frac{l^2}{6}\frac{\partial^3 u}{\partial t^3} - \frac{h^2}{12}\frac{\partial^4 u}{\partial x^4} + 2p(2\theta - 1)\frac{\partial u}{\partial t} + r^2\frac{\partial^2 u}{\partial t^2} + \ldots$$

   where $p = l/h^2$, $r = l/h$ and l, h are the time and space steps, respectively. Discuss the consistency of the scheme as θ varies.

4. If $u(x, t)$ satisfies the equation

   $$\frac{\partial u}{\partial t} = x\frac{\partial^2 u}{\partial x^2}; \ 0 < x < \tfrac{1}{2}$$

   with boundary conditions given by

   $$u(0, t) = 0, \ \frac{\partial u}{\partial x}(\tfrac{1}{2}, t) = -\tfrac{1}{2}u; \ t > 0$$

   and initial conditions given by

   $$u(x, 0) = x(1-x); \ 0 \leqslant x \leqslant \tfrac{1}{2},$$

show that approximating all derivatives by central difference replacements leads to the explicit scheme

$$U_m^{n+1} = mrU_{m-1}^n + (1-2mr)U_m^n + mrU_{m+1}^n; \quad m = 1, 2, \ldots, N-1$$

$$U_N^{n+1} = 2NrU_{N-1}^n + (1 - 2Nr - Nl)U_N^n,$$

where $r = l/h$ and h, l are the space and time increments, respectively. Show that the scheme is stable locally if $r < 2/(20 + 5h)$.

5. The function $u(x, t)$ satisfies the second order parabolic problem

$$\partial u/\partial t = \partial^2 u/\partial x^2; \quad -1 < x < 1$$
$$u(x, 0) = 1 - x^2; \quad -1 \leq x \leq 1$$
$$u(0, t) = u(1, t) = 0; \quad t > 0.$$

Use (i) the Crank–Nicolson method (ii) the extrapolated form of the fully implicit method based on the (1, 0) method (iii) the unextrapolated form of the fully implicit method based on the (2, 0) Padé approximant, to find the solution at $t = 1.0$ for $x = 0.9(0.1)0.9$. Take $h = 0.1$, $l = 0.01$.

6. The function $u(x, t)$ satisfies the simple diffusion problem

$$\partial u/\partial t = \partial^2 u/\partial x^2; \quad 0 < x < X$$
$$u(x, 0) = f(x); \quad 0 \leq x \leq X$$
$$u(0, t) = u(X, t) = 0; \quad t > 0$$

where $f(x)$ is a given continuous function of x. A rectangular mesh $(x_m, t_n) = (mh, nl)$, $m = 0, 1, \ldots, N+1$, $n = 0, 1, \ldots$ covers the region $[0 \leq x \leq X] \times [t \geq 0]$; l is the time step and h is the space step with $(N+1)h = X$.

The differential equation is replaced by the finite difference scheme

$$-\alpha p(U_{m-1}^{n+1} + U_{m+1}^{n+1}) + 2(1 + \alpha p)U_m^{n+1} = (2 - \alpha)p(U_{m-1}^n + U_{m+1}^n)$$

$$+ 2(1 - 2p + \alpha p)U_m^n,$$

where $p = l/h^2$ and α is a free parameter.

Show that the principal part of the local truncation error is

$$\left[\frac{l}{2}(1-\alpha) - \frac{h^2}{12}\right]\left(\frac{\partial^2 u}{\partial t^2}\right)_m^n + \frac{l^6}{6}\left(1 - \frac{3\alpha}{2}\right)\left(\frac{\partial^3 u}{\partial t^3}\right)_m^n$$

at the mesh point (mh, nl), $m = 1, \ldots, N$, $n = 1, 2, \ldots$.

Show that in the special case where $\alpha = 1$, the method is unconditionally stable (the method becomes the Crank-Nicolson method based on the (1, 1) Padé approximant).

7. The function $u(x, t)$ satisfies the diffusion problem

$$\partial u/\partial t = \partial^2 u/\partial x^2; \quad 0 < x < 1, \quad t > 0$$
$$u(x, 0) = g(x); \quad 0 \leq x \leq 1,$$
$$u(0, t) = u(1, t) = 0, \quad t > 0.$$

Using the notation used in Chapter 5, show that the explicit method based on the (0, 1) Padé approximant satisfies

$$\mathbf{u}^{n+1} - \mathbf{U}^{n+1} = l(\mathbf{T}^n + P\mathbf{T}^{n-1} + \ldots + P^n \mathbf{T}^0),$$

where $\mathbf{T}^n$ is the vector of local truncation errors at time $t = nl$ and P is the square matrix of order N given by

$$P = \begin{bmatrix} 1-2p & p & & & 0 \\ p & (1-2p) & p & & \\ & & \ddots & & \\ & & p & (1-2p) & p \\ 0 & & & p & (1-2p) \end{bmatrix}$$

with $p = l/h^2$ and $(N+1)h = 1$; h, l are the space and time increments.

Hence, using the conventional definition of stability outlined in subsection 5.3.2 of Chapter 5, prove that this finite difference scheme is convergent when it is consistent and stable.

8. Using the notation used in subsection 5.6.2 of Chapter 5, show that the (2, 1) Padé approximant yields the implicit scheme

$$\frac{1}{6}p^2(U_{m-2}^{n+1} + U_{m+2}^{n+1}) - \frac{2}{3}p(1+p)(U_{m-1}^{n+1} + U_{m+1}^{n+1}) + (1 + \frac{4}{3}p + p^2)U_m^{n+1}$$
$$= \frac{1}{3}p(U_{m-1}^n + U_{m+1}^n) + (1 - \frac{2}{3}p)U_m^n$$

for $m \neq 1, N$. Find the forms of this scheme when $m = 1, N$ and use the matrix method to prove that it is unconditionally stable.

9. The function $u(x, t)$ satisfies the parabolic equation

$$\frac{\partial u}{\partial t} = (1 - \alpha t)\frac{\partial^2 u}{\partial x^2}; \quad 0 < x < 1, \quad t > 0$$

with initial conditions

$$u(x, 0) = 1; \quad 0 \leq x \leq 1$$

and boundary conditions

$$\frac{\partial u}{\partial x}(0, t) = \tfrac{1}{2} u, \quad u(1, t) = 1; \quad t > 0.$$

The region $[0 < x < 1] \times [t > 0]$ is discretized in the way described in section 5.2 using step sizes h, l in space and time. The differential equation is approximated by the difference scheme

$$U_m^{n+1} = U_m^n + p(1 - \alpha t)(U_{m-1}^n - 2U_m^n + U_{m+1}^n)$$

and the derivative boundary condition by

$$(U_1^n - U_{-1}^n)/2h = \tfrac{1}{2} U_0^n; \quad n = 0, 1, \ldots$$

Show that the difference scheme may be written in matrix form as

$$\mathbf{U}^{n+1} = Q^n \mathbf{U}^n + \mathbf{q}^n; \quad n = 0, 1, \ldots$$

where Q^n is a square matrix of order $N+1$ of the form

$$Q^n = \begin{bmatrix} d & 2a & & & & \\ a & b & a & & 0 & \\ & \ddots & \ddots & \ddots & & \\ & 0 & & a & b & a \\ & & & & a & b \end{bmatrix}$$

with $a = p(1 - nl)$, $b = 1 - 2p(1 - nl)$, $d = 1 - p(1 - nl)(2 + h)$ and $p = l/h^2$; the vector $\mathbf{b}^n$ involves the boundary condition $u(1, t) = 1$, $t > 0$.

Given that α is a positive constant and that $\alpha nl < 1$, show that the method is stable provided $p(1 - \alpha t) \leq 2/(4 + h)$.

10. Show that any eigenvalue λ_s of the tridiagonal matrix A given by equation (5.10) corresponding to the eigenvector $\mathbf{v} = [v_1, v_2, \ldots, v_N]^T$ satisfies the difference equation

$$v_{s-1} - (2 + \lambda_s)v_s + v_{s+1} = 0; \quad s = 1, 2, \ldots, N$$

with $v_0 = v_{N+1} = 0$.
Hence show that

$$\lambda_s = -4h^{-2} \sin^2[s\pi/\{2(N+1)\}]; \quad s = 1, 2, \ldots, N.$$

11. For the variable coefficient problem

$$\frac{\partial u}{\partial t} = a(x, t) \frac{\partial^2 u}{\partial x^2}$$

with appropriate boundary conditions and initial conditions specified, verify that

(i) $\dfrac{1}{2h^2} \delta_x^2 (u_m^{n+1} + u_m^n) = \left(\dfrac{\partial^2 u}{\partial x^2}\right)_m^{n+\frac{1}{2}} + \dfrac{h^2}{12}\left(\dfrac{\partial^4 u}{\partial x^4}\right)_m^{n+\frac{1}{2}} + O(h^4 + l^2),$

(ii) $\dfrac{1}{2h^2} \delta_x^2 (u_m^{n+1} + u_m^n) = \left(\dfrac{1}{a}\dfrac{\partial u}{\partial t}\right)_m^{n+\frac{1}{2}} + \dfrac{h^2}{12}\dfrac{\partial^2}{\partial x^2}\left(\dfrac{1}{a}\dfrac{\partial u}{\partial t}\right)_m^{n+\frac{1}{2}} + O(h^4 + l^2),$

(iii) $\dfrac{1}{h^2 l} \delta_x^2 [(a_m^{n+\frac{1}{2}})^{-1} (u_m^{n+1} - u_m^n)] = \dfrac{\partial^2}{\partial x^2}\left(\dfrac{1}{a}\dfrac{\partial u}{\partial t}\right)_m^{n+\frac{1}{2}} + O(h^2 + l^2),$

where $\delta_x^2 u_m^n = u_{m-1}^n - 2u_m^n + u_{m+1}^n$.

12. Given the parabolic equation in two-space variables

$$\frac{\partial u}{\partial t} = \frac{\partial^2 u}{\partial x^2} + \frac{\partial^2 u}{\partial y^2}; \quad 0 < x, y < 1, t > 0$$

with appropriate initial and boundary conditions specified; it is approximated by the finite difference methods

(i) $U_{k,m}^{n+1} = [1 + p(\delta_x^2 + \delta_y^2)] U_{k,m}^n,$

(ii) $U_{k,m}^{n+1} = (1 + p\delta_x^2)(1 + p\delta_y^2) U_{k,m}^n,$

where $p = l/h^2$, l and h being the time and space increments, respectively, and $\delta_x^2 U_{k,m}^n = U_{k-1,m}^n - 2U_{k,m}^n + U_{k+1,m}^n$, $\delta_y^2 U_{k,m}^n = U_{k,m-1}^n - 2U_{k,m}^n + U_{k,m+1}^n$.

Show that the local truncation errors of each of these schemes is $O(l + h^2)$ but that, when $p = \frac{1}{6}$, the local truncation error of (ii) is $O(l^2 + h^4)$. Show that (i) is stable for $p \leq \frac{1}{4}$ and that (ii) is stable for $p \leq \frac{1}{2}$.

13. Given the parabolic equation of Problem 11, show that the implicit difference scheme

$$(1 - p\delta_x^2)(1 - p\delta_y^2) U_{k,m}^{n+1} = (1 + p^2 \delta_x^2 \delta_y^2) U_{k,m}^n$$

also has local truncation error $O(l + h^2)$.

Bibliography and References

Abarbanel, S., Gottlieb, D, and Turkel, E., 1975, 'Difference schemes with fourth order accuracy for hyperbolic equations', *SIAM J. Appl. Math.*, **29**, 329–351.

Abbott, M. B., 1966, *The Method of Characteristics* (Thames and Hudson).

Barker, V. (ed.) 1977, *Sparse Matrix Techniques* (Springer-Verlag).

Barrett, J. W. and Morton, K. W., 1980, 'Optimal finite element solutions to diffusion-convection problems in one dimension', *Int. J. Num. Meth. Engng,* **15**, 1457–1474.

Barrett, J. W. and Morton, K. W., 1981, 'Optimal Petrov-Galerkin methods through approximate symmetrization', *IMA J. Num. Anal.,* **1**, 439–468.

Beam, R. M. and Warming, R. F., 1980, 'Alternating direction implicit methods for parabolic equations with a mixed derivative', *SIAM J. Sci. Stat. Comput.,* **1**, 131–159.

Bieterman, M. and Babuska, I., 1982a, 'The finite element method for parabolic equations I. A posteriori error estimation', *Numer. Math.,* **40**, 339–371.

Bieterman, M. and Babuska, I., 1982b, 'The finite element method for parabolic equations II. A posteriori error estimation and adaptive approach', *Numer. Math.,* **40**, 373–406.

Bramble, J. H. and Sammon, P. H., 1980, 'Efficient higher order single step methods for parabolic equations: Part I', *Math. Comp.,* **35**, 655–677.

Burden, R. L., Faires, J. D. and Reynolds, A. C., 1981, *Numerical Analysis* (second edition) (Prindle, Weber and Schmidt).

Busch, W., Esser, R., Hackbusch, W. and Herrmann, U., 1975, 'Extrapolation applied to the method of characteristics for a first order system of two partial differential equations. Part One: The initial value problem', *Numer. Math.,* **24**, 331–353.

Clough, R. W. and Tocher, J. L., 1965, *Proceedings of the First Conference on Matrix Methods in Structural Mechanics* (Wright-Patterson Air Force Base, Ohio).

Craggs, J. W., 1973, *Calculus of Variations* (Allen & Unwin).

Crank, J. and Nicolson, P., 1947, 'A practical method for numerical integration of solutions of partial differential equations of heat conduction type', *Proc. Camb. Phil. Soc.,* **43,** 50.

Dittrich, S. and Hackbusch, W., 1980, 'A method of characteristics for solving the initial-boundary value problem of a hyperbolic differential equation of second order', *Numer. Math.,* **34,** 217–234.

Douglas, J. and Pearcey, C. M., 1963, 'On convergence of alternating direction procedures in the presence of singular operators', *Numer. Math.,* **5,** 175–184.

Dupuis, G. and Göel, J. J., 1970, 'Finite element with high degree of regularity', *Int. J. Num. Meth. Eng.,* **2,** 563–577.

Eiseman, P. R., 1982a, 'Coordinate generation with precise controls over mesh properties', *J. Comput. Phys.* **47,** 331–351.

Eiseman, P. R., 1982b, 'High level continuity for coordinate generation with precise controls', *J. Comput. Phys.,* **47,** 352–374.

Evans, D. J. and Danaee, A., 1982, 'A new group hopscotch method for the numerical solution of partial differential equations', *SIAM J. Numer. Anal.,* **19,** 588–598.

Finlayson, Isla E., 1982, *Numerical Modelling of the Otolith Membrane,* B. Sc. Dissertation (Brunel University).

Forsythe, G. E. and Moler, C. B., 1967, *Computer Solution of Linear Algebraic Systems* (Prentice-Hall).

George, J. A., 1977, 'Solution of linear systems of equations: direct methods for finite element problems', In Barker, V. A., ed., *Sparse Matrix Techniques* (Springer-Verlag).

George, A. and Liu, J. W. H., 1978, 'Algorithms for matrix partitioning and the numerical solution of finite element systems', *SIAM J. Numer. Anal.,* **15,** 297–327.

George, A. and Liu, J. W. H., 1981, *Computer Solution of Large Positive Definite Systems* (Prentice-Hall).

Gerald, C. F. and Wheatley, P. O., 1984, *Applied Numerical Analysis* (third edition) (Addison-Wesley).

Gladwell, I. and Thomas, R. M., 1983, 'Damping and phase analysis for some methods for solving second-order ordinary differential equations', *Int. J. Num. Meth. Engng.,* **19,** 495–503.

Goult, R. J., Hoskins, R. F., Milner, J. A. and Pratt, M. J., 1974, *Computational Methods in Linear Algebra* (Stanley Thornes).

Gourlay, A. R. and Morris, J. Ll., 1980, 'The extrapolation of first order methods for parabolic partial differential equations, II', *SIAM J. Numer. Anal.,* **17,** 641–655.

Gourlay, A. R. and Morris, J. Ll., 1981, 'Linear combinations of generalised Crank-Nicolson schemes', *IMA J. Numerical Analysis,* **1,** 347–357.

Graney, L. and Richardson, A. A., 1981, 'The numerical solution of non-linear partial differential equations by the method of lines', *J. Comp. Appl. Math.*, **7**, 229–236.

Greig, D. M., 1980, *Optimisation* (Longman).

Gustafsson, B., 1975, 'The convergence rate for difference approximations to mixed initial boundary value problems', *Math. Comp.*, **29**, 396–406.

Gustafsson, B., Kreiss, H. O. and Sundström, A., 1972, 'Stability theory of difference approximations for mixed initial boundary value problems. II', *Math. Comp.*, **26**, 649–686.

Hackbusch, W., 1977, 'Extrapolation applied to certain discretization methods solving the initial value problem for hyperbolic differential equations', *Numer. Math.*, **28**, 121–142.

Hackbusch, W., 1977, 'Extrapolation to the limit for numerical solutions of hyperbolic equations', *Numer. Math.*, **28**, 455–474.

Hackbusch, W., 1978, 'On a method of characteristics for solving a hyperbolic equation of second order', *Computing*, **20**, 47–60.

Hudetz, W. J., 1973, 'A computer simulation of the otolith membrane', *Comp. Biol. Med.*, **3**, 355–369.

Hughes, T. J. R. and Akin, J. E., 1980, 'Techniques for developing 'special' finite element shape functions with particular reference to singularities', *Int. J. Num. Meth. Engng.*, **15**, 733–751.

Hughes, T. J. R., Tezduyar, T. E. and Brooks, A. N., 1982, 'A Petrov-Galerkin finite element formulation for systems of conservation laws with special reference to the compressible Euler equations', in Morton, K. W. and Baines, M. J., eds, *Numerical Methods for Fluid Dynamics* (Academic Press), 97–125.

Irons, B. M., 1969, 'Economical computer techniques for numerically integrated finite elements', *Int. J. Num. Meth. Eng.*, **1**, 201–203.

Irons, B. and Ahmad, S., 1979, *Techniques of Finite Elements* (Ellis Horwood).

Irons, B. and Shrive, N., 1983, *A Finite Element Primer* (Ellis Horwood).

Iserles, A., 1981, 'Rational interpolation to $\exp(-x)$ with application to certain stiff systems', *SIAM J. Numer. Anal.*, **18**, 1–12.

Johnson, L. W. and Riess, R. D., 1982, *Numerical Analysis* (second edition) (Addison-Wesley).

Keast, P. and Mitchell, A. R., 1967, 'Finite difference solution of the third boundary problem in elliptic and parabolic equations', *Numer. Math.*, **10**, 67–75.

Khaliq, A. Q. M., 1983, *Numerical Methods for Ordinary Differential Equations with Applications to Partial Differential Equations*, Ph. D. thesis (Brunel University).

Khaliq, A. Q. M. and Twizell, E. H., 1982, 'The extrapolation of stable finite difference schemes for first order hyperbolic equations', *Intern. J. Computer Math.*, **11**, 155–167.

Khaliq, A. Q. M. and Twizell, E. H., 1984, 'Backward difference replacements of the space derivative in first order hyperbolic equations', *Comput. Meths. Appl. Mech. Engrg,* **43**, 45–56.

Kreiss, H. O., 1968, 'Stability theory for difference approximations of mixed initial boundary value problems, I', *Math. Comp.,* **22**, 703–714.

Lambert, J. D., 1973, *Computational Methods in Ordinary Differential Equations* (Wiley).

Lawson, J. D., 1972, 'Some numerical methods for stiff ordinary and partial differential equations', in *Proc. Second Manitoba Conference on Numerical Math.,* 27–34 (University of Manitoba).

Lawson, J. D. and Swayne, D. A., 1976, 'A simple efficient algorithm for the solution of heat conduction problems', in *Congressus Numerantium XVIII,* 239–250 (Utilitas Mathematica, University of Manitoba).

Lawson, J. D. and Morris, J. Ll., 1977, 'A note on the efficient implementation of splitting methods in two space variables', *BIT,* **17**, 492–493.

Lawson, J. D. and Morris, J. Ll., 1978, 'The extrapolation of first order methods for parabolic partial differential equations. I', *SIAM J. Numer. Anal.,* **15**(6), 1212–1224.

Layton, W. J., 1983, 'Stable Galerkin methods for hyperbolic systems', *SIAM J. Numer. Anal.,* **20**, 221–233.

LeTallec, P., 1980, 'A mixed finite element approximation to the Navier-Stokes equations', *Numer. Math.,* **35**, 381–404.

LeVeque, R. J. and Oliger, J., 1983, 'Numerical methods based on additive splittings for hyperbolic partial differential equations', *Math. Comp.,* **40**, 469–497.

Luskin, M., 1980, 'A finite element method for first order hyperbolic systems', *Math. Comp.,* **35**, 1093–1112.

McKee, S., 1983, 'Discretization methods and block isoclinal matrices', *IMA J. Num. Anal.,* **3**, 467–491.

Marszalek, W., 1984, 'Two-dimensional state-space discrete models for hyperbolic partial differential equations', *Appl. Math. Modelling,* **8**, 11–14.

Miller, J. V., Morton, K. W. and Baines, M. J., 1978, 'A finite element boundary computation with an adaptive mesh', *J. Inst. Maths Applics,* **22**, 467–477.

Mitchell, A. R., 1973, 'An introduction to the mathematics of the finite element method', in Whiteman, J. R., ed., *The Mathematics of Finite Elements and Applications* (Academic Press), 37–58.

Mitchell, A. R. and Phillips, G. M., 1972, 'Construction of basis functions in the finite element method', *BIT,* **12**, 81–89.

Mitchell, A. R. and Wait, R., 1977, *The Finite Element Method in Partial Differential Equations* (Wiley).

Mitchell, A. R. and Griffiths, D. F., 1980, *The Finite Difference Method in Partial Differential Equations* (Wiley).

Mitchell, A. R., Phillips, G. M. and Wachspress, E. L., 1971, 'Forbidden shapes in the finite element method', *J. Inst. Maths. Applics.*, **8**, 260–269.

Moore, P., 1978, 'Finite element multistep multiderivative schemes for parabolic equations', *J. Inst. Maths. Applics.*, **21**, 331–334.

Morris, J. Ll., 1983, *Computational Methods in Elementary Numerical Analysis* (John Wiley & Sons).

Morton, K. W., 1978, 'Numerical methods for hyperbolic equations; the evolution of finite element methods', University of Reading Numerical Analysis Report *1/78*.

Morton, K. W., 1980, 'Stability of finite difference approximations to a diffusion-convection equation', *Int. J. Num. Meth. Engng.*, **15**, 677–683.

Morton, K. W., 1982, 'Generalised Galerkin methods for steady and unsteady problems', in Morton, K. W. and Baines, M. J., eds, *Numerical Methods for Fluid Dynamics* (Academic Press), 1–32.

Morton, K. W. and Baines, M. J., eds, 1982, *Numerical Methods for Fluid Dynamics* (Academic Press).

Morton, K. W. and Parrott, A. K., 1980, 'Generalised Galerkin methods for first order hyperbolic equations', *J. Comput. Phys.*, **36**, 249–270.

Morton, K. W. and Stokes, A., 1982, 'Generalised Galerkin methods for hyperbolic equations', in Whiteman, J. R., ed., *The Mathematics of Finite Elements and Applications IV* (Academic Press).

National Physical Laboratory, 1961, *Modern Computing Methods*, Notes on Applied Science No. 16 (second edition) (HMSO).

Noble, B., 1973, 'Variational finite element methods for initial value problems', in Whiteman, J. R., ed., *The Mathematics of Finite Elements and Applications* (Academic Press).

Oliger, J., 1974, 'Fourth order difference methods for the initial boundary-value problem for hyperbolic equations', *Math Comp.*, **28**, 15–25.

Osher, S., 1982, 'Shock modelling in aeronautics', in Morton K. W. and Baines, M. J., eds, *Numerical Methods for Fluid Dynamics* (Academic Press), 179–217.

Osher, S. and Solomon, F., 1982, 'Upwind schemes for hyperbolic systems of conservation laws', *Math. Comp.*, **38**, 339–374.

Padé, M. H., 1892, 'Sur représentation approchée d'une fonction par des fractionelles', *Ann. de l'Ecole Normale Superieure*, **9** (Suppl.).

Peaceman, D. W. and Rachford, H. H., 1955, 'Numerical solution of parabolic and elliptic differential equations', *SIAM J.*, **3**, 28–41.

Ralston, A., 1965, *A First Course in Numerical Analysis* (McGraw-Hill).

Richards, T. H., 1977, *Energy Methods in Stress Analysis: With an Introduction to Finite Element Techniques* (Ellis Horwood).

Richtmeyer, R. D. and Morton, K. W., 1967, *Difference Methods for Initial Value Problems* (second edition) (Wiley).

Sermer, P., 1983, 'A Galerkin method for elliptic-hyperbolic type equations', *SIAM J. Numer. Anal.*, **20**, 471–484.

Siemieniuch, J. L. and Gladwell, I., 1978, 'Analysis of explicit difference methods for a diffusion-convection equation', *Int. J. Num. Meth. Engng.*, **12**, 899–916.

Sloan, D. M., 1980, 'On boundary conditions for the numerical solution of hyperbolic differential equations', *Int. J. Num. Meth. Engng.*, **15**, 1113–1127.

Smith, G. D., 1978, *Numerical Solution of Partial Differential Equations: Finite Difference Methods* (second edition) (Oxford).

Smith, I. M., Siemieniuch, J. L. and Gladwell, I., 1977, 'Evaluation of Nørsett methods for integrating differential equations in time', *Intern. J. Num. Anal. Meth. Geomechan.*, **1**, 57–74.

Smith, P. and Twizell, E. H., 1980, 'A finite element model of temperature distribution in the human torso', *Appl. Math. Modelling*, **4**(3), 146–154.

Smith, P., 1981, *Numerical Modelling of Human Thermoregulation*, Ph. D. Thesis (CNAA).

Starius, G., 1980, 'On composite mesh difference methods for hyperbolic differential equations', *Numer. Math.*, **35**, 241–255.

Strang, G. and Fix, G. J., 1973, *An Analysis of the Finite Element Method* (Prentice-Hall).

Strang, G. and Iserles, A., 1983, 'Barriers to stability', *SIAM J. Numer. Anal.*, **20**, 1251–1257.

Twizell, E. H., 1975, 'The numerical solution of second order hyperbolic partial differential equations with unequally spaced initial conditions', *Computer Journal*, **18**(3), 252–257.

Twizell, E. H., 1976, 'Some approximations of the characteristics of second order hyperbolic partial differential equations', *J. Inst. Maths. Applics.*, **17**(2), 209–218.

Twizell, E. H., 1977, *Some Approximations of the Characteristics of Second Order Hyperbolic Partial Differential Equations*, Ph. D. Thesis (CNAA).

Twizell, E. H., 1979, 'An explicit difference method for the wave equation with extended stability range' *BIT*, **19**(3), 378–383.

Twizell, E. H., 1980, 'A variable gravity model of the otolith membrane', *Appl. Math. Mod.*, **4**(2), 82–86.

Twizell, E. H., 1981, 'The numerical solution of the wave equation at the first time step', Brunel University, Department of Mathematics Technical Report *TR/13/81*.

Twizell, E. H. and Curran, D. A. S., 1977, 'A finite element model of the otolith membrane', *Comput. Biol. Med.*, **7**, 131–141.

Twizell, E. H. and Khaliq, A. Q. M., 1981, 'One-step multiderivative methods for first order ordinary differential equations', *BIT*, **21**(4), 518–527 (Also Brunel University Technical Reports *TR/02/81* and *TR/04/81*.)

Twizell, E. H. and Khaliq, A. Q. M., 1982, 'L_0-stable methods for parabolic partial differential equations', Brunel University Department of Mathematics Technical Report *TR/02/82* (Revised).

Twizell, E. H. and Khaliq, A. Q. M., 1984, 'Multiderivative methods for periodic initial value problems', *SIAM J. Numer. Anal.*, **21**(1), 111–122.

Twizell, E. H. and Smith, P., 1981, 'Heat flow in the human torso approximated by an elliptic cylinder', Brunel University Department of Mathematics Technical Report *TR/11/81*.

Twizell, E. H. and Tirmizi, S. I. A., 1984, 'Implicit methods for the simple wave equation', Brunel University Department of Mathematics Technical Report, *TR/09/84*.

Varah, J. M., 1980, 'Stability restrictions on second order three level finite difference schemes for parabolic equations', *SIAM J. Numer. Anal.*, **17**, 300–309.

Varga, R. S., 1962, *Matrix Iterative Analysis* (Prentice-Hall).

Wachspress, E. L., 1971, 'A rational basis for function approximation', *J. Inst. Maths Applics*, **8**(1), 57–68.

Wachspress, E. L., 1975, *A Rational Finite Element Basis* (Academic Press).

Whiteman, J. R. (ed.), 1973, *The Mathematics of Finite Elements and Applications* (Academic Press).

Whiteman, J. R. (ed.), 1975, *The Mathematics of Finite Elements and Applications II* (Academic Press).

Whiteman, J. R. (ed.), 1975, *A Bibliography for Finite Elements* (Academic Press).

Whiteman, J. R. (ed.), 1979, *The Mathematics of Finite Elements and Applications III* (Academic Press).

Whiteman, J. R. (ed.), 1982, *The Mathematics of Finite Elements and Applications IV* (Academic Press).

Zienkiewicz, O. C., 1967, *The Finite Element Method in Structural and Continuum Mechanics* (McGraw-Hill).

Zienkiewicz, O. C., 1971, *The Finite Element Method in Engineering Science* (McGraw-Hill).

Zienkiewicz, O. C., 1977, *The Finite Element Method* (3rd edition) (McGraw-Hill).

Index

A-stability, 214, 260
A_0-stability, 214, 217, 227, 240, 260
A(0)-stability, 215
A(α)-stability, 214
absolute stability, 214
acceleration parameter, 18
admissible function, 92
Alternating Direction Implicit (ADI) methods, 54
amplification factor, 142, 213, 223
amplification matrix, 144, 212
amplification symbol, 219, 230, 256
asymptotic error constant, 27

backward difference approximant, 33
backward difference method, 176
basis functions, 93
boundary conditions, 42, 46, 48, 132, 166, 202, 220
boundedness of a solution, 141
Brauer's Theorem, 12

Cauchy initial value problem, 117, 159, 201
Cauchy sequence, 84
Cayley-Hamilton Theorem, 12
central difference approximant, 34, 166, 253
characteristic, 117, 162
    cone, 154
    curve, 117, 162
    equation, 12
    mesh, 117
    polynomial, 12
    root, 12
Choleski method, 17
classification, 37
collocation, method of, 92
    points, 92
compatibility, condition of, 93

complete inner product space, 84
complex splitting, 229, 230
conforming condition, 93
conjugate gradient methods, 19
conjugate vectors, 19
conservation form, 161
consistency, 139, 207
convection parameter, 252, 259
convergence, 146, 215
convergent sequence, 84
corrective terms, 98
Courant-Friedrichs-Lewy (CFL) condition, 148, 178
Crank-Nicolson method, 204, 218
Crout method, 17
curved elements, 100

decomposition methods, 17
diffusion-convection equation, 200, 252
diffusion equation, 200
domain of dependence, 117
Doolittle method, 17, 20
direct methods, 17
Dirichlet problem, 42, 55, 70, 88, 91, 101
discontinuities, 164, 202, 214, 220, 227

eigenvalue problem, 12
elastic membrane, 28, 61
energy 141, 195
    inner product, 92
    method, 141, 209
    norm, 92
equilibrium problems, 42
Euclidean norm, 13
Euler corrector formula, 121
Euler-Lagrange equations, 87, 90, 91, 128, 130
Euler predictor formula, 121

Index

explicit method (scheme), 135, 156, 203, 218, 254
extrapolation, 172, 216, 225

finite difference methods, 44, 132
finite element method, 81, 128
finite elements, 81, 93
first time step, 138
five point formula, 44, 107
fixed point, 25
    uniqueness, 25
fixed point iteration, 26
forward difference approximant, 33
Fourier method (stability), 141, 179, 211, 255
f-type characteristics, 119
function space, 84

Galerkin method, 92
Gaussian elimination, 17
Gauss-Seidel method, 18
Geršgorin's Circle Theorem, 12
Geršgorin's First Theorem, 12
global accuracy, 140
grid, 43
g-type characteristics, 119

Hamilton's Principle, 128
hat functions, 129
heat equation, multi-dimensional, 223
    non-linear, 251
    one-dimensional, 200
    two-dimensional, 220, 240
Hilbert space, 84

ill set, 114
implicit method (scheme), 136, 158, 204, 218, 223
inflow problem, 160
initial-boundary value problem, 132
initial conditions, 117, 132, 166, 202, 220
inner product, 84, 85
    norm, 85
    space, 84
interpolating polynomial, 28
interval of stability, 215
interval of stiffness, 16
irregular boundaries, 61
iterative methods, 17

Jacobi method, 18

L-stability, 215, 256, 257
L_0-stability, 214, 217, 227, 235, 237, 240, 258
$L(0)$-stability, 215
$L(\alpha)$-stability, 215
L_1-norm, 13
L_2-norm, 13
L_∞-norm, 14
Lagrange interpolating polynomial, 29
Lagrangian interpolation, 28, 65, 66
Laplace's equation, 43, 101
Lax Equivalence Theorem, 148
Lax-Wendroff method, 175
least squares, method of, 93
linear multistep methods, 131
linear operator, 86
    adjoint, 87
    bounded, 86
    norm, 86
    positive definite, 87
    self-adjoint, 87
linear space, 83
linear system, 11
local discretization error, 209
local truncation error, 44, 139, 171, 207, 216, 222, 225
    principal part, 44, 140

Maclaurin expansion, 33, 135
Maclaurin formula, 32
    remainder term, 32
        Cauchy form, 32
        integral form, 32
        Lagrange form, 32
Maclaurin series, 33
matrix, 11
    block, 21, 221
    characteristic equation, 12
    diagonal, 22, 221
    eigenvalue, 12, 16
    eigenvector, 12
    fill, 17
    inverse, 11
    lower triangular, 20, 188
    non-singular, 11
    norm, 13
    quindiagonal, 22
    sparse, 17
    spectral radius, 13
    square, 17
    tridiagonal, 20, 221
    upper triangular, 20
matrix method (stability), 141, 212, 255
measurable function, 84
mesh, 43
Method of Characteristics, 116, 163

Index

modified Euler formula, 121
multiderivative methods, 131

nested dissection methods, 20
Neumann problem, 46, 57, 90
Newton-Raphson method, 27
    second order convergence, 28
Newton's method, 27
nodal point, 43
node, 43, 93
non-constant coefficients, 248, 249, 250
non-linear algebraic equations, 25
non-rectangular coordinates, 67
normalizing factors, 108
normed linear space, 83
normed vector space, 83
numerical eigenvalue problem, 12

one-point iteration formula, 25
one-step method, 126
one-way dissection methods, 20
order of accuracy, 140
order of convergence, 27
orthogonality, 19
oscillatory components, 213
outflow problem, 159, 177

Padé approximant, 134, 203
parallelogram elements, 100
partial differential equations
    classification, 37
    elliptic, 38, 42, 81, 88
    hyperbolic, 38, 116
    parabolic, 38, 200
    quasi-linear, 37, 116
patchwork, 101
Peaceman-Rachford method, 226, 240, 244, 246
Petrov-Galerkin methods, 131, 132
pivotal condensation, 17
Poisson's equation, 72, 88
Power method, 14
predictor-corrector method, 216
propagation of discontinuities, 126, 166, 213

quadrilateral elements, 98
quasi-variational principles, 128
quotient tree methods, 20

rational basis functions, 107
rectangular elements, 100

recurrence relation, 134, 167, 183, 203, 222, 254
residual vector, 19
Ritz method, 91
Robbin's problem, 48, 59, 90

Schwarz inequality, 13
semi-norm, 84
Sobolev norm, 85
Sobolev space, 85
spectral norm, 13
spectral radius, 13
splitting methods, 224, 240
    complex, 229, 230
stability, 141
    conditional, 223
    energy method, 141
    Fourier method, 141, 211
    interval, 215
    matrix method, 141, 212
    unconditional, 142
    von Neumann method, 141
standard triangle, 96
steady-state problems, 36, 42
steady-state solution, 16
stiff system, 16, 203
stiffness, 15, 203
stiffness ratio, 16, 203
subspace, 83
successive over-relaxation (SOR), 18
symbol, 214, 217

Taylor coefficients, 32
Taylor expansion, 33
Taylor formula, 31
    remainder term, 31
        Cauchy form, 31
        integral form, 31
        Lagrange form, 31
Taylor polynomial, 32
Taylor series, 33
time dependent problems, 36, 128, 200
tolerance, 14, 18, 28
transient solution, 16
trapezium elements, 100
triangle elements, 94
triangle inequality, 13

unconditional stability, 144
undetermined coefficients, methods, of, 29

variational formulation, 87
vector, length, 84
    magnitude, 84

vector space, 83
vectors, conjugate, 19
vibrating membrane, 154
von Neumann method (stability), 141
von Neumann stability criterion, 142, 212, 255

wedge functions, 108
weighting functions, 132
well set, 114

zero vector, 12